Comprehensive Virology 12

Comprehensive Virology

Edited by Heinz Fraenkel-Conrat
University of California at Berkeley

and Robert R. Wagner
University of Virginia

Editorial Board

Comprehensive

Edited by

Heinz Fraenkel-Conrat

Department of Molecular Biology and Virus Laboratory
University of California, Berkeley, California

and

Robert R. Wagner

Department of Microbiology
University of Virginia, Charlottesville, Virginia

Virology

12

Newly Characterized Protist and Invertebrate Viruses

PLENUM PRESS · NEW YORK AND LONDON

Library of Congress Cataloging in Publication Data

Main entry under title:

Newly characterized protist and invertebrate viruses.

  (Comprehensive virology; 12)
  Includes bibliographies and index.
  1. Viruses. 2. Host-virus relationships. 3. Unicellular organisms—Diseases. 4. Inverte-
brates—Diseases. I. Fraenkel-Conrat, Heinz, 1910-   II. Wagner, Robert R.,
1923-   III. Series: Fraenkel-Conrat, Heinz, 1910-   Comprehensive vir-
ology; 12. [DNLM: 1. Invertebrates—Microbiology. 2. Viruses. 3. Virus diseases—
Veterinary.
QW160 C737 v. 12]
QR357.F72 vol. 12 [QR325] 576'.64'08s [576'.64]
ISBN 0-306-35152-8   77-27552

© 1978 Plenum Press, New York
A Division of Plenum Publishing Corporation
227 West 17th Street, New York, N.Y. 10011

Printed in the United States of America

Foreword

The time seems ripe for a critical compendium of that segment of the biological universe we call viruses. Virology, as a science, having passed only recently through its descriptive phase of naming and numbering, has probably reached that stage at which relatively few new—truly new—viruses will be discovered. Triggered by the intellectual probes and techniques of molecular biology, genetics, biochemical cytology, and high resolution microscopy and spectroscopy, the field has experienced a genuine information explosion.

Few serious attempts have been made to chronicle these events. This comprehensive series, which will comprise some 6000 pages in a total of about 18 volumes, represents a commitment by a large group of active investigators to analyze, digest, and expostulate on the great mass of data relating to viruses, much of which is now amorphous and disjointed, and scattered throughout a wide literature. In this way, we hope to place the entire field in perspective, and to develop an invaluable reference and sourcebook for researchers and students at all levels.

This series is designed as a continuum that can be entered anywhere, but which also provides a logical progression of developing facts and integrated concepts.

Volume I contains an alphabetical catalogue of almost all viruses of vertebrates, insects, plants, and protists, describing them in general terms. Volumes 2–4 deal primarily, but not exclusively, with the processes of infection and reproduction of the major groups of viruses in their hosts. Volume 2 deals with the simple RNA viruses of bacteria, plants, and animals; the togaviruses (formerly called arboviruses), which share with these only the feature that the virion's RNA is able to act as messenger RNA in the host cell; and the reoviruses of animals and plants, which all share several structurally singular features, the most important being the double-strandedness of their multiple RNA molecules.

Volume 3 addresses itself to the reproduction of all DNA-containing viruses of vertebrates, encompassing the smallest and the largest viruses known. The reproduction of the larger and more complex RNA viruses is the subject matter of Volume 4. These viruses share the property of being enclosed in lipoprotein membranes, as do the togaviruses included in Volume 2. They share as a group, along with the reoviruses, the presence of polymerase enzymes in their virions to satisfy the need for their RNA to become transcribed before it can serve messenger functions.

Volumes 5 and 6 represent the first in a series that focuses primarily on the structure and assembly of virus particles. Volume 5 is devoted to general structural principles involving the relationship and specificity of interaction of viral capsid proteins and their nucleic acids, or host nucleic acids. It deals primarily with helical and the simpler isometric viruses, as well as with the relationship of nucleic acid to protein shell in the T-even phages. Volume 6 is concerned with the structure of the picornaviruses, and with the reconstitution of plant and bacterial RNA viruses.

Volumes 7 and 8 deal with the DNA bacteriophages. Volume 7 concludes the series of volumes on the reproduction of viruses (Volumes 2–4 and Volume 7) and deals particularly with the single- and double-stranded virulent bacteriophages.

Volume 8, the first of the series on regulation and genetics of viruses, covers the biological properties of the lysogenic and defective phages, the phage-satellite system P 2–P 4, and in-depth discussion of the regulatory principles governing the development of selected lytic phages.

Volume 9 provides a truly comprehensive analysis of the genetics of all animal viruses that have been studied to date. These chapters cover the principles and methodology of mutant selection, complementation analysis, gene mapping with restriction endonucleases, etc. Volume 10 also deals with animal cells, covering transcriptional and translational regulation of viral gene expression, defective virions, and integration of tumor virus genomes into host chromosomes.

Volume 11 covers the considerable advances in the molecular understanding of new aspects of virology which have been revealed in recent years through the study of plant viruses. It covers particularly the mode of replication and translation of the multicomponent viruses and others that carry or utilize subdivided genomes; the use of protoplasts in such studies is authoritatively reviewed, as well as the nature of viroids, the smallest replicatable pathogens.

The present volume deals with several special groups of viruses showing properties that set them apart from the main virus families. These are the lipid-containing phages and the viruses of prokaryotic and eukaryotic algae, fungi, and insects. This will be followed in Volume 14 by the special groups of vertebrate virus, such as the arena-, corona-, and bunyaviruses, while Volume 13 concludes our survey of structural aspects and assembly of viruses with chapters on the primary structure of viral nucleic acids and proteins, and the architecture and assembly of complex phages, viral membranes, and animal virions.

Contents

Chapter 2

Viruses in Fungi

K. N. Saksena and P. A. Lemke

Chapter 3

Cyanophages and Viruses of Eukaryotic Algae

Louis A. Sherman and R. Malcolm Brown, Jr.

Chapter 4

Viruses of Fungi Capable of Replication in Bacteria (PB Viruses)

T. I. Tikchonenko

Chapter 5

Bacteriophages That Contain Lipid

Leonard Mindich

CHAPTER 1

Viruses of Invertebrates

T. W. Tinsley and K. A. Harrap

Natural Environment Research Council
Unit of Invertebrate Virology
Oxford OX1 3UB, England

1. INTRODUCTION

Invertebrates constitute about 80% of the known animal species in the world, so it is not surprising to find that they harbor large numbers of a wide range of virus types. Unfortunately, our present knowledge of viruses that are pathogenic to invertebrates is largely centered on those reported from the class Insecta, but even here the available information is often fragmentary. However, this is more a reflection of the scarcity of virologists working in this field than of any intrinsic difficulty in studying the particular viruses (Tinsley and Melnick, 1974). It is encouraging that in recent years reports have appeared on virus diseases of invertebrates other than insects and mites. Very little work has been done to characterize such viruses, and some of the reports must be regarded with caution until the viral nature of the causative agent has been fully established, including the infectivity of the isolated viruses to uninfected members of the recorded host.

One of the inherent problems associated with virus infections of insects under natural conditions is the occurrence of subclinical or latent infections which, with present insensitive methods, are difficult to detect. When natural populations of insects are subjected to a physical or chemical stress, these low-level infections can be induced or activated to frank expression, often resulting in death of the host (Aruga, 1963). It is quite possible that such subclinical infections represent the

1

usual and most frequent host–virus interaction in the natural environment. This would suggest that invertebrates in general and insects in particular may be very efficient in "containing" virus infections and so preventing virus replication from reaching a harmful level. Salt (1970) commented that although the defensive mechanisms in insects are probably different from those exhibited by vertebrate animals it does not follow that they are in any way less efficient.

Viruses and viruslike conditions reported from invertebrates other than insects and mites are shown in Table 1. The wide gaps in our knowledge are obvious. The groups of viruses known to infect insects are shown in Table 2. At least six of these groups have physical and chemical similarities to viruses found in vertebrates and, to a lesser degree, to certain plant viruses. The importance of diseases in populations of beneficial insects such as the silkworm, *Bombyx mori*, and the honeybee, *Apis mellifera*, was recognized by the early workers in this field. Detailed observations on diseases now known to have a viral etiology are to be found in the writings of Nysten (1808), Cornalia (1856), Maestri (1856), Langstroth (1857), Bolle (1894), von Prowazek (1907), and White (1917). Reviews of much of this early work have been given by McKinley (1929) and Steinhaus (1949). Modern studies on viruses of invertebrates were stimulated by Bergold's demonstration of virus particles in preparations isolated from diseased insects using electron microscopy and other methods (Bergold, 1943, 1947). Much of the early work on insect viruses was concerned with descriptive pathology, and few attempts were made to isolate and characterize the causal agents. The concept that, because the viruses were isolated from different host species, they must in turn be different and unrelated gave rise to the erroneous belief that insect viruses are essentially host specific. A further complication arose from the inability to test host specificity with any degree of certainty because of the lack of characterization data and the absence of methods whereby unequivocal identification of the viruses could be made. The report by Smith *et al.* (1961) of an iridescent virus from the crane fly, *Tipula paludosa*, capable of infecting not only related *Tipula* spp. but also members of other natural orders in the Insecta provided experimental evidence that the concept of narrow host specificity could not be applied universally to viruses and their host species. Cunningham and Longworth (1968) showed, using serological methods, that seven insect cytoplasmic polyhedrosis viruses previously thought to be distinct were, in fact, very closely related if not one and the same virus.

The system of naming viruses of invertebrates has led to considerable confusion in the literature largely because it was common practice

TABLE 1
Viruses and Suspected Viruses of Invertebrates Other Than Insects and Mites

Host	Size of virus (nm)	Possible virus group suggested by authors	Reference
Entamoeba histolytica (amoeba)	1. 40 2. 70	—	1. Miller and Schwartzwelder (1960) 2. Diamond *et al.* (1972), Mattern *et al.* (1972)
Plasmodium gallinaceum (protozoon)	35–55	—	Terzakis (1969)
Paramoecium sp. (protozoon)	80	—	Preer and Preer (1967), Preer and Jurand (1968), Grimes and Preer (1971)
Naegleria gruberi (amoeba)	100	—	Schuster (1969), Dunnebacke and Schuster (1971), Schuster and Dunnebacke (1971)
Ignotocoma sabellarium (ancistrocomid ciliate)	75 × 50 250 × 50	—	Lom and Kozloff (1969)
Hydra vulgaris (coelenterate)	70–75	Adenovirus	Bonnefoy *et al.* (1972)
Enchytraeus fragmentosus (microannelid)	270 × 50	—	Dougherty *et al.* (1963)
Gyratrix hermaphroditis (platyhelminth)	70	—	Reuter (1975)
Sepia officinalis (squid)	75	Reovirus	Devauchelle and Vago (1971)
Octopus vulgaris (octopus)	120 × 140	ICDV	Rungger *et al.* (1971)
Trichosomoides crassicauda (nematode)	15	—	Foor (1972)
Ostrea edulis (oyster)	125–170	—	Bonami *et al.* (1971)
Crassostrea virginica (oyster)	70–90	Herpesvirus	Farley *et al.* (1972)
Carcinas maenas (crab)			Bang (1971)
Macropipius depurator (crab)	1. 50–60 2. 150–300	—	1. Vago (1966) 2. Bonami and Vago (1971)
Buthus occitanus (scorpion)		Reovirus	Morel (1975)
Penaeus duorarum (shrimp)	288 × 59 (occluded virions)	Baculovirus	Couch (1974a)

TABLE 2
Groups of Insect Pathogenic Viruses

Group	Insect host order	Nucleic acid	Inclusion body	Symmetry of particle	Similarities to viruses of vertebrates and plants based on morphological and biochemical criteria	
					Vertebrate	Plant
Baculoviruses	Lepidoptera Hymenoptera Diptera	DNA	+	Helical	None—apparently restricted to Invertebrates	None
Cytoplasmic polyhedrosis viruses	Lepidoptera Hymenoptera Diptera	RNA (ds)	+	Spherical	Reoviruses, bluetongue	Wound tumor, rice dwarf, viruses of fungi
Entomopox viruses	Lepidoptera Orthoptera Coleoptera Diptera	DNA	+	"Brick"	Poxviruses	None
Icosahedral cytoplasmic deoxriboviruses (iridoviruses)	Lepidoptera Hymenoptera Coleoptera Diptera Ephemeroptera Hemiptera	DNA	−	Spherical	Viruses, isolated from fish, frogs, lizards; African swine fever	Viruses of fungi and algae
Parvoviruses	Lepidoptera	DNA (ss)	−	Spherical	Parvoviruses	None
Picornaviruses	Lepidoptera Hymenoptera Coleoptera Orthoptera	RNA	−	Spherical	Enteroviruses	Small RNA viruses
Rhabdoviruses	Diptera	RNA	−	Bacilliform bullet shape	Rabies, vesicular stomatitis virus, viruses of fish	Potato yellow dwarf, lettuce necrotic yellows, etc.

to couple the virus group name with the Latin binomial of the host insect, e.g., "nuclear polyhedrosis virus of *Heliothis zea*." Such a system in common usage could only lend further credibility to the supposed host specificity of the viruses. Holmes (1948) introduced the names "Borrelina" (polyhedrosis viruses) and "Morator" (nonoccluded viruses), which did little to dispel the growing confusion in the literature. This was also the case with the further series of personalized names proposed by Bergold *et al.* (1960), e.g., "Borrelinavirus," "Bergoldiavirus," "Smithiavirus," and "Vagoiavirus." These systems reflected a serious disadvantage of insect virology in that it was developing in isolation from other branches of virology and was often unaware of modern concepts. However, in 1966, the International Committee on Nomenclature of Viruses (ICNV) was founded, and this organization established subcommittees to consider the whole field of nomenclature of viruses, first independently and later in concert. Various ideas and systems were also proposed to assist in bringing some order to the growing problem of describing viruses of invertebrates in different laboratories in an acceptable and cohesive form (Tinsley and Kelly, 1970, Harrap and Tinsley, 1971). Eventually, the Subcommittee on Invertebrate Viruses presented proposals to the ICNV, and these were incorporated in the general report of that body (Wildy, 1971). The facility to propose amendments and additions to the system was provided for, and the recent situation has been summarized by Vago *et al.* (1974). In essence, the names adopted for vertebrate virus groups are used wherever possible, and the only two present exceptions are baculovirus (nuclear polyhedrosis and granulosis viruses) and iridovirus (icosahedral cytoplasmic deoxyriboviruses). It is interesting that the officially encouraged use of vertebrate virus group names for similar viruses found in invertebrate animals has done much to draw together these two branches of animal virology. There is no doubt that further research into the similarities between viruses found in vertebrates and invertebrates could do much to encourage the growth of comparative virology.

A further significant factor in the need to develop knowledge of insect viruses is their potential in the biological control of insect pests (see Section 5.3). The use of viruses as pesticidal agents is not a new concept, but economic and safety factors have been serious obstacles to wide-scale development. The lack of basic virological data has also hindered development of these ideas, for, without systems of identification, the usefulness of viruses on the one hand and their ecological hazards on the other cannot be tested satisfactorily.

Another area of potential importance is the discovery of virus diseases in marine invertebrates of economic importance, although here the problem is prophylaxis and not biological control. The development of intensive growth systems of aquaculture or mariculture, i.e., intensive farming of marine resources, indicates that physical and/or chemical stresses on marine organisms will increase, and there are reports on the influence of toxic chemicals on the progress of a virus disease of the pink shrimp, *Penaeus duorarum* (Couch, 1974a,b). It is to be hoped that technical expertise in the virology of marine invertebrates will develop rapidly, because, with the mass culture systems presently being tested, dire consequences could result if basic virological information were not available.

2. GROUPS OF VIRUSES FOUND IN INVERTEBRATES

2.1. Nuclear Polyhedrosis and Granulosis Viruses

Many of the properties of nuclear polyhedrosis and granulosis viruses (NPV and GV) are similar. It is for this reason that the two groups have been combined under the title of "baculoviruses" in the First Report of the International Committee on the Nomenclature of Viruses (Wildy, 1971). Viruses of this type have been isolated mainly from infections in Lepidoptera, Hymenoptera, and Diptera, but of particular interest are the recent reports of baculovirus infections in a mosquito, *Aedes triseriatus* (Federici and Lowe, 1972), in a caddis fly larva, *Neophylax* sp. (Hall and Hazard, 1973), and in the pink shrimp, *Penaeus duorarum* (Couch, 1974a,b) (see Fig. 1). Virus infection usually occurs and is more pronounced in the larva, probably as a result of the food sources being contaminated with virus. The incubation period varies according to such factors as the age of the larva, temperature, food availability, population density, and incidence of other pathogens. The initial signs of infection are not marked, but, in time, behavioral changes may be observed that are probably indicative of metabolic disturbances. For example, larvae may cease feeding and leave the food plant entirely or may distribute themselves on it in some abnormal manner. Smirnoff (1965) describes how nun moth larvae gather characteristically at the tops of trees, hence the original name "Wipfelkrankheit" virus (treetop disease). Infected larvae are often sluggish, and eventually a change of color of the integument may be noticeable. In time, the integument becomes very fragile, and the larvae may be found hanging in an inverted position. Death soon follows and

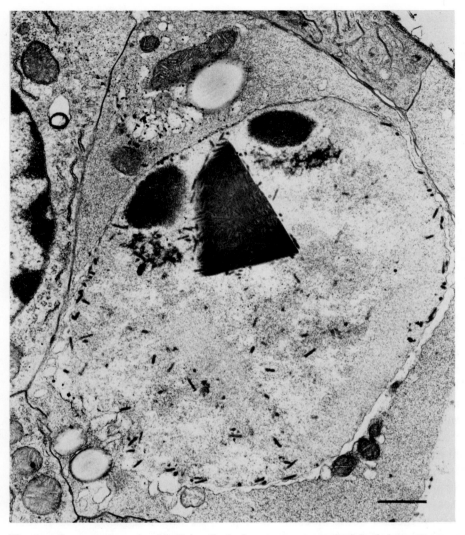

Fig. 1. Cross-section of epithelial cells in hepatopancreas of pink shrimp, *Penaeus duorarum*, showing an infected cell with virus particles and an inclusion body. Bar equals 1 μm. Courtesy of Dr. J. Crouch, U.S. EPA Laboratory, Gulf Breeze, Florida.

rupture of the epidermis will often occur, resulting in the liberation of masses of virus inclusion bodies (polyhedra or granules according to virus type) which at death account for the greater proportion of body weight. The polyhedra or granules released in this way are disseminated by such climatic factors as wind and rain, leading to contamination of adjacent food plants (Bird, 1961b; Vago and Bergoin, 1968). Other factors such as the behavior of parasites, predators, and scavengers are implicated in the spread of these viruses.

Nuclear polyhedrosis viruses as such were observed first in the mid-nineteenth century, but the associated disease condition was described in a poem by Vida in 1527 (Smith, 1971). The virus inclusion bodies are either polyhedral or capsular in shape according to type, and they range from 0.5 μm to 15 μm in diameter. They are very refractile and the polyhedral inclusion bodies may be observed easily under the light microscope. "Strains" of nuclear polyhedrosis and granulosis viruses with inclusion bodies of specific and consistent shape have been reported (Stairs, 1964, 1968). These reports support the view that inclusion body formation is virus directed. The inclusion bodies are composed of the virus particles (or an individual virus particle in the case of granulosis viruses) surrounded by a paracrystalline matrix of protein arranged as a cubic lattice (Bergold, 1963a,b). The virus particles are occluded in an apparently random manner and do not interfere with the regularity of the crystalline protein matrix (Engström and Kilkson, 1968). The way in which such a lattice could assemble and be organized has been considered by Harrap (1972a). Chemical analysis of polyhedra has shown them to contain protein, DNA, lipid, and a number of trace metals (Wellington, 1954; Shapiro and Ignoffo, 1971b). Silicon has been claimed to have an essential function in the structure of the cubic lattice (Estes and Faust, 1966; Faust and Adams, 1966), but the relative amounts of silicon to protein make this unlikely. There have been a number of reports of RNA associated with the protein matrix of the polyhedron (Faulkner, 1962; Aizawa and Iida, 1963). Gershenson *et al.* (1963) claim that such polyhedron RNA is infective and that the infectivity is destroyed by RNase. These findings have not been confirmed, and we therefore are doubtful of their validity. It is of course possible, on purely hypothetical grounds, that some virus-directed RNA species could become incorporated into polyhedra during their formation in the infected cell.

Mature inclusion bodies often show a peripheral electron-dense layer in ultrathin section. Treatment with alkaline solutions to disrupt the inclusion bodies can result in collapsed "baglike" ghosts of polyhedra being isolated by centrifugation. Such structures have been called polyhedron membranes. Evidence for the existence of these "membranes" has been a subject of dispute (Nordin and Maddox, 1971). However, negative staining has revealed distinctive subunit tessellation on these structures (Harrap, 1972a). The virus particles embedded in the polyhedral or capsular inclusion bodies are rod shaped and may be enveloped singly (as in GV and some NPV) or in groups or "bundles" (see Figs. 2 and 3). The size of the enveloped virus particles ranges from 40 to 140 nm in width × 300 to 330 nm in length depending on the number of rod-shaped virus particles within the envelope. Singly enveloped virus particles are typically 60–70 × 260–300 nm (Harrap, 1972b). The virus envelope, often referred to as the outer or developmental membrane, has a typical unit membrane trilaminar appearance in ultrathin section even though it is usually acquired actually within the nucleus during virus morphogenesis (see Section 4). Regular subunit tessellation can be demonstrated on the virus envelope by negative staining methods. The fine structure of the naked rod-shaped virus particle (or nucleocapsid) has been examined by a number of workers (Harrap and Juniper, 1966; Kozlov and Alexeenko, 1967; Himeno et al., 1968; Khosaka et al., 1971; Hughes, 1972; Harrap, 1972b). Regular repeating peripheral subunits can be resolved, and

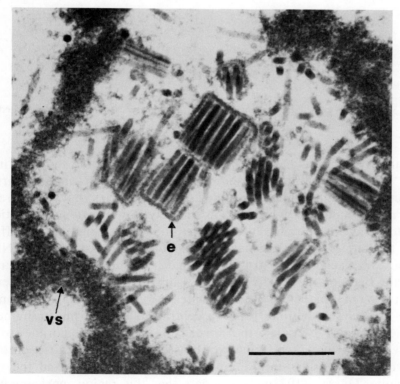

Fig. 2. Nuclear polyhedrosis virus particles in a fat body cell nucleus. Some naked particles are associated with virogenic stromal strands (vs) and groups of particles are acquiring an envelope (e). Bar equals 500 nm.

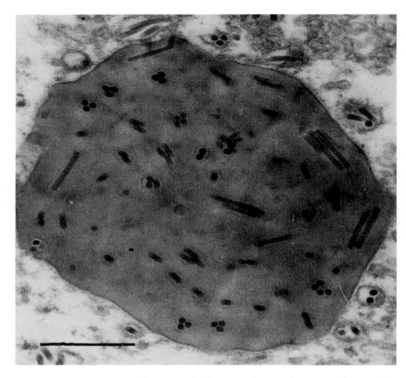

Fig. 3.   Mature polyhedron inclusion body. *Porthetria dispar* nuclear polyhedrosis virus. Bar equals 500 nm.

apparently "empty" capsids, probably equivalent to the so-called inner or "intimate" membrane, show a more complex ringlike surface structural arrangement. Disrupted virus particles have a core of internal component which stains densely in uranyl salt solutions. Beaton and Filshie (cited by Bellett *et al.,* 1973), using optical diffraction techniques, found the naked virus particles of a number of NPVs to have a stacked disk structure with 12–13 subunits per disk and a spacing between the disks of around 4 nm. The structure repeats every third row, making it a multistart helix with a pitch of 55°. Sections of virus particles sometimes reveal a central hole 10–15 nm in diameter. Some workers consider the inner or intimate membrane to be equivalent to that area which can be seen in sectioned virus particles between the densely stained inner "core" of the virus and the virus envelope. Harrap (1972*b,c*) considers that this region may simply be condensed nucleoplasm of the virus-infected cell as it cannot be explained satisfactorily by the results of observation of negatively stained virus preparations. It is possible that empty virus particles regarded as equivalent to the inner or intimate membrane (Bergold, 1963b) are not the virus capsids, as suggested by Harrap (1972b), but are some sort of matrix protein sheath outside the nucleocapsid proper. One thing is clear: the structure is not a membrane within the generally understood meaning of the term.

Large quantities of inclusion bodies can be isolated from dead or moribund larvae and purified by velocity and equilibrium sucrose gradient methods. On a large scale, this can be achieved most conveniently using zonal-type rotors (Cline et al., 1970; Martignoni et al. 1971; Harrap and Longworth, 1974). The inclusion body preserves the infectivity of the virus particles very effectively, and purified inclusion body preparations can be stored either as dry powders or as aqueous suspensions over a wide temperature range. Inclusion bodies are resistant to bacterial putrefaction and to a wide range of chemical treatments (Bergold, 1963b), but they are susceptible to treatment at high pH with solutions such as 0.1 M sodium carbonate. Under these conditions, the paracrystalline protein structure of the inclusion body is solubilized and the virus particles can be pelleted by centrifugation. The inclusion body protein can be precipitated from the alkaline solution by pH adjustment to its isoelectric point of pH 5.8. There is one report of successful recrystallization of the protein (Shigematsu and Suzuki, 1971), and latticelike aggregates of the protein can be observed especially when Mg^{2+} is present in the alkaline solution (Harrap, 1972a). Longworth et al. (1972) examined the properties of Pieris brassicae GV inclusion body protein (see Fig. 4). They isolated two proteins chromatographically and showed, by serological methods, that both were present at the capsule surface and in the solubilized protein obtained by treatment with alkali. One of the proteins, however, was also found to be common to the surface of the virus particle. The two proteins had different amino acid compositions and different isoelectric points. Egawa and Summers (1972) and Kawanishi et al. (1972b) studied the kinetics and ultrastructure of capsule solubilization using Trichoplusia ni granulosis virus. The pH was found to be the most influential factor in dissolution of the capsule, and, at near neutral pH, guanidine-HCl in propanol would cause dissolution, but virus particles could not be recovered. Three phases of the dissolution process were identified: hydrophobic and hydrogen-bonding forces were considered to be of major importance in the stability of the structure, while ionic bonding forces were considered to be of minor importance. This view contrasts with that of Estes and Faust (1966) and Faust and Adams

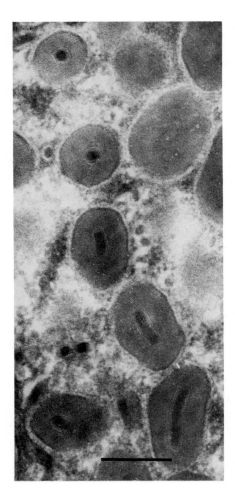

Fig. 4.   Capsules of *Pieris brassicae* granulosis virus. Bar equals 250 nm.

(1966), who implicated silicon as having a key role in inclusion body lattice structure. In ultrastructural studies, all the solvent systems except guanidine-HCl produced early longitudinal fracture of the capsule. The guanidine-HCl produced cavities in the capsule structure. Summers and Egawa (1973) studied the physical and chemical properties of *Trichoplusia ni* GV capsule protein using a variety of techniques. They found the major protein component of the inclusion body to have a molecular weight of 180,000 and a sedimentation coefficient of 7.2 S. Previous studies have reported 11 S as a typical value (Scott *et al.*, 1971; Harrap, 1972a). Koslov *et al.* (1975a,b) compared alkaline and acidic dissolution of two NPV inclusion bodies and concluded that the basic protein component had a molecular weight of 28,000 and a sedimentation coefficient around 2 S. Normal cleavage was thought likely to be a result of alkaline proteinase activity associated with the inclusion bodies themselves. Amino acid analyses of the inclusion body proteins of the two viruses were similar, and tryptic peptide maps differed in only a few peptides. Unfortunately, the purification criteria for the inclusion bodies were not clearly stated. A number of workers have studied methods for the disruption of NPV and GV inclusion bodies (Summers and Paschke, 1970; Nordin and Maddox, 1971). Investigation of the effects of varying conditions in which inclusion bodies are alkali treated, such as duration of treatment and temperature and salt effects, resulted in an attempt to standardize the treatment methods that produce maximum yields of virus particles from a number of NPVs and GVs (Harrap and Longworth, 1974). The virus particles can be purified, for example, using two cycles of gradient centrifugation in linear 30–60% (w/w) sucrose at 50,000g for 2 hr. Either a single band or multiple bands of virus particles will be obtained under these conditions depending on whether or not the virus envelope surrounds a single rod-shaped particle or several rod-shaped particles. These profiles are consistent even if centrifugation is prolonged. Similar results are obtained in cesium chloride, although this procedure is less satisfactory because the virus particles can be somewhat unstable in high salt concentrations. The buoyant density of singly enveloped virus particles is 1.32 g/ml. Separation of the different bands of multiply enveloped virus can also be achieved in 10–50% (w/v) sucrose velocity gradients. The different bands obtained from the multiply enveloped viruses reflect the differing numbers of naked virus particles (or nucleocapsids) surrounded by one envelope (Kawanishi and Paschke, 1970). We have found the number and distribution of the bands in both gradient systems to be characteristic of certain virus isolates. Virus particles purified in this way give consistent profiles in polyacrylamide gel electrophoresis; consistent precipitation lines are obtained when detergent-treated virus is reacted in immunodiffusion tests against antisera to both inclusion bodies and virus particles, and they appear similar structurally when examined in the electron microscope. Other workers have reported analogous findings in other hosts infected with certain NPVs and GVs (Summers and Paschke, 1970; Young and Lovell, 1973; Scott and Young, 1973). Scott *et al.* (1971) reported a sedimentation coefficient of 1530 S for virus particles isolated and purified from polyhedra of the cabbage looper, *Trichoplusia ni*. Work in our laboratory on the NPV of *Spodoptera* sp. has shown that treatment of the virus particles with a detergent such as Nonidet P40 results in profiles on polyacrylamide gel lacking many of the high molecular weight polypeptides. These are presumably envelope proteins, because electron microscope examination of such treated virus preparations reveals only naked rod-shaped particles with no evidence of envelope structure. The size of such naked particles is 40–50 nm $\times$ 200–300 nm depending on the isolate, and their buoyant density is 1.47 g/ml in cesium chloride. Two principal polypeptides are retained in these naked particles, those of molecular weight 15,000 and molecular weight 30,000. It seems then that different isolates of these viruses (by which we mean morphologically similar NPVs or GVs isolated from different host species) may have different envelope and nucleocapsid polypeptides. The structural polypeptides of a number of nuclear polyhedrosis viruses of Lepidoptera and Hymenoptera are currently under investigation in our laboratory. Three precipitation lines could be detected in immunodiffusion tests using Nonidet P40-treated virus particles. In the limited number of isolates tested, some lines were common to different isolates, some showed partial relationship, and some were specific. In complement fixation tests, significant differences in titer could be found between homologous and heterologous reactions involving different isolates. It is possible that such distinctions could form the basis for a typing method for these viruses.

Nuclear polyhedrosis and granulosis viruses contain DNA. Closed circular ("supercoiled"), open circular ("nicked") and linear forms of DNA can be identified if virus treated with sarcosine is centrifuged on cesium chloride gradients containing the intercalating dye ethidium bromide (Summers and Anderson, 1972a,b, 1973). Approximately 30% of the yield can be obtained in the covalently closed form and 70% in the open circular and linear forms. A surprising finding was that lyophilization of the polyhedra destroyed the supercoiled species. There is evidence from the work of D. L. Knudson in this laboratory that the alkaline treatment procedure for the extraction of virus from polyhedra and subsequent storage of the virus may cause degradation of the supercoiled species. Using virus grown in cell culture and purified directly as nonoccluded (i.e., non-polyhedron-bound) virus from the cells, thus avoiding alkali treatment procedures, as much as 90% of the extracted DNA was in the supercoiled form. Guanine plus cytosine content calculated from buoyant density determinations in cesium chloride ranged from 41% to 47% depending on the virus isolate under investigation. In neutral sucrose gradients, GV DNA has a molecular weight of about 100×10^6 and NPV DNA is slightly smaller. In a comparison of the DNAs of NPVs and GVs, Bellett (1969) found that a classification according to the base composition of the viral DNA was supported by available data on serological cross-reactions and amino acid composition of viral proteins. His results did not support the continuance of NPVs and GVs as separate groups, a factor significant in the present grouping and nomenclature of these viruses as baculoviruses. Kislev *et al.* (1972) compared *Spodoptera littoralis* virus DNA with that of the host insect and demonstrated a difference in the buoyant densities. Earlier studies on baculovirus DNAs reported molecular weight values as low as 2×10^6 (Onodera *et al.*, 1965). Molecular weight values of around $90-100 \times 10^6$ were obtained by Summers and Anderson (1972*a,b,* 1973), while values of $70-80 \times 10^6$ have been measured by C. C. Payne in this laboratory using sedimentation in neutral sucrose.

A subgroup of the baculovirus genus has been proposed to include a virus isolated from the rhinoceros beetle, *Oryctes rhinoceros* (Huger, 1966). Three other viruses are possible members of this subgroup on morphological grounds. One has been observed in the midgut cells of the adult whirligig beetle, *Gyrinus natator* (Gouranton, 1972). The other two cause diseases of the European red mite, *Panonychus ulmi* (Bird, 1967), and the citrus red mite, *Panonychus citri* (Reed and Hall, 1972). Only the *Oryctes* virus has been purified and characterized (Monsarrat *et al.,* 1973*a,b*; Revet and Monsarrat, 1974; Payne, 1974). The virus can be purified from both adults and larvae. The virus particles are rod shaped and enveloped, typically measure 220×120 nm, and have a density in sucrose of 1.18 g/ml. Eleven structural polypeptides have been resolved on 10% polyacrylamide gel. The virus genome is double-stranded DNA with a guanine plus cytosine content of 43–44% as measured from its density in cesium chloride (Revet and Monsarrat, 1974; Payne, 1974). A proportion of the DNA is in a covalently closed circular form and on neutral sucrose gradients the major component has a sedimentation coefficient of 57.2 S and an estimated molecular weight of 87×10^6. Although the virus is not found occluded in crystalline proteinaceous inclusions, its morphological and biophysical properties seem to justify its inclusion with baculoviruses.

2.2. Cytoplasmic Polyhedrosis Viruses

Cytoplasmic polyhedrosis viruses occur commonly in Lepidoptera and have also been recorded in Hymenoptera (Longworth and Spilling, 1970) and Diptera (mosquito, Anthony *et al.,* 1973; chironomid midge,

Federici *et al.*, 1973). Federici and Hazard (1975) have recorded cytoplasmic polyhedrosis (CPV) infections in a crustacean (*Daphnia* sp.).

The virus infection causes cessation of feeding and a reduction in
the size of the larva. Larvae often show signs of lassitude and debility
once infected and may change color as the disease progresses. Pupation
may occur, although the pupae are often small, and adults may fail to
emerge or be deformed or sterile. The infection was traditionally
thought to be confined to the midgut, and infected larvae often regur-
gitate large masses of inclusion bodies (polyhedra) or void them with
the feces. However, Kellen *et al.* (1966) reported CPV infections both
in the hypodermal cells and in the developing wing buds of *Culex
tarsalis*, and Stoltz and Hilsenhoff (1969) found CPV in the fat body of
Chironomus plumosus. These findings would suggest that the tissue tro-
pism of this group of viruses is wider than originally thought.

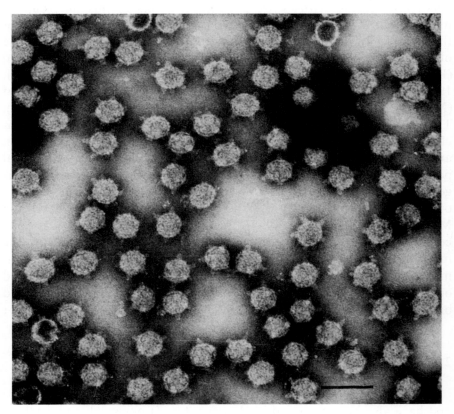

Fig. 5.   Virus particles isolated from *Phalera bucephala* cytoplasmic polyhedrosis virus
inclusion bodies. Bar equals 100 nm.

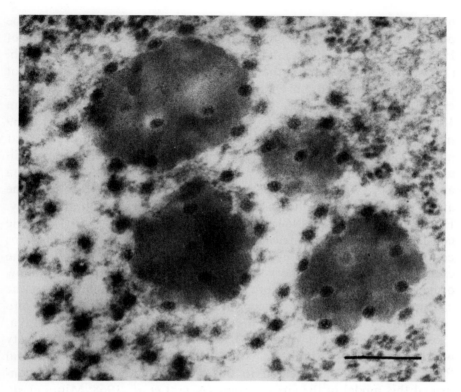

Fig. 6. Developing polyhedral inclusion bodies. Cytoplasmic polyhedrosis virus of
Bupalus piniarius. Bar equals 250 nm.

The polyhedra usually appear spherical in the light microscope and
are stained readily in hot Giemsa stain, in contrast to nuclear
polyhedra. "Strains" of virus with polyhedra of characteristic shape
have been reported (Hukuhara and Hashimoto, 1966). Polyhedra can
range widely in size up to several micrometers in diameter. Procedures
for the extraction and purification of virus inclusion bodies and the
virus particles contained within them have been described by several
workers (Cunningham and Longworth, 1968; Hayashi and Bird, 1968*a*,
1970; Lewandowski *et al.* 1969; Miyajima *et al.* 1969; Hayashi *et al.*,
1970) (see Figs. 5 and 6). In general, polyhedra can be extracted from
cooled, virus-infected midguts after removal of food by trituration in a
tris buffer, pH 7.6, followed by filtration and centrifugation through
sucrose to separate the polyhedra from cell debris. Virus particles can
be released from the purified polyhedra by treatment with alkali,
usually for periods up to an hour, and pelleted from the inclusion body
protein solution. Centrifugation on a linear sucrose gradient results in a

band of purified virus. The free, nonoccluded virus particles can also be purified directly from infected midgut tissue using the procedures of Hayashi and Bird (1968a). This method is valuable because Hayashi (1970a) found that more than 70% of the virus particles were not occluded into polyhedra even at 10 days after infection. The virus particles are spherical, with a fairly wide range of reported sizes, e.g., from 48 nm (*Nymphalis io* CPV, Cunningham and Longworth, 1968) to 69 nm (*Bombyx mori* CPV, Hosaka and Aizawa, 1964). The particles bear large apical "spikes" or projections on their surfaces. They will agglutinate mammalian and especially chick erythrocytes (Miyajima and Kawase, 1969). Asai *et al.* (1972) described terminal knobs on the apical "spikes," which they associated with the hemagglutinin activity. Hosaka and Aizawa (1964) proposed a model of the virus particle consisting of two concentric capsid layers with the apical projections contiguous with hollow cylinders leading into the virus core. The existence of a hollow tube vertex capsomer with its attached hollow projection connecting the core content of the virus particle to the exterior is intrinsic to the concept of the release of the viral RNA into cells through this structure (Hosaka, 1964; Nishimura and Hosaka, 1969). However, if the virus particle contains a transcriptase activity which functions only in intact virions, then it is unlikely that the viral RNA would ever be released in this way. Other workers have confirmed this model of the virus particle (Arnott *et al.*, 1968; Miura *et al.*, 1969). More recently, however, the validity of the model has been challenged (Lewandowski and Traynor, 1972; Payne and Tinsley, 1974). In particular, the existence of two concentric capsid shells has been criticized, and these workers consider the CPV particle to contain a single capsid analogous to the core or subviral particle which can be obtained from reovirus. Lewandowski and Traynor (1972) detected 20 surface capsomers on the circumference of the virus particle. Six projections or "spikes" could be seen, but the existence of 10 or more likely 12 was predicted. The "spikes" seemed to originate deep in the core region.

Not surprisingly, the CPV of the silkworm, *Bombyx mori,* has been studied most thoroughly. Virus particle sedimentation coefficients between 371 S and 440 S and buoyant densities of between 1.37 and 1.435 g/ml have been reported. The particle weight of the virus has been estimated at two widely differing values, 28.7×10^6 and $(54 \pm 4) \times 10^6$; the particles appear to be ether resistant (Kalmakoff *et al.,* 1969; Miyajima *et al.,* 1969; Nishimura and Hosaka, 1969; Miura *et al.,* 1968; Richards and Hayashi, 1971). A comparable sedimentation coefficient, 380 S, has been reported for another CPV—that of the white-marked tussock moth, *Orgyia leucostigma* (Hayashi and Bird, 1968*b*). Using iodination techniques, Lewandowski and Traynor (1972) found that *B. mori* CPV particles contained five structural polypeptides of molecular weights 151,000, 142,000, 130,000, 67,000, and 33,000. The inclusion body protein had two major polypeptides of molecular weight 30,000 and 20,000. Several minor proteins were observed. In contrast, Payne and Kalmakoff (1974), working with the same virus, reported slightly lower values for the three major polypeptides of the virion, and they were of the opinion that the polyhedral protein consisted of one glycosylated component. Lewandowski and Traynor (1972) found that, as well as the purified virus, two other species of virus particle could be isolated from the sucrose gradients during purification: a "top component," presumably devoid of RNA and of buoyant density 1.30 g/ml, and a shoulder to the main virus peak called "satellite virus" of buoyant density 1.425 g/ml, which contained virus particles deficient in some of the RNA genome segments (see below). This loss of genome segments appeared to be accompanied by a breakdown of the largest structural polypeptide. From the rate at which iodination proceeded before dissociation of the virus for electrophoresis, it was deduced that the 151,000 and 130,000 molecular weight polypeptides were externally located and the other three polypeptides were internally located. The molecular weights of the polypeptides are consistent with the hypothesis that they are coded for by monocistronic messenger RNA (mRNA) molecules transcribed from the distinct segments of the double-stranded RNA genome. Similarly, the two major and three minor polypeptides of the inclusion body protein could be correlated with the five remaining CPV RNA cistrons. This is to be expected if the inclusion body shape is characteristic for a given virus "strain" and thus is virus coded. Other minor polypeptides found in inclusion body protein could be explained by trapping of certain gene products in the protein matrix during occlusion of the virus particles in the cell. Working with another CPV isolate, that from *Nymphalis io*, Payne and Tinsley (1974) found three structural polypeptides in the virus particle with molecular weights of 116,000, 109,000, and 30,000 and a single major inclusion body polypeptide of molecular weight 37,000, which was glycosylated. All four structural polypeptides could be correlated with certain of the viral RNA cistrons. Three enzyme activities are associated with *B. mori* CPV particles: an RNA polymerase (transcriptase), a nucleotide phosphohydrolase (NTPase), and a nuclease that is ATP dependent (Lewandowski *et al.*, 1969; Storer *et al.*,

1973*a,b*). The NTPase and nuclease activity can be detected even after the virus particle is structurally disrupted, whereas the RNA polymerase can be detected only in the intact virus particles. This finding correlates well with the stability of the nuclease and NTPase as compared with that of the RNA polymerase in kinetics of inactivation studies. All three enzymes require magnesium ions for optimal activity. The localization of the enzymes is unknown, but Lewandowski and Traynor (1972) suggest that each surface capsomer might be a multicomponent aggregate with polymerase activity. If the virus possesses RNA polymerase activity, the viral RNA should not be infectious as claimed by Kawase and Miyajima (1968). However, these authors did state that some variability of the RNA infectivity was found. If the report were correct, the polymerase should be a host-specified enzyme. Lewandowski *et al.* (1969) could not substantiate the infectivity of the viral RNA. They found the polymerase associated with the virus particle to be capable of transcribing single-stranded RNA *in vitro* which would specifically hybridize with heat-denatured CPV genome RNA. This result was emphasized in later work (Lewandowski and Traynor, 1972).

There is a great deal of information on CPV RNA. Nishimura and Hosaka (1969), working with *B. mori* CPV, found about 30% of the virus particle to be RNA. They examined the RNA electron microscopically using the Kleinschmidt technique and found it to consist largely of two size classes of 0.36 μm and 1.12 μm. The longest molecule observed was 6.8 μm in length, equivalent to between 14 and 18 $\times$ 10^6 daltons. CPV RNA was first inferred to be double stranded from the base-pairing data of Hayashi and Kawase (1964). Miura *et al.* (1968) examined *B. mori* CPV RNA after phenol extraction and confirmed the two size classes electron microscopically. They found base pairing of G+C and A+U, and a G+C content of 43%. Features such as a sharp transition on heating, great hyperchromicity on degradation, nonreactivity with formaldehyde, and RNase resistance were all indicative of a double-stranded molecule. However, their calculated molecular weight of 4.7 $\times$ 10^6 for the viral RNA must be regarded as unusually low; a more realistic value is 15.7 $\times$ 10^6 (Fujii-Kawata and Miura, 1970). Payne and Tinsley (1974) reported *Nymphalis io* CPV RNA as having a molecular weight of 14 $\times$ 10^6. *Malacosoma disstria* CPV RNA has a reported molecular weight of 20 $\times$ 10^6 (Hayashi and Krywienczyk, 1972), although the large number (16) of RNA segments isolated electrophoretically and their apparent differences in staining intensity on the gels make one wonder whether this isolate might constitute a mixture of two CPVs. Several workers have demonstrated the segmented nature of the CPV genome on polyacrylamide gel electrophoresis. With *B. mori* CPV, usually ten RNA segments are found, of two size classes (L and S) ranging in molecular weight from around 2.5×10^6 to 0.35×10^6 (Kalmakoff *et al.,* 1969; Lewandowski and Millward, 1971; Fujii-Kawata and Miura, 1970; Hayashi and Krywienczyk, 1972). Other CPV RNAs with two size classes have been reported for *Orgyia leucostigma* (Hayashi, 1970*b*), *Malacosoma disstria* (Hayashi and Krywienczyk, 1972), and *Nymphalis io* (Payne and Tinsley, 1974). As mentioned above, each RNA segment appears to be monocistronic, and on theoretical grounds individual structural proteins could be coded for by individual RNA segments (Lewandowski and Traynor, 1972; Payne and Tinsley, 1974). Lewandowski and Millward (1971) showed that it was possible to oxidize *in situ* the 3′-terminals of the viral genome still contained in the structurally intact virus particle. The number of 3′-terminals subsequently found by reduction with tritiated sodium borohydride was the same as the number observed in RNA isolated from the virus before oxidation and reduction. Thus the viral RNA apparently occurs in polynucleotide segments within the virion or, more correctly, has as many oxidizable 3′-terminals. How or whether these pieces are linked in the virion is not clear. Furuichi and Miura (1972) and Lewandowski and Leppla (1972) extended the periodate oxidation–borohydride reduction technique and found that all 3′-termini were labeled to the same extent and none was phosphorylated as would be expected if a single large genome were cleaved by nuclease action. From analysis of nucleoside trialcohols obtained from ribonuclease or alkali digests of the labeled RNA, they reported 50% to be cytidine and 50% to be uridine. As there is no homology between the various RNA segments, their production by random breakage is unlikely. Cleavage probably occurs at specific weak points. There are three possibilities for the distribution of the uridine and cytidine nucleosides: a random distribution, five segments of U and five of C, or one strand of U and one strand of C for each of the segments. The data obtained so far suggest that the last possibility is the correct one, but unequivocal confirmation will have to await successful physical separation of the strands.

The nucleotide structure of the 3′ and 5′ ends of *B. mori* CPV genome RNA strands implies that the molecules are perfect double strands with no single-stranded tails. Each of the ten separated segments has been found to have the same terminal structure. Of the two genome strands, one chain carrying uracil at the 3′-terminus is selected as a template and transcription starts from the side where the 5'-terminus carries a methylated nucleotide (Shimotohno and Miura, 1974; Miura *et al.*, 1974). The RNA product of transcription has the same size distribution as the genome segments, and each synthesized mRNA fraction hybridizes specifically with the corresponding fraction of denatured RNA genome. The single-stranded mRNA product leaves the virus particle only after transcription has been carried out over the whole length of each genome segment (Shimotohno and Miura, 1973). *S*-Adenosyl-L-methionine (SAM) greatly stimulates the synthesis of the single-stranded mRNA. When it is present in the *in vitro* synthesis, a ribose moiety of adenylic acid in the 5'-terminal region of the mRNA is methylated at a very early stage of transcription and 7-methyl guanylic acid is found combined with the 5'-terminus of the mRNA by a 5'-5'-pyrophosphate linkage. The viral transcription therefore seems to be a "methylation-coupled" reaction, and the modified 5'-terminal nucleotide of mRNA may be necessary for its successful completion (Furuichi, 1974; Furuichi and Miura, 1975). The presence of such methylated residues attached to adenosine by a pyrophosphate linkage has been found in a number of RNAs and may be important for cellular processing of mRNA (Rottman *et al.*, 1974).

The segment profiles of CPV RNA seem to be distinctive and characteristic (C. C. Payne, personal communication) and could perhaps be used as a method of identification or indeed detection of mixtures of these viruses currently considered as individual isolates. The difference in RNA profiles between, for example, *B. mori* CPV and *N. io* CPV (Payne and Tinsley, 1974) is also apparent in serological comparisons of CPVs. Cunningham and Longworth (1968) examined a number of CPVs in complement fixation tests and found *B. mori* CPV to be unrelated to the others tested, the CPVs of *N. io, Aglais urticae, Vanessa cardui, Arctia caja, Porthetria dispar, Phalera bucephala,* and *Euproctis chryssorhoea.* Krywienczyk *et al.* compared the CPVs of *M. disstria, O. leucostigma,* and *B. mori* in double-diffusion and immunoelectrophoresis tests on cellulose acetate. They found that the former two were very closely related, sharing at least five antigens, only one of which was shared with *B. mori* CPV. It is also worth noting that in a comparison of the proteins of some CPVs differences in amino acid analysis were found (Hayashi and Durzan, 1971).

Until quite recently, all cytoplasmic polyhedrosis viruses had been found, as the name implies, developing in the cytoplasm of infected cells. However, Kawase *et al.* (1973) have reported a virus resembling

CPV in which polyhedra are formed in the nucleus of the midgut cells of the silkworm, *B. mori*. Purification of the virus was described, and the form of the virion was typically CPV-like, having a diameter of 62 nm and a sedimentation coefficient of 430 S. Subsequently, the virus was shown to contain a segmented genome of double-stranded RNA (Kawase and Yamaguchi, 1974). These reports make the already clumsy nomenclature of this group of viruses even more unsatisfactory.

2.3.   Poxviruses

The first insect poxvirus was described by Vago (1963) in the larva of the common cockchafer, *Melolontha melolontha*. The infected larvae are white in appearance, and the hemocytes and the fat body are the susceptible tissues. Many other insect poxviruses have since been found; those infecting the Coleoptera are the largest (400 × 250 nm) and have an eccentrically positioned core (Bergoin *et al.*, 1971; Goodwin and Filshie, 1969; Vago *et al.*, 1968*a,b* 1969*b*). There is a single lateral body in the concavity of the core. Poxviruses isolated from Lepidoptera and Orthoptera are smaller (350 × 250 nm) and have a symmetrical core and two lateral bodies (Meynadier *et al.*, 1968; Henry *et al.*, 1969; Granados and Roberts, 1970) (see Fig. 7). Those poxviruses found in Diptera are the smallest so far recorded in insects (320 × 230 nm) and are cuboidal (Götz *et al.*, 1969; Weiser, 1969). In general, all insect poxviruses show a structural similarity to vertebrate poxviruses, having a "mulberry" appearance on their surface, lateral body or bodies, and a flexible "ropelike" structure inside the core folded into four or five segments. Insect poxviruses are occluded into proteinaceous crystalline inclusion bodies analogous to those of baculoviruses. The inclusion bodies are purified by dissecting the fat body from paralyzed or moribund larvae into tris-HCl, pH 7.6, containing 2% ascorbate. The inclusion bodies are released from the tissue by sonication, and, after low-speed centrifugation to remove cell debris, the sample is layered on 50–65% (w/w) sucrose gradients and centrifuged at 75,000*g* for 1 hr. The pure band of inclusion bodies can be collected and the virus particles released from them after dilution and repeated washing. The release of the virus particles from the inclusion bodies is accomplished in solutions of 0.08 M sodium carbonate and 0.02 M sodium thioglycollate at pH 10.9 for 2 min. The reaction can be slowed by the addition of sucrose (60% w/w) and stopped by lowering the pH to 8.5–9.0. The virus particles are purified on 40–60%

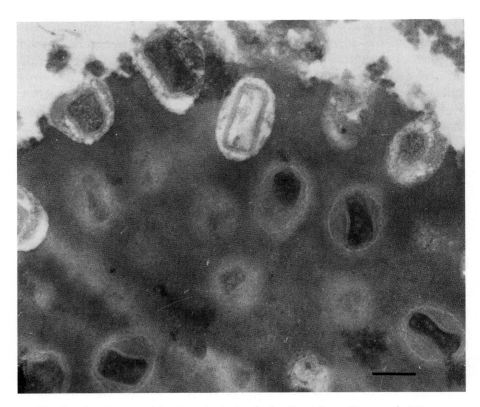

Fig. 7.   Poxvirus particles in an inclusion body. *Hepialis* sp. Bar equals 250 nm.

(w/w) linear sucrose gradients centrifuged for 2 hr at 85,000g. The resulting virus band is harvested and dialyzed against 10^{-3} M tris-HCl, pH 8.6, for 48 hr (Pogo *et al.,* 1971). Four enzyme activities are found associated with virus particles purified in this way: a nucleotide phosphohydrolase, a DNA-dependent RNA polymerase, and a neutral and an acidic DNase. These four enzymes are also found in vaccinia, but they may in fact be common to all DNA viruses that replicate in the cytoplasm. The enzymes are located in the core of the virus particles, as is the case in vertebrate poxviruses (Kates and McAuslan, 1967; Munyon *et al.,* 1967, 1968; Gold and Dales, 1968; Pogo and Dales, 1969). The DNases are less active than those found in vertebrate poxviruses. This is possibly due to inactivation during extraction and purification, as prolonged alkali treatment results in a total loss of activity. A uniform coat or "halo" has been observed around virus particles isolated from inclusion bodies by alkaline treatment (Bergoin *et al.,* 1968*a*; McCarthy *et al.,* 1974). The halo could be removed by trypsin treatment, resulting in a change in buoyant density in cesium chloride from 1.282 g/ml to 1.262 g/ml. Virus particles without the halo were 15–45 times more infective. Halo formation may be a step in the process of occlusion of the virus particle. Polyacrylamide gel electrophoresis of *Amsacta moorei* poxvirus particles in the presence of SDS has revealed 13–17 polypeptides depending on the percentage gel employed. The major bands reportedly have molecular weights of 17,500, 60,000, 84,000, and 93,000 (Roberts and McCarthy, 1972). A DNA content of 1.4%, somewhat lower than the average figure for vertebrate poxviruses, has also been reported, with a molecular weight tentatively put at 160×10^6 (McCarthy *et al.,* 1974). However, McCarthy *et al.* (1975) found that this estimate was too low and that the DNA content was nearer to 5% and thus would approximate the DNA content of vertebrate poxviruses.

No common antigens have been reported between poxviruses of vertebrates and invertebrates. Suckling mice injected intracerebrally and intraperitoneally with poxvirus particles of three insect species were apparently unaffected after 1 year. Similarly, pocks did not form on the chorioallantoic membranes of 11-day-old chicken eggs at either 31°C or 37°C after the introduction of insect poxvirus particles (Roberts and McCarthy, 1972).

The properties of poxviruses of vertebrates and invertebrates have been reviewed and compared by Bergoin and Dales (1971).

2.4. Iridescent Viruses

Possibly the best known insect virus is *Tipula* iridescent virus found in the larva of the crane fly, *Tipula paludosa,* and first reported by Xeros (1954) (see Fig. 8). It causes a disease condition characterized by the presence of a blue-green iridescent coloration in the tissues of the host. Since then, many so-called iridescent viruses have been described in, for example, beetles (*Sericesthis pruinosa*), a stem borer (*Chilo suppressalis*), mosquitoes (*Aedes taeniorhyncus, A. cantans, A. annulipes, A. stimulans, A. detritus*), midges (*Culicoides* sp.), a chironomid, and several grassland pests (*Wiseana cervinata, Witlesia sabulosella, Costelytra zealandica*), among others (Steinhaus and Leutenegger, 1963; Day and Mercer, 1964; Clark *et al.,* 1965; Weiser, 1965; Fukaya and Nasu, 1966; Chapman *et al.,* 1966; Stoltz *et al.,* 1968; Vago *et al.,* 1969*a*; Anderson, 1970; Kalmakoff and Robertson,

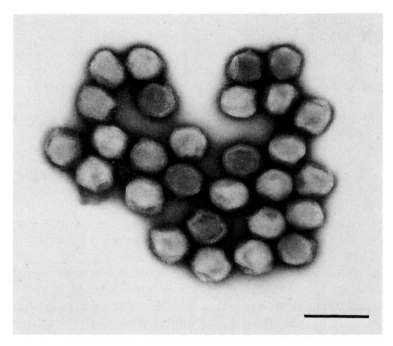

Fig. 8.   Particles of *Tipula* iridescent virus (type 1). Bar equals 250 nm.

1970; Tinsley *et al.*, 1971; Hall and Anthony, 1971; Fowler and Robertson, 1972; Kalmakoff *et al.*, 1972). Indeed, the number of isolates was becoming so large and potentially so confusing that Tinsley and Kelly (1970) suggested an interim nomenclature analogous to that used for adenoviruses, in which *Tipula* iridescent virus (TIV) became iridescent virus type I, *Sericesthis* iridescent virus (SIV) became type 2, and so on. An extended list was provided by Kelly and Robertson (1973) in their review of icosahedral cytoplasmic deoxyriboviruses, and a further updated version is provided in Table 3. This method of nomenclature avoids the confusing implication that each iridescent virus isolate is a distinct and separate virus specific to a given host.

    In nature, acquisition of most insect viruses is usually thought to occur by mouth, but in the case of iridescent viruses such a route of entry has been difficult to substantiate (Smith, 1967; Vago *et al.*, 1969a; Stoltz and Summers, 1971; Carter, 1974). It may be that accidental injury and predatorial behavior represent ways that the virus gains access to susceptible tissues, but cannibalism or feeding on infected cadavers seems to be the most important method (Linley and Nielsen,

1968; Carter, 1973*a*). In the laboratory, intrahemocoelic injection is the routine infection technique, although mosquito larvae are infected from an aqueous virus-contaminated environment (Wagner *et al.*, 1973).

Several workers have reported extraction and purification procedures for iridescent viruses (Glitz *et al.*, 1968; Kalmakoff and Tremaine, 1968; Cunningham and Hayashi, 1970; Kelly and Tinsley, 1972; Yule and Lee, 1973). They usually involve trituration of dead and dying larvae in neutral phosphate buffer, followed by cycles of differential centrifugation and sucrose density gradient (e.g., 5–40%) centrifugation at speeds between 10,000 and 15,000 rev/min for periods up to 1 hr. Reported variations on this general method include use of zonal rotors, use of Sepharose 2B columns, and use of enzymes, surface-active agents, or organic solvents. As many as three virus-containing

TABLE 3
Types of Iridescent Viruses[a]

Type	Invertebrate host	Reference
1	*Tipula paludosa*	Xeros (1954)
2	*Sericesthis pruinosa*	Steinhaus and Leutenegger (1963)
3	*Aedes taeniorhyncus*	Matta and Lowe (1970)
4	*Aedes cantans*	Weiser (1965)
5	*Aedes annulipes*	Weiser (1965)
6	*Chilo suppressalis*	Fukaya and Nasu (1966)
7	*Simulium ornatum*	Weiser (1968)
8	*Culicoides* sp.	Chapman *et al.* (1968)
9	*Wiseana cervinata*	Fowler and Robertson (1972)
10	*Witlesia sabulosella*	Fowler and Robertson (1972)
11	*Aedes stimulans*	Anderson (1970)
12	*Aedes cantans*	Tinsley *et al.* (1971)
13	*Corethrella brakeleyi*	Chapman *et al.* (1971)
14	*Aedes detritus*	Hasan *et al.* (1970)
15	*Aedes detritus*	Vago *et al.* (1969*a*)
16	*Costelytra zealandica*	Kalmakoff *et al.* (1972)
17	*Pterostichus madidus*	J. S. Robertson (unpublished)
18	*Opogonia* sp.	Kelly and Avery (1974*a*)
19	*Odontria striata*	J. Kalmakoff (unpublished)
20	*Simocephalus expinosus*	Federici and Hazard (1975)
21	*Heliothis armigera*	D. C. Kelly (unpublished)
22	*Simulium* sp.	B. S. Batson and D. C. Kelly (unpublished)
23	*Heteronycus arator*	D. C. Kelly (unpublished)
24	*Apis* sp.	L. Bailey (unpublished)
25	*Tipula* sp.	R. M. Elliott and D. C. Kelly (unpublished)
26	*Ephemeropteran* sp.	B. Federici (unpublished)
27	*Nereis diversicolor*	Devauchelle and Durchon (1973)

[a] Courtesy of Dr. D. C. Kelly.

bands can be found on sucrose gradients: a main band of discrete virus particles, a lower band of aggregated virus, and an upper or "top component" band. Some workers find this top band to be of particles morphologically distinguishable from complete virus particles in terms of stain penetration and shape under the electron microscope. Other workers find no obvious differences.

The viruses can be divided into two main groups on the basis of size: those around 130 nm in diameter and those approximately 180 nm in diameter. The *Chironomus plumosus* virus (Stoltz, 1971) is intermediate at 165–170 nm in diameter. *Tipula* iridescent virus (TIV) (type I) has been shown to be icosahedral by a double shadow-casting method (Williams and Smith, 1958), and others can be observed in negatively stained preparations on each of the three axes of symmetry, 2, 3, and 5, from which it may be inferred that they have icosahedral symmetry. TIV (type I) was originally reported as having 812 surface subunits (Smith and Hills, 1962), but Wrigley (1969, 1970) showed that a detailed substructure could be revealed on both TIV and *Sericesthis* iridescent virus (SIV) (type 2) but not on *Chilo* iridescent virus (CIV) (type 6) after treatment with a nasal decongestant ("Afrin"). Prolonged storage of virus in the cold led to dissociation into triangular, pentagonal, and linear fragments made up of subunits of the same size as those visible on the treated intact virus particles. From his micrographs and by use of the Goldberg diagram, Wrigley was able to postulate that SIV (type 2) probably had 1562 morphological subunits, although 1292 and 1472 could not be excluded, and that TIV (type 1) probably had 1472 subunits, although 1292 and 1562 could not be excluded. Stoltz (1971, 1973) found similar structural arrangements in the *Chironomus plumosus* virus and in TIV using pronase treatment, and in the "T" strain of an iridescent virus infecting the mosquito *Aedes taeniorhyncus* (type 3). In this case, treatment with "Afrin" gave negative results. At least 1560 morphological subunits were proposed for the *Chironomus* virus. Fibrils were observed attached to the intact *Chironomus* virus particle and to the trisymmetrons produced by dissociation. Fibrils or a peripheral fringe of projections has been observed in other iridescent viruses such as TMIV (Stoltz, 1973; Wrigley, quoted by Stoltz, 1973). Willison and Cocking (1972) also reported numerous fibrous projections on TIV in freeze-fractured preparations. Such structures might account for the large interparticle spacing found in quasicrystalline arrays of these viruses.

The position of a lipid layer or lipid-containing structure in iridescent viruses is a matter of dispute. Reports of small amounts of

lipid (around 5% or less) in these viruses were originally dismissed as
being due to host cell contamination (Thomas, 1961; Glitz et al., 1968;
Bellett, 1968; Kalmakoff and Tremaine, 1968). However, Kelly and
Vance (1973) extracted lipid from purified SIV (type 2) and CIV (type
6) and analyzed it by thin-layer chromatography. The lipid pattern was
very different from that of the host insect, Galleria mellonella, and
sphingomyelin, a lipid in which plasma membrane is usually quite rich,
was only a minor lipid component of the viruses. They found that the
virus particles contained 9% by weight of lipid, equal to 59×10^6
daltons per virus particle, and calculated that there was enough lipid to
form a continuous bilayer at an internal site within the virus particle.
Such a site would be at a 106-nm-diameter position as compared with
an external diameter position of 130 nm. They claimed that the position
of the lipid bilayer was discernible at a 106-nm-diameter position in
sectioned and negatively stained virus preparations. Wagner et al.
(1973) found only $3.9 \pm 1\%$ and $3.4 \pm 1\%$ lipid, respectively, in each of
the two strains ("Regular" and "Turquoise") of the mosquito (Aedes
taeniorhyncus) iridescent virus (type 3) designated as RMIV and TMIV
(Matta and Lowe, 1970). Both of these viruses are in the larger size
category, with diameters in section of 187 nm and 181 nm, respectively.
Despite this, Stoltz (1971), in a technically excellent electron microscope study, demonstrated two closely appressed trilaminar membranes
in section at the periphery of the mature intact RMIV particle. When
the virus particles were treated with excreted digestive enzymes from
first instar larvae, the outer membrane retained its original angular
contour, whereas the inner membrane was apparently more flexible and
associated with the inner nucleoprotein core. The inner membrane was
also clearly resolved as a trilaminar structure in TIV and Chironomus
virus but the outer membrane was more electron dense. The hypothesis
was proposed that the virus particle shell was a unit membrane
modified by an icosahedral lattice of morphological subunits in the
manner also described by Wrigley (1969, 1970). The virus core or nucleoid (as defined by Stoltz) was surrounded by a second internal
membrane. Solvent resistance as reported for similar virus particles
(Day and Mercer, 1964) could be explained by nonaccessibility of the
membranes. However, in a later paper Stoltz (1973) virtually retracted
this hypothesis in favor of a different model. Working largely with "top
component" preparations of TMIV (which yields a greater proportion
of top component than RMIV) but also with other isolates such as the
iridescent virus of Corethrella brakeleyi (type 13), he concluded that
the only consistent trilaminar structure observed is internal in relation

to the shell and represents the core membrane. The outer trilaminar structure seen unequivocally only in MIV must represent sectioned outer morphological subunits which for some reason are less electron dense in MIV. Thus there now seems to be agreement that the lipid-containing structure in these viruses is internally located somewhere within the virus shell, possibly surrounding the virus core. Conclusive evidence could no doubt be provided by X-ray diffraction studies such as those undertaken with bacteriophage PM2 and Sindbis virus (Harrison *et al.*, 1971*a,b*).

Iridescent viruses of the smaller size group with diameters around 130 nm have sedimentation coefficients of around 2200 S and buoyant densities of the order of 1.33 g/ml. TIV top component has a sedimentation coefficient of 1400 S. Diffusion coefficients of 1.03×10^{-8} cm²/sec and 1.6×10^{-8} cm²/sec have been reported for TIV. Particle weights for both TIV and SIV have been calculated in independent reports at $11.9–13.0 \times 10^8$ and 5.51×10^8 (Day and Mercer, 1964; Glitz *et al.*, 1968; Kalmakoff and Tremaine, 1968). The larger group of iridescent viruses of diameter around 180 nm such as the mosquito iridescent viruses (type 3, R and T strains) have reported sedimentation coefficients of 4458 S and 4041 S (R strain) and 3318 S (T strain), and buoyant densities in cesium chloride of 1.354 g/ml and 1.319 g/ml (R strain) and 1.311 g/ml (T strain). Particle weights have been calculated at 2.486×10^9 and 2.75×10^9 for RMIV and 2.10×10^9 for TMIV (Matta, 1970; Wagner *et al.*, 1973). Estimates of the DNA content of the iridescent viruses vary between 10.5% and 19% depending on individual reports and the virus investigated (Bellett, 1968; Kalmakoff and Tremaine, 1968; Faust *et al.*, 1968; Matta, 1970; Wagner *et al.*, 1973). The virus contains double-stranded DNA which appears to be linear. It ranges in molecular weight from 114 to 160×10^6 in the smaller size group of iridescent viruses. Of the larger size group, one virus, MIV (type 3), has a reported viral DNA molecular weight of 4.64×10^8 for the R strain and 2.71×10^8 for the T strain based on protein–DNA ratio data—a somewhat inexact method. The base composition of the different viral DNAs ranges from 28–29% G+C to 41% G+C (Bellett and Inman, 1967; Bellett, 1968; Kelly and Robertson, 1973; Wagner *et al.*, 1973; Kelly and Avery, 1974*a*). Evidence of base sequence homology between members of the iridescent virus group was first demonstrated by Bellett and Fenner (1968) with TIV, SIV, and CIV (types 1, 2, and 6), although the SIV isolate used was found to be CIV at least in major proportion as a result of this work. Kelly and Avery (1974*a*) grew four iridescent viruses, SIV, CIV, *Wiseana cervinata* iridescent virus (WIV) (see Fig. 9), and *Opogonia* sp. iridescent virus (types 2, 6, 9, and 18) in *Galleria mellonella* larvae, and using fragmented DNA found no common sequences between CIV and any of the other three viruses but complete homology between WIV and the *Opogonia* iridescent virus and 45% of WIV homologous to SIV. In reciprocal experiments, 25% of SIV was common to WIV and the *Opogonia* virus. SIV and CIV contained repeated sequences in the genome.

Tryptic peptide analyses of SIV and TIV are similar; the virus top components have a reduced arginine content. Matta (1970) found the amino acid analysis of RMIV to be similar to those of other iridescent viruses that have been investigated, and Wagner *et al.* (1974*a*) found similar amino acid analyses for the R and T strains of MIV with the exception of phenylalanine and serine. Analysis of the virus structural proteins by polyacrylamide gel electrophoresis has been reported for four iridescent viruses. Kelly and Tinsley (1972) found 20 and 19

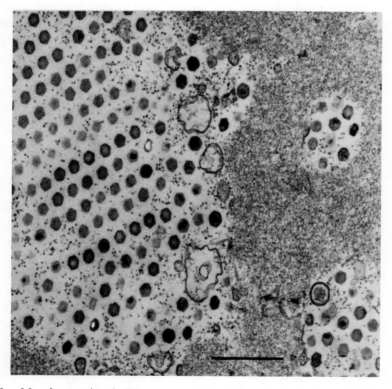

Fig. 9. Morphogenesis of *Wiseana cervinata* iridescent virus (type 9). Bar equals 1 μm.

polypeptides, respectively in SIV and CIV, ranging in molecular weight from 213,000 to 10,000 or 11,000. The total molecular weight of the proteins added up to around 2×10^6 for both viruses, and they calculated that about 30% of the genome directs the synthesis of these viral structural proteins. Krell and Lee (1974) identified as many as 28 structural polypeptides in TIV, ranging in molecular weight from 17,500 to 300,000. The smallest polypeptide was identified as a lipoprotein by Sudan black B and Oil red O staining. Similar analysis of virus top component showed the absence of two of the polypeptides and a reduction in three others. The authors calculated the total molecular weight of the proteins to be 1.72×10^6, which would require 25% of the viral genome for their synthesis—a figure in reasonable agreement with that for SIV and CIV. Wagner *et al.* (1974a) examined the proteins of the R and T strains of MIV (type 3) using similar techniques and found only nine bands, ranging in molecular weight from 16,000 to 98,000. The profiles were consistent for gels of different concentration, for different concentrations of SDS and 2-mercaptoethanol, and for different sample volumes. According to Kelly and Robertson (1973), further comparative work has shown that those viruses which are serologically related differ in their profiles of structural proteins on polyacrylamide gels less than those that are unrelated serologically.

A complete study of the serological interrelationships of iridescent viruses has not been made and the present situation is rather confused. Cunningham and Tinsley (1968) found TIV and SIV to be serologically related in complement fixation tests, tube precipitation tests, and immunodiffusion tests. They found TIV grown in three species of Lepidoptera to be identical, although distinct from but related to the virus isolated from *Tipula paludosa* larvae, and they speculated that host antigens were incorporated into the virus particle. No serological relationship with MIV was found in complement fixation tests. Bellett (1968) records CIV as being related to TIV and SIV, but this is almost certainly due to the fact that the "Canberra" stock of SIV contains CIV (Bellett and Fenner, 1968). Day and Mercer (1964) recorded TIV and SIV as not being identical serologically, and Glitz *et al.* (1968) found quantitative differences between the two viruses in precipitin reactions. Fowler and Robertson (1972) reported WIV and *Witlesia sabulosella* iridescent virus (types 9 and 10) as possessing at least one antigen in common and being related to TIV. Kalmakoff *et al.* (1972) considered WIV (type 9) to be closely related to TIV, SIV, and *Costelytra zealandica* iridescent virus (type 16), and CIV to be serologically distinct. However, the *C. zealandica* virus antiserum was specific to homologous reactions only and in this respect appeared to be serologically distinct. Cross-reaction between this virus, TIV, SIV, and WIV could be explained by postulating that WIV is a mixture of all three. From these and other unpublished observations quoted by Kelly and Robertson (1973), the situation can be summarized as follows: types 1, 2, 9, 10, 16, 17, 18, and 19 have antigens in common but are not identical; types 3 and 12 are serologically related but do not have antigens in common with types 1, 2, 6, 9, and 10; type 6 is not serologically related to types 1, 2, 3, 9, 10, 12, and 16.

SIV and CIV (types 2 and 6) possess RNA polymerase activity (Kelly and Tinsley, 1973), and preliminary data also suggest the presence of triphosphatase activity (Kelly and Robertson, 1973).

The apparent morphological similarity of insect iridescent viruses is misleading, and there is undoubtedly real diversity in their properties. Whether this diversity is sufficiently marked to jeopardize their grouping cannot be resolved at this time. Possible interrelationships of these viruses with morphologically similar viruses isolated from a variety of other hosts and their inclusion in a "supergroup" of icosahedral cytoplasmic deoxyriboviruses have been carefully evaluated in a review (Kelly and Robertson, 1973). More recently, Kelly and Avery (1974b) have demonstrated that there is no sequence homology between frog virus 3 DNA and the DNA extracted from either SIV (type 2), CIV (type 6), or *Wiseana* iridescent virus (type 9).

2.5. Small Isometric Viruses

Some 10 years ago, comparatively few nonoccluded viruses smaller than iridescent viruses had been recorded in invertebrates. There are now a large number of reports of such viruses, and it is likely that many more will be made in the future when one considers the large number of invertebrate species. Only a few of these viruses have been characterized in any detail; on the others, only the barest details can be given.

One of the best known viruses of this group is the densonucleosis virus (DNV) of the wax moth, *Galleria mellonella,* first described by Meynadier et al. (1964) (see Fig. 10). The virus causes a fatal disease, killing larvae in 4–6 days at 28°C. It was shown by Truffaut et al. (1967) to contain DNA, apparently of a double-stranded configuration. Longworth et al. (1968) purified the virus by organic solvent extraction and sucrose gradient centrifugation and found the proportion of virus

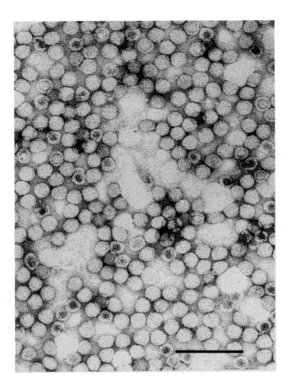

Fig. 10. The virus of *Galleria mellonella*. Bar equals 100 nm.

to top component in their preparations to be 1:3, with sedimentation coefficients of 119 S and 58 S, respectively. The serological properties of the virus and the top component were identical in immunodiffusion tests, and their amino acid compositions were similar. They confirmed that the virus contained DNA, and they found a DNA–protein ratio of 37:63. Barwise and Walker (1970) examined the structure of the DNA. When isolated in solutions of high ionic strength or by thermal denaturation, it had a sedimentation coefficient of 16 S and a molecular weight of 4×10^6 and the characteristics of a double-stranded molecule. However, the particle weight of the virus was determined as 5.7×10^6 and that of the top component as 3.5×10^6, leaving a theoretical molecular weight for the DNA of 2.2×10^6. Further investigation showed that the DNA in the virus was, at least in part, single stranded, and it was postulated therefore that the individual DNA molecules from different virus particles were complementary, or, in other words, the + and − strands were separately encapsidated and annealed on extraction to give a double-stranded structure. This result was confirmed by Kurstak *et al.* (1971), who obtained a comparable DNA molecular weight of 1.6×10^6. A similar situation has been reported with adeno-associated viruses, and indeed DNV shares a number of other properties such as size, heat stability, solvent resistance, and buoyant density in cesium chloride (1.44 g/ml) with members of the parvoviruses as a whole (Tinsley and Longworth, 1973). DNV and another virus with similar properties isolated from the lepidopteran *Junonia coenia* (Rivers and Longworth, 1972) show four structural polypeptides on 10% polyacrylamide gel with molecular weights of 72,000, 57,000, 53,000, and 46,000, whereas only three polypeptides can be resolved on 5% gels, a result analogous to those with other parvoviruses (Tinsley and Longworth, 1973). Kurstak and Côté (1969) suggested, however, that DNV particles had 42 capsomers and that a new family should therefore be erected for the group, as the most likely number of capsomers for parvoviruses is 32. However, DNV has many properties in common with other parvoviruses, and therefore we believe that a possible discrepancy of capsomer number should not be given undue weight when one considers the difficulties of negative staining techniques with so small a virus. DNV and *Junonia coenia* virus are closely related serologically, although they differ in host range and histopathology. Hoggan (1971) could not demonstrate any serological relationship between DNV and other parvoviruses.

Longworth and Harrap (1968) isolated isometric virus particles 37–42 nm in diameter from four species of saturniid moth larvae with flaccidity and diarrhea. One of the isolates was shown to contain DNA and have a sedimentation coefficient of 200 S. Recently however reexamination of one of these isolates indicated that it contained RNA.

There are several isometric RNA viruses of invertebrates. Such viruses have been isolated from three species of Lepidoptera, two of which are saturniids, *Antherea eucalypti* (Grace and Mercer, 1965), (see Fig. 11) and *Nudaurelia cytherea capensis* (Tripconey, 1970), and the other a lasiocampid, *Gonometa podocarpi* (Harrap *et al.,* 1966) (see Fig. 12). The *Antherea* virus was originally stated to be 50 nm in diameter with cores 30 nm in diameter, but in later work two sizes of particles were found of 32 nm and 14 nm diameter, the smaller possibly being a type of "associated" virus. RNA was extracted from preparations containing both sizes of particle and was found to be infective either by feeding to or injection into larvae (Brzostowski and Grace, 1970). *Gonometa* virus was purified by Longworth *et al.* (1973a). The virus is 32 nm in diameter, has a sedimentation coefficient of 180 S, has a buoyant density in cesium chloride of 1.35 g/ml, and contains 37% RNA. There are four principal polypeptides of molecular weight 36,500, 32,000, 29,000, and 12,000, and a minor one of 47,500.

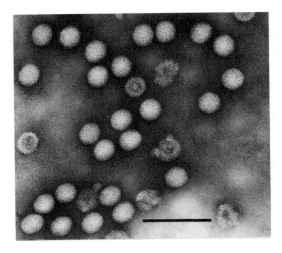

Fig. 11. The virus of *Antherea eucalypti*. Bar equals 100 nm.

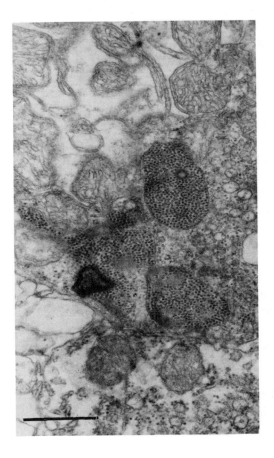

Fig. 12. *Gonometa podocarpi* virus in a section of a gut cell. Bar equals 1 μm.

Naturally occurring antibodies to *Gonometa* virus can be found in the sera of cattle, horses, sheep, dogs, and deer (Longworth *et al.*, 1973*b*). Tripconey (1970) purified the *Nudaurelia* virus and adapted it to grow in three other species and in a cell culture. The virus has a diameter of 35 nm, a sedimentation coefficient of 210 S, and a buoyant density of 1.298 g/ml. Polson *et al.* (1970) calculated the particle weight to be 16.3×10^6 using an electron microscope counting technique with hemocyanin as a standard and calculations from sedimentation and diffusion coefficients. Struthers and Hendry (1974) found the molecular weight of the coat protein to be between 60,000 and 62,000. Only a single polypeptide could be isolated on polyacrylamide gels despite prolonged treatment with SDS. In the analytical ultracentrifuge the protein had a sedimentation coefficient of 14 S, but there was some heterogeneity in the profile which was difficult to reconcile with the polyacrylamide gel result. The virus contained 11% RNA which was single stranded and had a base composition of 28% G, 23% U, 25% C, and 24% A and a molecular weight of 1.8×10^6. The virus was relatively stable, resisting treatment with mild alkali, cold and warm salt, and 67% acetic acid. Formic acid treatment was used to dissociate the virus and prepare virus protein. It was proposed that the capsid had 234–241 subunits organized into 42 capsomers. Finch *et al.* (1974) confirmed this value in three-dimensional reconstruction work from electron micrographs. The 240 protein subunits are clustered in trimers with four trimers per icosahedral face. The virus is the first record of a $T = 4$ type. *Nudaurelia* β virus, as it has been called more recently, is one of five viruses isolated from this species.

At least two isometric viruses have been isolated from *Drosophila* sp. P virus (Plus and Duthoit, 1969) is 25 nm in diameter, and ι (iota) virus (Jousset, 1970) is 30 nm in diameter. The isolation of the viruses has been described by Plus *et al.* (1972). P virus causes sterility in female flies and a 50% reduction in life span. Signs of infection are suggestive of malfunctioning of the Malpighian tubules, and virus can be found both there and in the cytoplasm of ovarian and gut cells. The fact that the virus contains RNA was determined using an electron microscope staining technique (Teninges and Plus, 1972).

The subject of isometric viruses isolated from mites is confusing. Steinhaus (1959) described a possible virus disease of the European red mite, *Panonychus ulmi,* thought to be caused by a spherical virus, but Bird (1967) showed the virus to be rod shaped and to have a morphology similar to the virus of *Oryctes rhinoceros* (see Section 2.1). An isometric virus was first reported in the citrus red mite, *Panonychus citri,* by Smith *et al.* (1959). Diarrhea and paralysis of the legs occurred in diseased individuals, and a spherical virus 35 nm in diameter was isolated. Estes and Faust (1965) claimed that the virus contained RNA, but their purification of the virus was only rudimentary. Reed *et al.* (1972) found birefringent crystals associated with the disease, but transmission of virus to other mites could be accomplished from mites in which no crystals were present, so, although they were useful in diagnosis, the association remained equivocal. However, Reed and Hall (1972) found that diseased citrus red mites contained rod-shaped particles in the midgut epithelium similar to those reported by Bird (1967) in the European red mite. No such particles were found in healthy mites so it seems that the disease could be due to a virus of the *Oryctes* type. Two types of spherical viruslike particles, one 24–25 nm and one 34–35 nm, were also detected in this work, but they could be found in both healthy and diseased mites and did not produce disease symptoms.

Juckes *et al.* (1973) compared serologically in immunodiffusion tests *Antherea* virus, *Gonometa* virus, a *Panonychus* virus, and two *Nudaurelia* viruses β and ε. The only cross-reaction was between the two saturniid viruses *Antherea* and *Nudaurelia* β.

Isometric viruses have been isolated from three species of Orthoptera. Meynadier (1966) isolated an isometric virus 34 nm in diameter from the cricket *Gryllus bimaculatus.* Reinganum *et al.* (1970) isolated a virus from the field crickets *Teleogryllus oceanicus* and *Teleogryllus commodus* which caused rapid and fatal paralysis in the early nymphs. The virus particles were purified using organic solvent extraction and sucrose gradient centrifugation. They were 27 nm in diameter and contained RNA. Crystalline aggregates of virus particles were found in the cytoplasm of cells of the epidermis, alimentary tract, and nervous tissue.

Jutila *et al.* (1970) reported a virus which infected and killed several species of grasshopper. The virus was observed in paracrystalline array in the cytoplasm of muscle, tracheal epithelium, and pericardial cells. The virus is only 13 nm in diameter and is probably the smallest virus ever isolated. It was purified by sucrose gradient centrifugation and contains 18–22% RNA and has a sedimentation coefficient of 42 S.

Six viruses known to contain RNA are associated with honey bees. Chronic bee paralysis virus (CBPV) is not isometric but ellipsoidal and is included here only for convenience. CBPV is isolated from diseased bees with common "bee paralysis," and reinoculation of the virus into healthy bees produces similar symptoms (Bailey *et al.,* 1963). The virus can be purified from aqueous extracts of paralyzed bees using organic solvents and centrifugation or salt precipitation methods. Ellipsoidal virus particles of three sizes have been found, all 22 nm wide but 41 nm, 54 nm, and 64 nm long. They have sedimentation coefficients of 97 S, 110 S, and 125 S, respectively. The shortest particles were found to contain the least RNA, and preparations predominant in such particles were least infective. Determination of the amount of RNA in the virus particle was inconclusive, but the RNA was found to have a base composition of 20% G, 24% A, 28% C, and 28% U (Bailey *et al.*, 1968). Acute bee paralysis virus (ABPV) is an isometric virus 28 nm in diameter. As with CBPV when fed or injected the virus caused "trembling" within a few days, but bees infected with ABPV died quickly (Bailey *et al.*, 1963). Furgala and Lee (1966) could not find paralysis associated with ABPV. The virus is stable at pH 3–7, and has a sedimentation coefficient of 160 S and a buoyant density of 1.34 g/ml. It contains 30% RNA, which is single stranded, sediments at 30 S, and has a base composition of 30.3% A, 20.5% C, 18.8% G, and 30.4% U. The sedimentation behavior of the RNA in the presence of formaldehyde suggests that there are many hidden breaks in the chain (Newman *et al.*, 1973). Sacbrood virus is a similar although serologically unrelated virus which affects the larva of the bee causing failure to pupate after the larvae have been sealed in their brood comb cells by worker bees. Fluid accumulates between the body of the larva and the unshed skin and the larvae changes color from pearly white to dark brown on death. North American and European strains of sacbrood virus are identical. The virus particles are stable at pH 5–7, are 28 nm in diameter, and have a sedimentation coefficient of 157 S and a buoyant density of 1.33 g/ml. They contain single-stranded RNA which sediments at 35 S and has a base composition of 32.1% A, 17.9% C, 19.1% G, and 30.9% U. In the presence of formaldehyde the RNA sediments at 15–16 S, rather lower than RNA from mammalian picornaviruses. The molecular weight of the RNA is said to be in the range 2.1–2.6 × 10⁶ (Bailey *et al.*, 1964; Lee and Furgala, 1965; Newman *et al.*, 1973). Bailey and Woods (1974) described three further serologically unrelated RNA-containing isometric bee viruses: Arkansas bee virus (30 nm diameter, 128 S, 1.37 g/ml), bee virus X (35 nm, 187 S, 1.36 g/ml), and slow paralysis virus (30 nm, 176 S, 1.35 g/ml). Bee virus X was not in itself associated with signs of infection or mortality but killed bees when injected together with sacbrood virus.

Three possible paralysis viruses of another social insect, the termite, have been described by Gibbs *et al.* (1970). A virus about

30 nm in diameter was found in each of three different species. The nucleic acid type has still to be determined. The virus infection caused lack of coordination and death in 4–5 days. One isolate was found to have a sedimentation coefficient of 155 S. The viruses are serologically distinct from ABPV.

An intriguing virus, Nodamura virus, was isolated from the mosquito *Culex tritaeniorhyncus* in the village outside Tokyo from which it takes its name. It was originally thought to be a togavirus but is resistant to organic solvents. It has an unusually wide host range originally being shown to multiply in moth larvae (*Plodia interpunctella*), ticks (*Ornithodorus savigni*), and mosquitoes (*Culex tarsalis*), and possibly a beetle, *Attagenus piceus* (Scherer and Hurlbut, 1967). Pigs in the same locality were found to have antibodies to the virus. Murphy *et al.* (1970) showed that the virus would multiply in infant mice, where paracrystalline arrays of particles could be found in limb muscle, causing severe structural disruption of the tissue. Kupffer liver cells also contained large aggregates of virus. The virus particles were shown to be acid-stable, 28–29 nm in diameter, and of buoyant density 1.34 g/ml, and therefore characteristic of a picornavirus, although multiplying in and transmitted by arthropods. Bailey and Scott (1973) found that homogenized infected mouse brain could be used to transmit the virus by injection to adult bees and wax moth (*Galleria mellonella*) larvae. The moth larvae died mainly as pupae, but the bees showed paralysis of the anterior two pairs of legs shortly before death. Harvested virus reacted strongly with homologous antiserum but not with ABPV antiserum. Newman and Brown (1973) purified the virus from infected mouse muscle using a trituration and salt precipitation procedure followed by sucrose gradient centrifugation in the presence of SDS. They achieved a recovery of infectious virus of the order of 10^7 LD_{50}/mg muscle tissue. They confirmed the size of the virus as 29 nm and found that it sedimented at 135 S. Two species of RNA were isolated, sedimenting at 22 S and 15 S, with molecular weights of 1.0×10^6 and 0.5×10^6, respectively. The base compositions of the two RNA species differed. Each RNA had low infectivity, but this could be enhanced a hundredfold by mixing the two. One major polypeptide of molecular weight 35,000 and two minor ones were resolved on polyacrylamide gel. The buoyant density of freshly purified virus was confirmed as 1.34 g/ml, but purified virus stored at 4°C for several days showed most radioactivity in peaks corresponding to 1.50 g/ml and 1.60 g/ml when centrifuged to equilibrium in cesium chloride. It seemed then that protein was detaching from the virus particle. Particles banding at 1.50 g/ml were found to contain 22 S RNA, and those banding at 1.60 g/ml contained 15 S RNA. It appears therefore that the two RNA species are separately encapsidated, with some particles containing one molecule of 22 S RNA and others containing two molecules of 15 S RNA. Bailey *et al.* (1975) have shown that Noda-mura virus will multiply in the mosquito cell lines of *Aedes aegypti* and *Aedes albopictus* and in the hamster BHK cell line. Virus yields were determined by infectivity assay in adult bees and wax moth larvae.

Wagner *et al.* (1974*b*) isolated an isometric virus from the mosquito *Aedes taeniorhyncus* and found that it contained 20% RNA. The virus particle was around 30 nm in diameter and was resistant to ether. Virus could be purified from apparently healthy larvae, but infection with an iridescent virus was enhanced in terms of yield when the picornavirus also was present in the host.

Just as two of the small DNA-containing viruses of invertebrates can be placed in the parvovirus group on the basis of their biophysical and biochemical properties, so it seems that several of the RNA-containing viruses of invertebrates can be thought of as picornaviruses.* It is for this reason that a number of the RNA viruses described above were included in a review of picornaviruses by Brown and Hull (1973).

2.6. Sigma Virus of *Drosophila*

Sigma (σ) virus is a hereditary infectious agent often present in populations of *Drosophila* spp. which can be detected by the sensitivity of the infected flies to carbon dioxide. Flies exposed to an atmosphere rich in carbon dioxide become anesthetized but in normal air will recover and are then quite normal in behavior. Infected flies reawaken but remain paralyzed and die after a few hours. Berkaloff *et al.* (1965) demonstrated a rhabdoviruslike particle in sections of tissues of infected flies and in negative stain preparation which they considered could probably be the σ agent. Teninges (1968) found similar particles associated with germinal cells, and particles were observed budding from the cell surface. The particles measured 140–180 × 70nm, and a cross-striation was visible in negatively stained preparations. Printz (1967) found that another rhabdovirus, vesicular stomatitis virus (VSV), would grow in *D. melanogaster,* resulting in the selection of a variant which would grow in the insect and induce carbon dioxide

* *Editor's comment:* Others appear to resemble plant coviruses (see Volume 11 of *Comprehensive Virology*).

sensitivity. Ohanessian and Echalier (1967) infected embryonic cells of *Drosophila* maintained in microdrop culture with σ virus. No cytopathic effect was observed, but the culture medium was assayed for infectious virus on populations of flies by end-point titration using carbon dioxide sensitivity, and multiplication was demonstrated. Subsequently, multiplication of σ virus has been demonstrated in continuous cultures of *Drosophila* cells (Ohanessian, 1971). Bussereau (1970*a,b*) found that carbon dioxide sensitivity appeared more rapidly if σ virus preparations were injected intrathoracically rather than abdominally and that such an injection route resulted in rapid increase of virus in nerve ganglia, although maximal yields in digestive and genital organs were obtained more slowly. It was concluded that the site of injection determined the manner of virus multiplication and that carbon dioxide sensitivity was due to nervous center infection. The σ virus has not yet been purified, and we know from experience in our own laboratory that this is very difficult to achieve. Nevertheless, it seems clear that evidence exists that there is an agent which multiples, is transmissible, and will cause sensitivity to carbon dioxide in fruit flies, that the route of administration of the agent affects the signs of infection or the speed with which they are acquired, that these signs of infection can be inherited, and that rhabdoviruslike particles can be observed in infected flies in tissues which include genital organs. All these features have been reviewed by Printz (1973). However, that the particles in themselves are responsible for the manifestation of carbon dioxide sensitivity cannot be proved unequivocally without purification of the virus. This fact still arouses doubts as to whether σ is a virus at all.

2.7. Other Possible Viruses of Invertebrates

Many other viruses and viruslike particles have been reported in invertebrates. Little is known of the properties of most of them, and the reports are largely descriptions of disease conditions and suspect viruslike causal agents visible in tissue section (see Table 1). However, Richardson *et al.* (1974) have produced evidence of two inapparent virus infections in *Culex tarsalis*. Two sizes of particles were observed, of about 56 nm and 35 nm, and the experimental results strongly suggested that both viruses replicated on serial passage. There are also a number of reports of similar particles in *Drosophila* sp. (Akai *et al.*, 1967; Philpott *et al.*, 1969, Felluga *et al.*, 1971; Gartner, 1972; Tandler, 1972).

3. THE MECHANISM AND PATHWAYS OF INFECTION

The means by which viruses infect invertebrates are not fully understood. Only recently have the early stages and pathway of infection become known and then largely for one group of viruses, the nuclear polyhedrosis and granulosis viruses (baculoviruses). This is a fruitful area of study for those interested in virus specificity and host susceptibility, and it is to be hoped that the recent progress with virus growth in insect cell cultures will add impetus to work in this area. The scope for novel comparative work on virus–cell interaction in the natural animal host and in cultured cells is large and deserves to be exploited. Virus infection in invertebrates has been reviewed recently by Harrap (1973).

3.1. Routes of Entry

Virus may be acquired either by mouth, transovarially, or as a result of injury, e.g., through parasitism, and the larval stages are the most susceptible to infection. Oral infection is the most common by far of these three modes of acquisition. Jaques (1962) considers the most usual source of infection of occluded insect viruses to be the contamination of foliage with the inclusion bodies from insect cadavers. In addition, healthy larvae are often attracted to infected cadavers and become infected by feeding on them. It is known that dispersal of these viruses is usually assisted by wind and rain (Bird, 1961b; Vago and Bergoin, 1968). Other viruses are probably acquired by their host in a similar way, although their dispersal can be influenced by a number of factors such as movement of healthy carriers and infected hosts, the behavior of predators such as birds, and environmental factors such as population density, food availability, temperature, and the incidence of other pathogens (see Section 5).

There are many reports of transovarial transmission of insect viruses. The subject has been critically reviewed by Longworth (1973). He concluded that unequivocal transovarial transmission of a virus in insects has been demonstrated in two cases only. This is because in many reports the real possibility of contamination of the egg surface by either meconium or feces has not been excluded. A prerequisite of transovarial transmission is penetration of the oocytes by virus, presumably as a result of active ovariole infection. It is difficult to see how this can occur in exclusively enteric infections such as cytoplasmic polyhedroses. Transovum transmission or surface contamination of the egg by virus accounting for infection of the emerging larvae has been demonstrated many times. A good example is provided by the work of Hamm and Young (1974).

Parasites have been shown to transmit virus from larva to larva. David (1965) demonstrated the transmission of granulosis viruses from infected to healthy larvae of *Pieris brassicae* by the hymenopterous parasite *Apanteles glomeratus*. It is not known how common virus acquisition is as a result of parasite activity in natural conditions.

3.2. Attachment and Early Infection

The common route of virus entry is oral, and it is reasonable to suppose therefore that virus attachment and cell invasion will occur initially in the gut. However, there are few recorded attempts to establish how this initial infection proceeds. There are reports of virus particles in association with midgut cell microvilli (Stoltz and Hilsenhoff, 1969), but in general the early events of virus association with gut cells have been disregarded. Indeed, the histopathology of the gut has been studied usually only in those virus infections where signs of gut infection are obvious, such as cytoplasmic polyhedroses, nuclear polyhedroses of *Hymenoptera,* and some picornaviruses.

Stoltz and Summers (1971) tried to follow the pathway of infection of an iridescent virus in second and third instar *Aedes taeniorhyncus* larvae. The efficiency of infection was low despite the large amounts of virus ingested, and they found that most if not all of the virus particles observed in electron microscope thin sections were degraded shortly after entering the midgut. The virus was unable to penetrate the peritrophic membrane, and the virus particles therefore could not be found in contact with the gut cells. This contrasts markedly with the situation in Lepidoptera, for example, where the virus probably penetrates the peritrophic membrane through discontinuities within it. These do not seem to occur in mosquitoes. We have noted elsewhere in this chapter that the natural route of infection of iridescent viruses may not be an oral one, although mosquito larvae are usually infected in this way in the laboratory.

In nuclear polyhedroses of Lepidoptera, virus is usually acquired orally, and, in time, large quantities of virus inclusion bodies accumulate in many body tissues. The gut of the larva appears to be unaffected by this otherwise acute and systemic infection despite its implication as the route of entry. This apparent paradox was investigated by Harrap and Robertson (1968) and Harrap (1969, 1970), and they demonstrated that nuclear polyhedrosis virus in the larvae of the small tortoiseshell butterfly, *Aglais urticae,* infected midgut columnar cells and multiplied in an unusual way. These findings were confirmed and extended by others (Summers, 1969, 1971; Tanada and Leutenegger, 1970; Kawanishi *et al.* 1972*a*) in GV and other NPV infections so that it was possible to propose a pathway of infection for baculoviruses in Lepidoptera which had some experimental support.

It is clear that polyhedra and granulosis virus capsules must dissociate in the gut lumen. Faust and Adams (1966) have suggested that gut secretions solubilize the inclusion bodies primarily by a pH effect, together with some enzymatic degradation. The released, enveloped virus particles (or bundles of virus particles—see Section 2.1) can be found in association with columnar cell microvilli. The virus envelope apparently interacts and eventually fuses with the plasma membrane of the microvillus so that the now naked virus particle (or nucleocapsid) is within the cell cytoplasm. Such naked virus particles can be found within microvilli early in the infection of the host. A similar situation has been reported by Granados (1973) with an insect poxvirus. The virus envelope fuses with the plasma membrane of the microvillus at or near the tip and the core structure, and lateral bodies are released from the virion and "migrate" along the microvillus toward the body of the cell. The microvillus is usually enlarged in the region of the virus core. In granulosis virus infection, naked nucleocapsids have been found in an end-on association with nuclear pores (Summers, 1969, 1971). Full, empty, and partly empty nucleocapsids were observed, which suggested the explanation that the DNA content of the nucleocapsid is injected into the nucleus through the nuclear pore. Similar findings have been reported in nuclear polyhedrosis virus infections of a line of cells in culture (Raghow and Grace, 1974). Viral morphogenesis in the nuclei of the midgut columnar cell is atypical (*cf.* Section 4.1). Histopathological effects are less dramatic than in other tissues, and the infected cells cannot be detected by light microscopy. The region of infected cells is quite localized, and this also complicates their detection. Both naked and enveloped virus particles can be seen in infected columnar cell nuclei, together with small masses of what is apparently polymerized inclusion body protein as it possesses the characteristic lattice appearance and dimensions. Vesicular structures, electron-dense spherical bodies, and granular nucleoplasm are commonly present in such infected nuclei, but an obvious network of viroplasm is absent. The nuclear membrane is usually found to be ruptured, sometimes so extensively that it is difficult to find. However, in Lepidoptera the most striking difference between viral morphogenesis in columnar cells and in cells of other tissues is the fact that virus particles are not occluded by polymerizing polyhedron protein and true inclusion bodies do not form. Virus particles can also be found in the cell cytoplasm in large numbers or more specifically in that region of the cell cytoplasm between the nucleus and the basal lamina on which the epithelial cells are situated. It is not clear how the virus particles are confined to one region of the cell cytoplasm, although they may move along some type of gradient existing within the cells, perhaps as a result of their likely absorptive function. This region of the columnar cell cytoplasm is very convoluted, however, and it is not easy to decide whether the enveloped virus particles seen there are inside or outside the cell membrane. In either case, clear indications of how virus particles leave cells have not been obtained, although these particles can be found obviously outside the cell, within the matrix of the basal lamina, and adjacent to the tracheal epithelium cells. Early stages of viral morphogenesis can be found in some of these tracheal epithelium cells, and sections of tissue taken at later stages of infection show such cells in an advanced state of infection, having within them inclusion bodies containing many virus particles.

Despite the dangers inherent in this type of electron microscopic investigation, it seems reasonable to regard the above events as representing a pathway of infection, especially as virus particles cannot be found in other tissues of the insect larva at the same time as the events in the gut are observed. Virus particles that do not invade through the columnar cell microvilli are probably destroyed in the gut lumen. The localized nature of the infected area may be a reflection of the lability of the virus particles in the gut lumen. On the other hand, it could equally well occur because of the activity of some host defense mechanism (see Section 3.3). Lack of development of virus-containing inclusion bodies means that all infective virus produced in gut columnar cells is potentially capable of infecting cells in that same host. However, in other tissues, a large proportion of the infective virus particles are bound into inclusion bodies and hence effectively deprived of a further infective function in that host, being preserved for future infection of other host individuals. The advantage of precluding this feature of the virus morphogenesis in the primary invasion site is obvious. The infected midgut columnar cell can be regarded as a primary site responsible for furthering the early stages of infection in a given host. Whether or not this function continues throughout the course of the disease is not known. If it does, then the number of infected gut cells and the level of virus replication within them may determine the progress of the disease. Otherwise, disease progression must be dependent on those virus particles which fail to become incorporated into inclusion bodies during normal morphogenesis in such susceptible tissues as fat body, hypodermis, and hemocytes. It is not known how such virus particles are released from cells, although it may result simply from cell lysis which occurs late in infection. It should be noted, however, that Robertson *et al.* (1974) observed synhymenosis or "budding" of naked GV particles through the gut cell membrane, so it is possible that more than one method of virus release and more than one method of envelope acquisition may occur.

The function of the virus envelope is not yet resolved. Recent work in this laboratory seems to indicate that it is the site of group-specific antigens. In the insect host, it is clear that the envelope has a function in the infection of gut cells. Kawarabata (1974) has shown that in the hemolymph of silkworms infected with NPV the highly infectious fractions from density gradients contained naked virus particles. It is easy to imagine that the active hemocytes would readily acquire virus particles by viropexis and that possible envelope attachment sites may be superfluous. Certainly with insect poxviruses, viropexis by hemocytes of purified virus particles injected into the hemocoel is readily detected, and engulfed virus particles can be found in vesicles. The outer coat of the virus is lost, the core is released into the cytoplasm by disruption of the vesicle, and larval infection results. This contrasts with the apparently functional role of the virus envelope in fusing with the gut epithelial cell microvillar membrane in oral routes of entry (Granados, 1973).

The nature of the infectious agent in NPV-infected cell cultures is in dispute. Henderson *et al.* (1974), working with *Trichoplusia ni* NPV in a *T. ni* cell line, found that the infective agent sedimented at 830 S and had a buoyant density of 1.26–1.35 g/ml. After treatment with deoxycholate or Tween-ether, the values were 325 S and 1.42 g/ml respectively. They concluded therefore that they were dealing with a fragile enveloped particle. Raghow and Grace (1974), in an electron microscopic study of *B. mori* NPV in cell culture, found that all the virus particles in association with cells very early in infection were enveloped. No naked virus particles were seen penetrating the plasma membrane. However, virus entry did appear to be one of active invagination (viropexis), presumably of those virus particles that had attached to the cell. Knudson and Tinsley (1974) using ³H-labeled

Spodoptera frugiperda NPV in a *S. frugiperda* cell line, found that infectious virus not incorporated into polyhedra (unbound) formed, upon harvesting from the cell cultures, a homogeneous peak in a sucrose gradient. In contrast, polyhedra harvested from infected cell cultures and treated with alkali to liberate the enclosed virus particles gave a heterogeneous profile of several peaks, reflecting the different numbers of virus particles enveloped, as would be expected with this "multiply enveloped" type of NPV. Because the normal sucrose gradient profile of enveloped virus particles of an NPV of this type is one of several bands, the presence of a single band corresponding to infectious virus seems to indicate that a population of "unbundled" and therefore unenveloped virus particles was isolated, a situation that is analogous to that reported for insect hemocytes, as mentioned above. The banding density of the infectious virus also supports this interpretation. In a subsequent electron microscope study using the same virus–cell system, Knudson and Harrap (1976) demonstrated that both enveloped and naked virus particles could be found in association with the cells early in the infection cycle. This was so with both cell-culture-grown and insect-grown virus inoculum. If the mode of virus entry is viropexis in cell culture and hemocyte infection, it is conceivable that it may not matter whether the infecting virus is enveloped or naked; such a situation would contrast strongly with that of the host insect oral infection route.

Viropexis also seems to be the prime method of uptake of iridescent viruses by hemocytes (Leutenegger, 1967; Younghusband and Lee, 1969, 1970). The input virus partices then seem to become localized in lysosomelike structures. Viropexis has also been demonstrated in cell cultures infected with iridescent viruses, although subsequent association of the virus with lysosomes has not been found (Kelly and Tinsley, 1974a).

There are no factual reports of the early events of CPV infection. The "spike" of the CPV particle has been claimed to be a structure important for attachment, and it has been proposed that the virus core material is released through the hollow tubular structure, a part of which forms the "spike" after attachment (Hosaka, 1964; Nishimura and Hosaka, 1969).

In conclusion, mention should be made of those reports and postulations in the literature on NPVs and GVs suggesting subviral particles or free DNA as major factors involved in virus entry and early infection (Aizawa, 1967; Barefield and Stairs, 1969; Zherebtsova *et al.*, 1972). In the opinion of the authors, there is no compelling evidence that mechanisms involving such entities are real either from electron microscopic examination of high-speed centrifugation pellets or from dynamic infectivity assays in insect populations. The proposed mechanisms would have to be considered as either contrary to accepted theories of virus structure and function in the case of the "infective subviral particle" or highly improbable in the case of "infective free DNA," especially with a viral DNA of such high molecular weight. We mention them only for the sake of completeness; in our view, the weight of experimental evidence is against both concepts.

3.3. Possible Mechanisms of Resistance to Infection

The existence of some form of defensive reaction is clear from the undoubted success of insects, as shown by the enormous number of species, their countless individuals, and their ubiquity. The cellular defense reactions of insects are probably simpler than those of vertebrates but not necessarily in any way makeshift or less effective. Salt (1970) pointed out that insects could not have become so dominant if their defense reactions had been anything but fully effective. There is an obvious need to evaluate the nature and mode of action of these defensive reactions. Salt also commented that insects live in almost all terrestrial and freshwater habitats and consume almost every type of organic matter as food. It is thus likely that insects are exposed to a greater range of diseases than any other group of animals. Insects also possess certain structural and physiological qualities that could be useful in reducing the risks of infection. The exoskeleton, made up of chitin and tanned proteins, constitutes a chemically resistant covering for the softer and more vulnerable tissues of the body. Further, the chemical composition of the gut fluids is such that many microorganisms would be quickly destroyed after ingestion. However, such factors would not be adequate in themselves to protect insects against infectious virus taken in by mouth; therefore, the existence of additional systems must be considered.

The search for immune responses in insects comparable to the vertebrate antigen/antibody reactions has not met with any success (Bernheimer et al., 1952; Briggs, 1958; Stephens, 1959; Phillips, 1960; Kamon and Shulov, 1965; Stephens Chadwick, 1967a). Insects apparently do not produce vertebrate-type immunoglobulins, which is not too surprising as they do not possess systems analogous to the lymphatic or glandular organization of vertebrates. The phylum

Arthropoda represents a climax group, and, in the middle Paleozoic era, vertebrates and arthropods were already so widely divergent as to provide ample opportunity for the development of widely different but nonetheless effective systems of defense. The existence of systems in arthropods either not connected with or widely divergent from those present in vertebrates will demand new techniques for their elucidation and a new terminology for their description (Stephens Chadwick, 1967*b*; Salt, 1970).

As previously mentioned, the mouth is the significant portal of entry for viruses into insects and so initial defensive mechanisms could be expected at any point along the route from the mouth to the gut tissues. Aizawa (1962) investigated the inhibitory or antiviral properties of gut juices on the nuclear polyhedrosis virus of the silkworm, *Bombyx mori*. Untreated material and gut juices adjusted to pH 7.0 with buffer both reduced virus infectivity by about 50% after incubation at 28°C for 60 min, and Aizawa concluded that an antiviral substance existed in the gut juices. However, Aruga and Watanabe (1964) included a control treatment in their later experiments whereby virus was incubated with pH 10.0 buffer to simulate the pH of lepidopteran gut juices and only slight antiviral activity was recorded in the gut juice. Aruga and Watanabe suggested that gut juices constituted only a minor factor in the general system of resistance. Similarly, Rehaček (1965) found no evidence of digestive enzymes inhibiting or destroying the infectivity of arboviruses when the gut juices of ticks were incubated with virus for 30 min at 22°C. More precise experimentation is clearly needed to resolve these differing opinions, probably by incubating gut juices with purified virus preparations and using insect cell cultures as a test system.

The permeability of the peritrophic membrane has been suggested as a determinant in the penetration of virus particles. Chamberlain and Sudia (1961) thought, given two viruses of equal size, that equal rates of either retention or diffusion across the peritrophic membrane could be expected, but these authors drew attention to experience with arboviruses which provided no support for this idea. It is also possible that the nature of electrical charge on this membrane, when related to the charge of the virus particles, may have some relevance to permeability, but little information is available on this aspect.

The epithelial cells of the insect midgut seem to be the focal entry point for all virus groups associated with insects (Chamberlain and Sudia, 1961; Sinha, 1967; Harrap and Robertson, 1968; Harrap, 1970). Chamberlain and Sudia (1961) discussed the possibility of specific receptor sites on the midgut epithelial cells acting as determinants of virus attachment. However, the range of virus types capable of infecting insects would make it necessary to postulate the existence of a wide range of specific receptor sites. Little is known about the mode of entry of virions into the epithelial cells, and this might well determine the type and specificity of such receptor sites. Chamberlain and Sudia suggested that the receptor site hypothesis could be tested by first feeding "killed" virus to occupy the essential receptor sites and following this at a suitable interval with infectious virus. Further refinements would be to use virus irradiated by ultraviolet light as well as infectious virus labeled with radioactive tracers.

The possibility that the "gut barrier" might be simply a manifestation of a surface inhibitory substance secreted by or associated with the gut epithelial cells was put forward as an alternative hypothesis by Chamberlain and Sudia (1961). This idea would also lend itself to critical experimentation using both inactivated and infectious virus preparations, with the interesting possibility that the activity of the inhibitor could be deflected by attachment to the noninfectious virus, leaving the way clear for infectious virus. Some support for this hypothesis is provided by the numerous examples in the literature of successful infections being brought about in apparently resistant hosts simply by increasing the ingested dose. However, the "infection threshold" of arboviruses in mosquitoes varies widely with different viruses, and very often critical dosages are required for systemic infections to occur (Chamberlain and Sudia, 1961; Chamberlain, 1968). Therefore, if a nonspecific surface inhibitor exists, either a quantal response results or it functions with varying degrees of efficiency, depending on the type of ingested virus. On the other hand, if the inhibitors are virus specific, then it follows that a wide range of types exists. This would be a very complex situation and, at the present time, would be very difficult to test with existing techniques. Therefore, it would be more meaningful to investigate the presence or absence of preformed, nonspecific inhibitors on the surface of the midgut epithelium.

At the same time, it is equally possible that the critical stages in the determination of successful virus infection occur after the entry of virus particles into the midgut epithelium. Therefore, the "gut barrier" may not arise from the processes already discussed but may result from a tissue-mediated response manifested after the entry of infectious virus into such tissues. McLean (1955) suggested that the factor determining whether Murray Valley encephalitis was capable of undergoing a "biological" cycle in the vector was the capacity of the midgut to support virus replication. Conditions which operated against the liberation of nucleic acid following virus entry would effectively ensure that infection did not take place, so the gut cells would represent, to all intents and purposes, a barrier of nonsusceptible tissue. Whether this state can be changed through stress factors or changes in host metabolism is not known. Alternatively, even if the viral nuclei acid were liberated, its activity could be repressed to such an extent that only very low levels of virus replication occurred, resulting in subclinical infections, and again these would be very difficult to detect and measure with present insensitive methods. On the other hand, assuming that conditions for virus replication were optimal in the gut cells, movement of virions into the hemocoel would not occur if the progeny virus could be contained within the gut tissue in some way. Such a system would effectively prevent systemic invasion of the host. The effects of bypassing the gut through intrahemocoelic infection of virus are relevant to all of these hypotheses. The original concept of the "gut barrier" developed from comparisons of the level of dosages necessary to cause infection by ingestion with the minimal quantities required to cause infection by injection into the hemocoel (McLean, 1955; Bailey and Gibbs, 1964; Peers, 1972). Nonvectors of plant viruses and of arboviruses can be converted into efficient vectors simply by puncturing the gut after ingestion of the virus. This operation evidently allows systemic invasion, together with passage into or infection of the salivary glands (Storey, 1932, 1939; Merrill and ten Broeck, 1935; McLean, 1955). The general susceptibility of mosquitoes to all biologically transmitted arboviruses can be demonstrated by intrahemocoelic injection (Hurlbut, 1951, 1953, 1956; McLean, 1955). Further, nonhemophagous insects such as various members of the Coleoptera and Lepidoptera prove to be susceptible to infection with a wide range of arboviruses by intrahemocoelic injection (Hurlbut, 1956; Hurlbut and Thomas, 1960, 1969). Scherer and Hurlbut (1967) have shown that Nodamura virus, an insect-borne picornavirus, will replicate in larvae of the Indian meal moth, *Plodia interpunctella,* and Bailey and Scott (1973) found that Nodamura virus caused fatal results in the wax moth, *Galleria mellonella* L., and in honey bees following injection into the hemocoel. Whether Nodamura virus can infect these hosts after oral ingestion was not tested. Viruses that are pathogenic for their insect host need only be injected in very small doses to produce fatal results, whereas very large doses are needed *per os* to produce the same level of mortality. Therefore, there is the implication that if defensive mechanisms do exist in the hemocoel, i.e., hemolymph factors, then they are not immediately operative. These observations lead to the suggestion that the defensive reactions of insects are so organized as to prevent entry of virus particles from the midgut tissue into the hemocoele.

An observation of considerable significance was made by Whitfield *et al.* (1973), who found that after successful infection of the midgut an equilibrium was reached in that the number of infected cells and number of virus particles appeared to be reasonably constant. This was in agreement with an earlier suggestion of Chamberlain and Sudia (1961) that an "infectivity plateau" occurred in the midgut before other tissues were or could be invaded. A failure in the rate of virus replication to reach the requisite infectivity plateau and therefore the consequent mobility to invade the blood system could be classed as a "containment" of infection. Rehaček (1965) was of the opinion that the threshold phenomenon played a major role in the development of infection in arthropods and was most likely to be influenced by receptors or inhibitors in the gut cells, less by the action of digestive enzymes, and least by the permeability of the peritrophic membrane.

Insect pathogenic viruses are often associated with latent or subclinical infections in natural host populations. These can be activated by a wide range of physical and chemical stimuli which put the host under stress, and such frank infections necessarily involve the invasion of the hemocoel (Aruga, 1963). The mode of action of these stress factors is not understood, but it is possible that they either remove the barriers against systemic infection or stimulate a higher rate of virus replication, so that the infectivity plateau is reached. A possible explanation of both the gut barrier reaction and of latent infections could be the activities of a tissue-mediated viral inhibitor which depresses virus replication. This inhibitor could be a normal constituent of the midgut and related tissues, because production of an inhibitor as a response to viral particles would not be swift enough to be effective. Varma (1972) suggested that the disappearance or elimination of virus from gut cells could be the result of destruction of the virus by antiviral substances already present in the cells. However, there is no evidence in the literature of the occurrence of such inhibitors in midgut tissues. This could be investigated by incubating highly purified virus with midgut cell extracts and then determining the degree of inhibition of virus infectivity, preferably in an insect cell culture system. The attachment of inhibitors to the virus surface, possibly leading to aggregation, could be tested by incorporating radioactively labeled markers into the living host some time before excision of the midgut tissue.

Tinsley (1975) found that aggregates of virus particles could be obtained when extracts of healthy insects, treated only with low-speed centrifugation, were incubated with highly purified virus particles at 4°C for 12 hr, using the densonucleosis virus (DNV) of the wax moth, *Galleria mellonella* L. When the mixture of virus plus healthy insect extracts was run in a 2.5% polyacrylamide gel system (PAGE), two bands could be seen in the stained gel columns. One band (B) represented normal virus particles and corresponded in rate of migration to the purified virus sample; the other band (A) migrated much more slowly into the gel, indicating that a change in overall charge had occurred, possibly by the incorporation of host material. When this slow-running band was excised from a comparable unstained gel and examined with the electron microscope, it was shown to contain virus particles surrounded by an amorphous substance. When densonucleosis virus was treated with 1% sodium dodecylsulfate (SDS) and run in a 10% PAGE system incorporating β-mercaptoethanol, four polypeptide bands were obtained. When band A excised from the 2.5% gels was treated in a similar manner and run on the 10% gel system, an extra band was observed in addition to four virus polypeptide bands. When comparable gel columns were stained with the Schiff reagent, specific for carbohydrate moieties, this fifth band showed faint red coloration, indicating the presence of carbohydrate. A logical extension of these investigations was to determine if similar bands corresponding to coated particles could be found when extracts from infected insects were tested by PAGE. *G. mellonella* larvae infected *per os* with the densonucleosis virus were sampled at daily intervals after infection feeding. Samples were run on 2.5% gels, and on day 5 a strongly staining band in the position of band A was visible close to the top of the column, while a continued treatment of purified particles showed the normal position for virus much farther down the gel. Excision of band A revealed coated virus particles with the electron microscope, and treatment of this material with SDS followed by electrophoresis in 10% PAGE gels also showed the additional band in the normal virus profile. When virus incubated with healthy insect extract was co-run on similar gels with the 5-day sample from infected insects, band A was clearly visible at the top of the column and stained more intensely than bands in the A position produced by either of the separate treatments. These preliminary experiments indicate that substances found in both healthy and infected insects can be associated in some way with virus particles, although it is not known if the infectivity of such coated virus particles is affected. However, these attached substances were demonstrated in whole-insect extracts, and it remains now to determine if they are present in the midgut tissue. The chemical nature of the additional substance has not been fully established, but certainly protein and probably carbohydrate are involved, which is suggestive of a glycoprotein.

Longworth *et al.* (1968) noted a 17 S contaminant during the initial purification of the densonucleosis virus from *G. mellonella*. This substance was invariably associated with virus particles and was eliminated from virus preparations only by fairly extensive treatment. The 17 S fraction had a very different amino acid composition from virus protein, was highly immunogenic, and showed no relationship with virus protein in serological tests. The effect of the 17 S substance on the infectivity of the densonucleosis virus is not known. Bailey *et al.* (1970) criticized the methods used by other workers in the purification of insect viruses and urged the use of a less rigorous system: the data provided by these authors from experiments with the densonucleosis virus, albeit for other purposes, are relevant to a consideration of the nature and role of these contaminating fractions. Extracts made from younger (fifth instar) *Galleria* larvae had less of the contaminants, particularly the 17 S component, than those derived from older (eighth instar) larvae used by Longworth *et al.* (1968). It would be of interest to follow the development of the small S components in the course of insect development. More importantly, when virus was extracted from a mixture of healthy larvae labeled with ^{14}C and larvae infected with the densonucleosis virus, Bailey *et al.* (1970) found that both the small S material and the virus preparation had residual radioactivity. Indeed, the virus preparation possessed over 60% of the total residual activity of the whole mixed preparation. This is suggestive of the "contaminant" taking up a radioactive label and then attaching to or being associated with the purified or possibly semipurified preparation. The nature and role of the 17 S host protein component would repay further study, particularly in relation to virus infection and replication.

Salt (1970) was impressed by the key role played by mucopolysaccharides in the reactions of the insect host following the deposition of parasite eggs. He thought that if immunoglobulins could not be evoked in defense reactions, then mucopolysaccharides would head the list of possible candidates. The preliminary results reported by Tinsley (1975) suggest that glycoproteins other than immunoglobulins may also have a role to play, very probably in contributing to the gut barrier reactions. Salt also suggested that the involvement of polysaccharides, especially mucopolysaccharides, in both mammalian and insect cellular defense reactions, seemed too frequent to be a mere coincidence. It was possible that an original invertebrate system based on mucopolysaccharides was taken over by insects and vertebrates; the latter, however, found the system unsatisfactory, either because it was insufficiently specific or perhaps because it was too easily mimicked, and so developed a parallel system based on immunoglobulins. A similar case can be made for the retention and development of glycoproteins other than immunoglobulins as a defense reaction in the invertebrate, which in time became more sophisticated in their action and probably more specific. The insect has no requirement for a system of acquired immunity, principally because its life span is short, and yet, in relative terms, the virus infection can occupy a large proportion of the insect's life. Therefore, whatever defense system may be in operation, it is clear that it must be swift in appearance and is very probably a normal host constituent. There could also be a finite quantity of inhibitor, which would explain why the insect hosts can deal effectively with low virus dosages commonly met in nature but are in serious difficulties when exposed to high doses. It is significant that some vertebrate animals do have a preformed inhibitor system based on glycoproteins that is quite different from immunoglobulins. The best known of these is the α inhibitor, which is effective against influenza virus and is found in various organs, including the submaxillary glands (Cohen, 1963).

4. VIRUS REPLICATION

4.1. Assay and Growth Kinetics

Data on the growth kinetics of viruses of invertebrates are scanty. Multiple infection cycles occur in the host animal, and bioassay data which have usually been obtained on a population of laboratory-reared animals reflect this situation. Such assay systems are virtually useless for understanding the kinetics of virus replication and have value only as crude titrations of infectivity. Usually, available bioassay data measure only lethal doses (LD_{50}), that is, quantification of host death. Yet it is known that in many hosts viruses can replicate without causing death of the host. It is unfortunate that as yet neither detection nor assay systems are available for such sublethal infections. Reliable single-cycle growth kinetic data for viruses of invertebrates, as for viruses of vertebrates, will be derived only from *in vitro* systems of cell culture.

A number of assays of the infection time course have been made for baculoviruses in their host. The results suggest that a single cycle of growth occurs within 24 hr (Tanada and Leutenegger, 1970; Wäger and Benz, 1971; Summers; 1971). Aizawa (1959) examined the virus titer in the hemolymph of silkworm larvae kept at 25°C and found an initial drop in titer with a logarithmic increase from 10 hr to a maximum value around 50 hr. Quantitative growth kinetic data have also been obtained from work in cell cultures. Vaughn and Stanley (1970) described an end-point dilution method of assay using primary insect cultures, but it is clear that methods employing continuous cell lines hold out most promise. Faulkner and Henderson (1972) used a quantal response method for the assay of *Trichoplusia ni* NPV in a *T. ni* cell line. They scored the number of uninfected cells 24 hr after infection. However, the population doubling time of the cells was 16 hr and, as Knudson and Tinsley (1974) have pointed out, this is likely to result in conservative estimates of the virus titer as the number of uninfected cells would be increasing. Hink and Vail (1973) described a plaque assay method of titration in the same system in which 0.6% methylcellulose was used as an overlay. This overlay was the only one of several that were tried which was successful. Two types of plaque were observed—one with cells containing many polyhedra (MP) and one with cells containing only a few (FP). In a subsequent publication, it was claimed that the virus in the MP plaque was of the "multiply enveloped" type and that of the FP plaque was of the "singly enveloped" type (Ramoska and Hink, 1974). However, attempts to clone the two plaque types were not entirely successful; thus when the FP and MP types were grown individually in insects and then purified and reinoculated into cell culture, both types of plaque could be recovered from each despite the fact that the plaques remained true to type in cell culture. This led Knudson and Tinsley (1974) to question whether the existence of two plaque types was real or whether the FP plaque resulted from secondary lateral transmission from the MP plaque. These workers studied the growth characteristics and assay of *Spodoptera frugiperda* NPV in a continuous cell line of *S. frugiperda*. In this system, they found the optimal temperature for replication to be 27°C. A single particle could initiate infection, and particle–infectious unit calculations gave a ratio of 62–310 nonoccluded virus particles per $TCID_{50}$. Infectious virus was released from the cells from 12 hr after infection and approached maximum titer at 96 hr after infection. Two methods of assay were used—an end-point dilution method and a plaque method. For both assays, the polyhedron provided a readily observable CPE. It was found that cultures infected at high multiplicity did not attain confluency, which indicated that infected cells do not divide; and, as no CPE was observed when confluent cultures were inoculated, it must be assumed that cells need to be actively dividing for infection to proceed. Therefore in cultures at selected cell densities and infected at low multiplicity, foci or plaques of infected cells could be observed randomly distributed in the monolayer. This occurred because cells not infected by the input virus continued to divide to produce a confluent culture, incapable of generalized infection, before the release of virus from the first cycle of growth. Thus a plaque (or infection focus) assay titration method was provided in which no overlay was employed. Two plaque types were also found in these cultures, but from regression analysis of the real and expected dose response data using the relationship of 0.7 pfu per $TCID_{50}$, it was clear that the small plaques observed were derived from the large plaques as a result of lateral transmission of infectivity. The manipulation of cell density and virus multiplicity is perhaps unlikely to be as successful in preventing this as the conventional overlay methods.

Kawase and Miyajima (1969) could detect virus-specific proteins in the host insect *Bombyx mori* at 6 hr after infection in cytoplasmic polyhedrosis virus infection. Inclusion body formation in the gut cells could be detected at 15 hr reaching a maximum around 48 hr (Miyajima and Kawase, 1968). Successful hemagglutination and hemagglutination-inhibition tests have been achieved with *B. mori* CPV inclusion bodies, optimally with chick erythrocytes at 4°C, which could be used as an assay method (Miyajima and Kawase, 1969). A positive reaction was also claimed with nuclear polyhedrosis viruses but no details were given.

Iridescent viruses are usually detected initially by iridescence of the host tissues, and the presence of the virus can easily be demonstrated by electron microscopy. Bellett and Mercer (1964) and Bellett (1965a,b) reported an assay method for *Sericesthis* virus (type 2) multiplication in cell culture based on fluorescent antibody staining, and they attempted to quantify virus replication. Virus antigen and DNA accumulated in discrete foci in the cytoplasm which could first be seen 48 or 12 hr after infection, depending on the incubation temperature, 25°C or 21°C, respectively. The number of foci was maximal after 4 days (25°C) or 6 days (21°C). The assay measured virus titer in infective units estimated from the proportion of cells staining with fluorescent antibody 6.5 days after infection with various dilutions of virus. The relationship between virus titer and concentration was linear, with a particle–infectivity ratio of 80:1. In quantitative experiments, it was found that virus could not be dissociated from cells by repeated washing after 10 hr postinfection. The yield of virus per cell was 500 infectious units, of which 360 were released between 4 and 8 days postinfection.

A latex agglutination test has been described for the assay of *Tipula* iridescent virus in clarified larval extracts (Carter, 1973b). The test has a sensitivity equal to that of passive hemagglutination, in which it has previously been demonstrated that minimal detectable quantities are of the order of 0.26–0.015 μg (Cunningham et al., 1966).

4.2. Histopathological and Biochemical Events

4.2.1. Baculoviruses

The process of baculovirus assembly in susceptible cells of the host insect is well known (Xeros, 1956; Summers, 1971; Harrap, 1972c). In nuclear polyhedrosis virus infections, an enlarged nucleus is the first obvious histopathological event. A dense network of viroplasm, "virogenic stroma," is formed in the enlarged nucleus, and the assembly of rod-shaped virus particles can be observed from the substance of the stromal strands. As this process proceeds, the viroplasm network apparently contracts, or the nucleus enlarges further, giving the infected nuclei a characteristic appearance described as a "ring zone." In the region between the viroplasm network and the nuclear membrane, and in the lacunae of the network itself, the naked virus particles (nucleocapsids) acquire envelopes which are laid down either around individual virus particles or around groups or "bundles" of varying numbers of virus particles, depending on the type of baculovirus. In section, the virus envelope has a trilaminar structure similar to plasma membrane, and it is interesting that such a structure is acquired by the virus within the nucleoplasm rather than at a cell membrane site as is usually the case with enveloped viruses (Harrap, 1972c; Stoltz et al., 1973). However, it should be noted that granulosis virus particles have been observed budding from gut cells and acquiring an envelope as a result (Robertson et al., 1974) and that the situation can be different in cell culture systems (see below). Polymerizing inclusion body protein is deposited on the external surface of the virus envelope, perhaps because it carries some binding site on which protein deposition can proceed. Naked virus particles are never occluded by the polymerizing protein even though they may be present in the cell during inclusion body formation. The virus envelope which at first surrounds the nucleocapsid only loosely becomes closely apposed during inclusion body protein deposition. Increasing numbers of enveloped virus particles become occluded into the protein matrix, and eventually a densely stained surface layer, the so-called polyhedron membrane, can be observed at the periphery of the inclusion body. The nucleus, which now accounts for most of the cell volume, is packed with many mature polyhedra, each containing large numbers of virus particles at the terminal stage of virus multiplication. The nucleus often ruptures at this late stage of virus assembly, and those few virus particles not occluded into polyhedra may be responsible for the infection of adjacent cells. It is not clear what factors control the size or shape of the polyhedra. Inclusion body protein synthesis has been shown, by fluorescent antibody techniques, to take place on the ribosomes as would be expected (Krywienczyk, 1963). At the later stages of infection, each cell contains very little cytoplasm, and polyhedron growth may be limited simply by protein monomer depletion. Some ultrastructural features of NPV and GV inclusion body formation have been described by Arnott and Smith (1968*a,b*) and Summers and Arnott (1969).

These profound and complex morphogenetic events can also be observed in cell cultures infected with NPV. Ignoffo *et al.* (1971) attempted the multiplication of an NPV in a continuous cell line of *Heliothis zea* ovarian cells. No virus particles or inclusion bodies containing them could be found in the cells, but when infected cells were fed to larvae after seven passages the larvae died with typical NPV infection. Sohi and Cunningham (1972*a,b*) studied the multiplication of an NPV of the western oak looper, *Lambdina fiscellaria somnaria,* in two cell lines of hemocytes of *Malacosoma disstria.* Polyhedra were observed 3 days after inoculation. The infected cells of one cell line showed typical virus morphogenesis with an extensive virogenic stroma, nucleocapsid formation, and envelope acquisition followed by polyhedron formation. In the other cell line, few cells showed polyhedron formation and although virogenic stroma could be detected in the cells few virus particles seemed to be present. The few polyhedra that were detected contained no virus particles. Virus replication was therefore considered to be more efficient in one cell line than in the other. Vail *et al.* (1973) found that polyhedron formation occurred by 20 hr after infection when a virus isolated from the alfalfa looper, *Autographa californica,* was inoculated into *Trichoplusia ni* cells. The polyhedra were as infectious for larvae of several species as those produced in the insects themselves. However, MacKinnon *et al.*

(1974) examined *T. ni* cells electron microscopically, paying particular attention to morphogenetic changes occurring during long-term serial passage of the virus. They found that the capacity for *in vivo* replication of virus that had been repeatedly passaged in cell culture decreased with time. Virus morphogenesis in early passages was very similar to the process in insect host tissue. Virogenic stroma could be observed at 24 hr after infection and polyhedra at 48 hr. A few polyhedra contained a few or no virus particles. Sheets of fibrous material were observed containing electron-dense "spacers" similar to polyhedron membranes back to back. Such spacers were also observed by Summers and Arnott (1969) and have been noted in this laboratory. In the 24–28 hr postinfection period, nucleocapsids could be seen acquiring envelopes within infected nuclei. However, at the periphery of the nuclei, some nucleocapsids could be seen budding through the nuclear envelope into the cisterna of the endoplasmic reticulum. Budding through the plasma membrane was also seen between 48 and 72 hr when the nuclear envelope had ruptured and the nucleocapsids were released directly into the cytoplasm. After 15 serial passages, a reduction in polyhedron number occurred and continued progressively. After 20 passages, abnormalities of morphogenesis were noticeable and became pronounced from 40 passages onward. Alterations such as nuclei with virogenic stroma but no nucleocapsids, or virus particles and nuclei with virus present but no or few polyhedra, were obvious. Long cylindrical profiles possibly of capsid protein, sheets, and whorls interspersed between capsids of variable diameter (15–43 nm instead of 42 nm) were all unusual features in late-passage infected cells. Polyhedra from early-passage material contained the usual bundled virus particles, but later-passage material had polyhedra containing aberrant virions with putative nucleocapsids of about a third of the normal length, although a few normal nucleocapsids could be observed budding through the plasma membrane. These alterations were not caused by changes in the cell line as it remained fully susceptible to infection by NPV-infected insect hemolymph, resulting in normal NPV morphogenesis. The suggestion was advanced by these workers that repeated passage had resulted in the accumulation of defective interfering particles.

Raghow and Grace (1974) studied the multiplication of *Bombyx mori* NPV in a *B. mori* cell line and constructed a growth curve on the basis of an LD_{50} assay in the host insect of cell-associated and released virus in the cultures at various times. They examined the cells at intervals after infection by ultrathin sectioning and electron microscopy. Attachment of enveloped virus particles and their invagination into the cell by viropexis were observed up to 8 hr after infection. Disruption of the virus envelope in the cytoplasm and alignment of the transverse end of the naked virus rod with the nuclear membrane pore were observed between 4 and 8 hr after infection. Some of the naked virus rods associated with the nuclear pores were partially "empty." Nuclear changes were observed first at 16 hr after infection, and there was an obvious virogenic stroma with naked virus particles by 24 hr. Virus envelopes were apparent around individual and groups of virus particles by 36 hr and by 40 hr small polyhedra could be detected. Polyhedron formation was virtually complete at 72 hr. Knudson and Harrap (1976) studied *Spodoptera frugiperda* nuclear polyhedrosis virus replication in a *S. frugiperda* cell line, also using ultrathin sectioning techniques. The growth kinetics for this cell–virus system had been established previously (Knudson and Tinsley, 1974) and are described at the beginning of this section. By 1 hr after infection, enveloped naked virus particles could be found in contact with the outer surface of the plasma membrane; naked virus particles could be found in the cytoplasm; and enveloped, partially enveloped, and naked virus particles could be found in vacuoles in the cytoplasm. Using large quantities of insect-produced virus as an inoculum in a separate experiment, it was clear that all forms of virus particles can readily be engulfed by the cells. Isolated virus particles could be found in the nucleus as soon as 3 hr after infection. Virus could not be found in the cells from 4 to 8 hr but at 9 hr nuclear changes were apparent, with the development of a virogenic stroma and some naked virus rods. By 12 hr, the "ring zone" stage of development was obvious and virus envelope formation could be observed. At both 12 and 15 hr, large numbers of virus particles could be seen leaving the cell, both by budding from the plasma membrane and by extrusion of vacuoles containing naked and enveloped virus rods. It was not clear how the virus left the nucleus. Although suggestions of a budding process into the cisterna of the endoplasmic reticulum were found, naked virus rods could also be observed in the cytoplasm, possibly after rupture of the nuclear envelope. Early polyhedron formation was detected at 18 hr and was well advanced by 24 hr. Virus particles could still be found leaving the cell at this stage. The beginnings of the deposition of a dense peripheral layer around the polyhedra were seen at 36 hr, and this process was largely complete by 48 hr. It seems clear then that the morphogenesis of an NPV in cell culture is analogous to if not always identical with the situation in the insect host, but there is a real possibility of aberrant virus formation occurring if the virus is repeatedly passaged.

At the present time, there is little information on the biochemistry of NPV replication. What work has been done reports the situation in the host insect and therefore represents a cumulative effect of infection in the host rather than specific events in a single virus growth cycle. Morris (1968a,b) found that DNA synthesis increased until polyhedron formation occurred, and in Lepidoptera both nuclear and cytoplasmic RNA increased initially, then decreased, whereas in Hymenoptera the reverse was true. In later work, an autoradiographic study of protein changes showed an increase in total cell protein at an intermediate stage of infection (Morris, 1971). Benz and Wäger (1971) studied a granulosis virus infection and found two peaks of RNA synthesis and a large increase in DNA synthesis. Young and Lovell (1971) reported changes in hemolymph proteins in NPV-infected *T. ni* larvae. A much lower concentration of total protein was found in late instar infected larvae than in uninfected larvae. In NPV-infected *H. zea* larvae, Shapiro and Ignoffo (1971a) reported elevated protein levels early in infection, although these levels subsequently declined. Young and Johnson (1972) studied the virus-specific soluble antigens in the fat body of *T. ni* larvae infected with NPV and could demonstrate both inclusion body and virus particle antigens in immunodiffusion tests. Morgante *et al.* (1974) examined the patterns of DNA synthesis in NPV-infected caecal cells of the dipteron *Rhynosciara angelae* and concluded that the virus infection induced increased host DNA and RNA synthesis and that the bulk of viral DNA synthesis was evident only after this amplification of host synthesis. Watanabe (1972) showed that the viral DNA synthesis was associated with the virogenic stroma.

With some successful NPV–cell culture systems now available in which assay is possible and virus growth kinetics have been established, one might expect an explosion of information on the biochemistry of NPV replication. Knudson (personal communication) has recent information on the biochemical events of *S. frugiperda* NPV replication in *S. frugiperda* cells. He finds an early suppression of host DNA and a marked increase in viral DNA at a maximum at 12 hr, two peaks of viral RNA at around 8 and 24 hr, with protein peaks at a maximum a little later at around 10 and 26 hr. Cellular RNA declines steadily after 12 hr.

4.2.2. Cytoplasmic Polyhedrosis Viruses

As the name implies, cytoplasmic polyhedrosis viruses mature in regions of the cell cytoplasm. The virogenic areas are devoid of the usual cell organelles and contain only dense reticulate areas of viroplasm in which the spherical virus particles can be seen (Stoltz and Hilsenhoff, 1969; Longworth and Spilling, 1970). The virus particles are incorporated into masses of inclusion body protein to form polyhedra. Stoltz and Hilsenhoff (1969) suggested that individual virus particles were coated by protein first and then such coated particles fused to form a polyhedron which was "rounded off" by further protein deposition. However, this process does not seem to be common to all CPV infections. Arnott *et al.* (1968) distinguished between an early virogenic stroma with incompletely formed virus particles and a later fibrillar "crystallogenic matrix" containing mature virus particles. Kawarabata and Hayashi (1972) reported the possible development of a CPV in a mosquito cell line. No polyhedra were observed, but a labeled RNA species could be isolated from the infected cells similar to CPV viral RNA. There have been other reports of successful CPV growth in primary cell cultures and of the spontaneous appearance of a CPV in a continuous cell line (Grace, 1962; Vago and Bergoin, 1963; Sohi *et al.,* 1970). Recently, however, Granados *et al.* (1974) reported the complete replication of a *T. ni* CPV in *T. ni* cells. The cells were inoculated immediately following subculture using either extracts of diseased larval guts or virus particles released by alkaline treatment of polyhedra. Both methods were successful. Large masses of virogenic matrix and virus particles were observed in the electron microscope, and both spherical and cuboidal polyhedra were found. The virus was passaged twice using triturated cells as a source of inoculum. Cell lysis was not commonly observed and cell-free medium did not prove to be a good inoculum.

There is some information on the biochemistry of CPV replication, all of which has so far been made on infected insects. Kawase and Kawamori (1968) found incorporation of isotope into viral RNA 24 hr after infection of *B. mori* larvae with CPV. At 72 hr, incorporation into viral RNA was at least 3 times that of ribosomal RNA. Watanabe (1967) suggested that viral RNA was synthesized in the nucleus, but Hayashi and Retnakaran (1970) implicated both nucleus and cytoplasm in viral RNA synthesis. Several workers have shown that synthesis of both single-stranded and double-stranded RNA continues in CPV infection in the presence of actinomycin D (Hayashi, 1970c; Hayashi and Kawarabata, 1970; Furusawa and Kawase, 1971; Kawase and Furusawa, 1971; Payne, 1972). Richards and Hayashi (1972) showed that satisfactory separation of the single-stranded and double-stranded RNAs could be achieved if use was made of their differential aggregation in the presence of magnesium ions and their different solubilities in 2 M lithium chloride. Also, Hayashi and Donaghue (1971) showed that, whereas both RNA forms were produced *in vivo,* only the single-stranded RNA was produced in an *in vitro* system in which dissected infected midguts of larvae previously treated with actinomycin D were triturated in buffer and incubated. Eventually, the single-stranded RNA was shown to be capable of reannealing with heat-denatured viral RNA, indicating that it was transcribed from the viral genome and was probably virus-specific messenger RNA (Furusawa and Kawase, 1973; Payne and Kalmakoff, 1973). Other information on RNA species produced in *in vitro* systems is considered in Section 2.2.

4.2.3. Poxviruses

Invertebrate poxviruses have been described developing in the cytoplasm of hemocytes or fat bodies of beetles (Bergoin *et al.,* 1968*b,* 1969; Bergoin and Devauchelle, 1972; Devauchelle *et al.,* 1971; Goodwin and Filshie, 1969, 1975), in a grasshopper (Henry *et al.,* 1969), in chironomids (Huger *et al.,* 1970; Stoltz and Summers, 1972), and in Lepidoptera (Granados and Roberts, 1970; Bird, 1974). The subject was reviewed by Bergoin and Devauchelle (1972). The viral genome is released after disruption of the core coat, and foci of virus synthesis can be observed in adjacent areas. These foci seem to give rise to virogenic areas, portions of which are sequestered into developing virus particles. These immature particles are spherical or nearly spherical and bear a double-layered envelope, at least part of which is a unit membrane. So-called intermediate forms with a dense eccentric nucleoid inside the envelope can be found, so it seems that differentiation from immature to mature particles occurs within the envelope, sometimes even during the occlusion of the virus particle into the inclusion body. Progeny virus particles migrate from the foci on maturation. In hemocytes, they are released from the cell by exocytosis, acquiring an external coat from modified cell membrane as they do so. This process also occurs in some fat body cells. However, the majority of the virus particles are occluded into spheroidal or ovoid crystalline proteinaceous inclusion bodies analogous to those of baculoviruses. These inclusion bodies develop in areas away from the original morphogenetic foci. Spindle-shaped or fusiform inclusions which do not contain virus particles are also found in some poxvirus infections (Bird, 1974).

4.2.4. Iridescent Viruses

The morphogenesis of iridescent viruses has been studied in both insect tissues and cell cultures. The virus is assembled in the cytoplasm in virogenic areas which can be detected by fluorescent staining and fluorescent antibody techniques (Bellett and Mercer, 1964) and by electron microscopy. In ultrathin section of the virogenic regions, the virus appears to be assembled by sequestration of portions of the virogenic matrix by developing virus capsids (Bird, 1961a; Harrap and Robertson, unpublished data). Yule and Lee (1973) proposed an alternative hypothesis in which the viral DNA core material is introduced into the intact preformed capsid through a hole. Such a system seems to be unlikely and is not supported by studies of the virus replication in cell culture where progeny viral DNA can be detected before the formation of intact virus particles (Kelly and Tinsley, 1974b). Virus assembly in cell culture systems is similar to that observed in host insect tissues, although virus-associated structures (microtubule formation and paracrystalline and amorphous aggregates) have been reported in some virus–cell combinations (Kelly and Tinsley, 1974a). In these systems, mature virus particles were detected 144 hr after infection, i.e., 48 hr after the detection of structural antigens, when incubation was at 21°C. Virus release was achieved by cell lysis and exocytosis, but budding processes involving acquisition of a cell membrane envelope have been reported with other systems (Bellett and Mercer, 1964; Hukuhara and Hashimoto, 1967). Using autoradiographic techniques, Morris (1970) reported a slight increase in nuclear DNA synthesis during infection, with a large increase in virus-induced DNA synthesis in the cytoplasm. The infected fat body cells were greatly enlarged, although the nuclei were only slightly bigger. Yule and Lee (1973), using immunoferritin techniques, detected viral protein at 12 hr after infection. However, Krell and Lee (1974) reported the detection of proteins on polyacrylamide gels with the same mobility as TIV polypeptides from hemocytes at 4 hr after infection. Six noncapsid viral polypeptides were also said to be present in infected cells. Inhibition of host protein synthesis apparently occurred within 1 hr of infection; because of this, it is claimed that the viral genome is released and activated within 1 hr of infection. By use of cell culture techniques with two different cell lines and SIV (type 2) and CIV (type 6), Kelly and Tinsley (1974b) found that overall DNA and RNA syntheses were depressed up to 48 hr after infection and then both were enhanced for the remainder of the infection. They used hybridization techniques to demonstrate that RNA complementary to viral DNA could be detected from 24 to 144 hr (the full extent of the time period examined), and viral DNA was detectable from 72 to 144 hr. Early virus-specific RNA synthesis is probably directed by input genome as virus-specific DNA synthesis was not detected until 72 hr after infection and the virus-associated DNA-dependent RNA polymerase was thought unlikely to account for all the RNA synthesis detected. It was postulated that a virus-specific DNA-dependent RNA polymerase which utilized the input genome might be synthesized early in infection. Virus-specific DNA synthesis occurred when total DNA synthesis in the infected cells was enhanced and at times when DNA-dependent DNA polymerase and thymidine kinase activities were stimulated. Also, viral structural proteins were synthesized concomitantly, although before mature virus particles were detected. This study indicates that host cell DNA synthesis was curtailed during infection, a result which contrasts with that of Bellett (1965*b*).

McIntosh and Kimura (1974) have reported the multiplication of iridescent virus type 6 in a vertebrate cell line, that of viper spleen.

4.2.5. Other Viruses

There is little information on the histopathology or biochemistry of replication of other viruses of invertebrates, either in host tissues or in cell cultures. In the main, a description of the site of mature virus particles in the tissues of the diseased host is all that is known (Lee and Furgala, 1965; Longworth and Harrap, 1968; Bailey and Milne, 1969; Henry and Omas, 1973). An exception is the densonucleosis virus (DNV) of the wax moth, *Galleria mellonella*. The virus was first reported by Meynadier *et al.* (1964). Larvae are killed by the virus after 4–6 days when incubation is at 28°C, and many if not all tissues are susceptible. Ovarian cells of *G. mellonella* and of the silkworm *B. mori* grown in culture have also been infected successfully. Using techniques such as acridine orange staining, immunofluorescence, and immunoperoxidase, virus lesions can be detected in the nucleus and later in the cytoplasm, and virus antigen can be detected first in the cytoplasm, then in the nucleus, and finally over the entire cell (Kurstak and Stanislawski-Birencwajg, 1968). Using electron microscopy, dense nuclear material can be recognized first, then virus particles can be seen by 8–12 hr. Paracrystalline inclusions of virus particles can be seen in some cells later in infection. Morris (1970), using autoradiographic methods, found a large increase in the size of the nucleus and a big increase in nuclear DNA synthesis and late in infection in nuclear RNA synthesis.

Mouse L cells are reportedly "transformed" by densonucleosis virus (Kurstak *et al.*, 1969), although to be accepted unequivocally this result needs confirmation, especially in view of its implications. The extensive work on densonucleosis virus and its replication by Kurstak and his co-workers has been reviewed by Kurstak (1972).

This survey of what is known about the multiplication of viruses of invertebrates shows that much remains to be done. Much of the information that is available depends on morphological observations, and in very few instances is there any understanding of the underlying biochemical events of the replicative process. This situation is a direct consequence of the lack of suitable virus-sensitive cell culture systems in which such processes can be studied—a situation that has frustrated workers in this field for many years. In some cases, suitable cell culture systems are now available, and it is to be hoped that knowledge of the multiplication of these viruses will accumulate rapidly as a result.

5. ECOLOGY AND EPIZOOTIOLOGY OF THE VIRUS DISEASES

5.1. Movement and Spread of Virus

The needs and constrictions of biological control systems have greatly influenced the investigations of the ecology of insect viruses in the natural environment. Consequently, most of the available information has been derived from epizootiological studies on a few pests of cultivated crops and forest trees, and, for this reason, very little is known about the role of viruses on the population densities of invertebrates. Further, the dispersal and stability of viruses of marine invertebrates in an aquatic environment have not been studied at all. Clearly, the studies and methodology used in the association of mammalian viruses, such as hepatitis virus, poliomyelitis virus, and coxsackievirus, with oysters and mussels will have direct relevance once the investigations into invertebrates begin.

Reports in the literature suggest that the effects of a virus disease on populations of terrestrial insects under natural conditions can be considerable; this is well illustrated by the case history of the spruce sawfly, *Gilpinia hercyniae*, in North America. This pest became established in eastern Canada and flourished in the absence of any natural enemies, including virus diseases. Some 2000 square miles of spruce in the Gaspé peninsula were already affected when the insect was first recorded in 1930, and this had increased to some 12,000 square miles of affected trees by 1938. At this time, a nuclear polyhedrosis virus, probably accidentally introduced, spread naturally from south to north within the next 4 years and was responsible for keeping the sawfly down to a minor pest level over most of its range (Balch and Bird, 1944). The same sawfly appeared in quantity a few years ago in the forests of mid-Wales in the United Kingdom and has now spread rapidly in this area. It is interesting that a nuclear polyhedrosis virus also appeared after the population of sawflies had developed significantly. It is still too early to say whether or not the Canadian experience of natural control will be repeated in Wales, but the situation is being studied intensively because a unique opportunity is provided to study the movement and dispersal of a virus under natural conditions.

Three factors affecting the association of viruses with their invertebrate hosts are of particular importance in the consequent ecological development of disease. These are the stability or retention of infectivity, the dispersal of such infective virus, and its eventual pathogenicity or virulence for the host.

The stability of the baculoviruses can be very great in the laboratory and, also, given certain conditions, in the field. The proteinaceous polyhedron in which the virions are occluded is markedly resistant to physical and chemical treatment. It is, for example, resistant to acid and alkali treatment over a range of pH 2–9. Desiccation is scarcely a problem, as the chemical nature of the polyhedron is a good protectant for the virus particles. In the laboratory and under refrigeration (4°C), the baculovirus of *B. mori* was still infective after 21 years (Steinhaus, 1960), whereas the baculovirus of *Gilpinia hercyniae* lost infectivity after storage for 11 years at 4.5°C (Nielson and Elgee, 1960). Aizawa (1954) found that infectivity of the silkworm nuclear polyhedrosis virus was lost after 37 years when the virus was stored at room temperature. A nuclear polyhedrosis virus of the cabbage looper moth, *Trichoplusia ni,* was still active in the soil after at least 4 years (Jaques, 1969).

The stability of the other groups of insect viruses outside their hosts is less well documented but the virus of the red mite *Panonychus ulmi* (Putman, 1970) and the iridescent virus of the mosquito *Aedes taeniorhynchus* (Linley and Nielsen, 1968) seem to be extremely unstable.

Ultraviolet light is considered to be a prime agent in reducing or destroying infectivity of baculoviruses (Gudankas and Canerday, 1968). However, work needs to be done on the critical wavelengths which are operative. David (1970) was of the opinion that wavelengths of 290–320 nm were most active, but he suggested that the effects of prolonged exposure to longer wavelengths, i.e., 320–380 nm, should be investigated. Naturally, the effects of solar radiation could be of higher magnitude if the polyhedra and virus particles plus envelopes were on exposed surfaces. There is no doubt that in soil or under vegetation the stability and therefore longevity of the virus would be considerably increased. It is also true that infectivity of the virus is adequately protected within the insect host body, whether living or not. The release of masses of active virus from a moribund or dead insect will lead to the food plant becoming extensively contaminated. The protective action of dried body fluids has not been actively investigated, but it is probably considerable.

The relationship of insect viruses with leaf surfaces could be one of the key factors in determining whether an epizootic develops or not. The main portal of entry of viruses into insects is the mouth. Therefore, if virus, protected by dried body fluids, adheres closely to the leaf surfaces, this association provides an important and highly stable source of infection. Liberation of virus from dying or dead insect larvae very probably constitutes an important source of infection for contiguous larvae and is therefore of greater significance for species that are gregarious and/or sedentary in habit. The state of the foliage of the host plant and the nature of any exudates that are secreted could have a profound effect on the retention of virus activity at the leaf surface. It is clear that virus–leaf surface interactions could be very significant in determining the stability of virus inocula under field conditions and certainly could be a fruitful area of research.

A high degree of virus persistence outside the host insect could greatly influence the enhancement of an epizootic, but persistence within the living host is also of great importance. Many viruses are able to persist in the insect for most if not all of its reproductive cycle. The level of virus replication can vary from induction of a fulminating infection and rapid death to the production of a subclinical condition which under normal circumstances is undetectable. The baculovirus of the spruce sawfly, *Gilpinia hercyniae,* replicates only in the midgut epithelium. Young larvae that acquire the virus during an active feeding period usually die within 5–10 days following ingestion. On the other hand, larvae acquiring virus within 3 days of the end of the last feeding instar usually survive to the prepupal stage. At this time, the infected gut cells slough off and come to lie in the lumen of the gut which has been re-formed from the small, undifferentiated cells arising from the surface of basal lamina separating the gut epithelium from the tracheal tissue. The prepupal gut made up of these undifferentiated cells is apparently immune or at least very resistant to infection, so the virus is "contained" in or near the gut tissue. The prepupal stage of this sawfly is part of the diapause stage and it can overwinter in this form. Virus replication recommences once these small cells differentiate and continues in the true gut cells of the emerging adult. This means that the adult may develop a high concentration of virus which would be available during oviposition to contaminate the conifer needles (Bird, 1953; Nielson and Elgee, 1968; Tinsley and Entwistle, 1974). A further example is provided by the iridescent virus infection of the mosquito *Aedes taeniorhynchus* (Linley and Nielson, 1968). Larvae acquiring virus at a late instar after feeding on adjacent larvae killed by an early infection usually survive and give rise to infected females which lay a fairly low percentage of infected eggs. The hatching larvae from these eggs usually die in about the fourth instar. The infection of adults by the iridescent virus seems to have no deleterious effects. In both *Gilpinia* and *Aedes*, the alteration of apparent pathogenic and benign stages of the virus infection is closely linked with the life cycle of the host. In the mosquito this allows virus persistence totally within the host, and in the sawfly persistence in the host broken only by a limited period of exposure between oviposition and ingestion by first instar larvae. The virus can persist between this and perhaps two further recycles before again entering a late final instar larva and persisting through to the eonymph. *Gilpinia* eonymphs can remain in a state of diapause for up to six winters (Balch, 1939). This would provide the opportunity for prolonged virus survival, particularly important in periods of low host density.

The cytoplasmic polyhedrosis viruses commonly survive in their lepidopterous hosts from one generation to the next. Eggs of the moth *Malacosoma fragile* laid on tree bark can become contaminated with nuclear polyhedrosis virus simply by being splashed with raindrops containing polyhedra originating from larval residues of the previous summer. The larvae hatching from such contaminated eggs ingested virus with the eggshell and so became infected (Clark, 1956). The nuclear polyhedrosis virus of the gypsy moth, *Porthetria dispar*, can also persist on the egg surfaces during winter months (Doane, 1969). Virus movement from contaminated vegetation is predominantly downward, and any virus which reaches the soil in an infective state is likely to retain its activity for long periods (Jaques, 1969). Rain splashing can take up virus from the soil and transfer it back onto lower foliage of either existing perennial or succeeding annual crops. Jaques (1964) has shown that such a system of recycling can be responsible for the infection on new host insect generations. Irrigation of alfalfa fields by controlled flooding, as is practiced, for example, in southern California, could also play a part in washing infective virus out of the soil and back onto foliage.

The occurrence of sublethal or "latent" infections could have significant effects on the appearance of epizootics. Such infections are very common and can be stimulated or activated to lead to rapid replication of the virus, with fatal results. The stimuli appear to be many and varied, and cover chemical, physical, and biological stress factors. This topic has been reviewed in detail by Aruga (1963). The state of virus in sublethal infections has not been investigated fully, but there are indications that the process is possibly linked to the "gut barrier phenomenon" and the defensive reactions of the insect host (Tinsley, 1975). Sublethal infections are probably much more common than has been realized hitherto, because if the insect host does not die after infection it is judged not to be infected. The development of sensitive methods to detect these low-level infections is an urgent necessity, not just for its own sake but also to permit meaningful host range investigations to be made using tested virus-free insects. Further, there is the probability that the exposure of such latently infected insects to a second and possibly unrelated virus is sufficient to bring about activation of the first (Krieg, 1957; Bailey and Gibbs, 1964; Smith, 1967; Longworth and Cunningham, 1968). Such events would obviously invalidate cross-transmission experiments in the absence of unequivocal methods of identification. Indeed, much of the published information on the host ranges of insect viruses and cross-transmission studies must therefore be regarded with caution, particularly where no checks on virus identity were made.

The occurrence of virus strains in populations of host insects is an unknown quantity because here again there is a lack of sensitive methods for their detection. However, it is encouraging that the conventional biochemical and biophysical techniques of mammalian virology are now being more widely used with insect viruses, and this can only be good for the general development of the subject.

The dispersal or dissemination of virus occurs in two directions. There can be vertical transmission between successive generations, either inside or on the outside of the egg, or simply depending on residual virus at the initial feeding level. Horizontal transmission occurs between individuals of the same or at least the current season's generation and could be influenced by both space and time. The flight and dispersal habits of the adult insects obviously play very important parts in the general pattern of virus movement from the initial focus of infection. Gregarious or weak-flying host insects may have a high local incidence of virus infection but it would be essentially discrete, whereas an active insect like *Gilpinia hercyniae* may fly a distance of at least 1 mile before oviposition (Balch, 1939) and this would account for what is known as "jump-spread." This has been exploited with the alfalfa caterpillar, *Colias eurytheme*, by applying a paste containing a nuclear polyhedrosis virus to the genitalia of adult females. This resulted in a very effective dispersal of this virus in the larval population of this species in the area of release (Martignoni and Milstead, 1962). The introduction of pupae arising from sublethal infections in larvae proved to be an effective way of introducing the virus into new larval populations of the sawfly, *Neodiprion swainei,* an effect which persisted into subsequent generations (Smirnoff, 1962).

It is not possible to study dispersal of insect viruses without a detailed knowledge of the biology and behavior of the host insect. However, such data are all too frequently either not available or too fragmentary to be of value. Two examples showing the importance of oviposition habits are shown by the conifer sawflies *N. sertifer* and *G. hercyniae*. *N. sertifer* lays eggs in groups, whereas *G. hercyniae* lays them singly and so contributes to much larger numbers of infection foci. This sort of information enabled Bird (1961*b*) to suggest that the egg-laying habits of *G. hercyniae* accounted for the much wider spread and distribution of its nuclear polyhedrosis virus in North America than of the similar virus of *N. sertifer*.

There is now considerable evidence of the important role of birds in the dissemination of baculoviruses in nature (Bird, 1955; Hostetter and Biever, 1970). The nuclear polyhedrosis virus of *G. hercyniae* is passed out with the feces in a highly infective state after infected larvae have been eaten by birds (Entwistle, 1974). Further surveys showed that 15 species of birds trapped in the spruce forests of mid-Wales were carriers of this virus. Half of a sample of 100 birds trapped in September 1973 were carrying the virus, and the coal tit, *Parus ater*, proved to be the most important and common carrier. A significant feature of these investigations was the record that feces containing infective virus could be obtained in the winter and spring, a period when living sawfly larvae

were absent from spruce. Entwistle concluded that the most plausible
explanation was that the birds were feeding on virus-containing corpses
of *G. hercyniae* larvae that had persisted on the trees from the previous
summer.

It is also probable that parasitic insects may have a role in virus
dispersal, particularly during oviposition of the parasite egg when the
ovipositor could become contaminated with virus particles after
penetration of the host tissues. It is remarkable that so little is known
about this biological transmission and the other obvious dispersal
factor, that of wind currents. Obviously much more work is needed
before the full pattern of spread is understood.

Susceptibility of an insect to infection with a virus naturally
depends on age and possible factors of natural resistance. In the
Lepidoptera, for example, the very young larvae can be much more susceptible than later instars, which can prove, on occasion, to be so
resistant as to be immune to infection for all practical purposes.
However, some of the observations that have been made do not account
for the relationship between dosage of virus and body weight of the
host.

Several workers have suggested that strains of insect viruses with
varying virulence can be found in natural populations and also that
particular individuals of a host species show differences in susceptibility
(Bird and Elgee, 1957; David, 1957; Rivers, 1959; Sidor, 1959;
Ossowski, 1958, 1960; David and Gardiner, 1960; Martignoni and
Schmid, 1961). It would indeed be remarkable if such phenomena did
not exist. However, since we know so little about the mode and
mechanism of virus replication in the host tissues, it is not possible to
distinguish clearly between actual rates of infection and the subsequent
replication of the virus. Therefore, until methods are available to
measure infection with virus, as well as the degree of local multiplication and the progress of the subsequent systemic invasion of the host, it
is of litte use to consider questions of pathogenicity, virulence, and
susceptibility.

The relationship between viruses found in insects and those occurring in vertebrates and plants could have obvious epidemiological
importance. This is well illustrated by the arboviruses, now termed
togaviridae, which have considerable importance to human health in
many areas of the world. The possible relationships of those viruses
that are pathogenic to insects (as presented in Table 2) to other groups
of viruses is not known, but considering the ubiquity of insects and the
close association they have with vertebrates as competitors for food this

is certainly a problem that should be investigated. No one appears to have examined the bird carriers of the nuclear polyhedrosis viruses for viremia or antibodies to these viruses in the blood. Birds must receive a high dose of virus after ingesting diseased insects, and this inoculum will contain not only occluded virus within the polyhedra but also all the stages of virus synthesis, possibly including infectious DNA. This situation would also apply to any small insectivorous mammal that had access to diseased larvae or pupae. Such infections in birds or animals could well be subclinical as commonly occurs with arbovirus infections; thus, without deliberate investigation, they would go undetected. Therefore, it is dangerous to claim that, as no serious effects were observed in certain wild animals, it follows that no cross-infections had taken place.

Few attempts have been made to test the susceptibility of vertebrate cell lines to infection with insect viruses. Himeno *et al.* (1967) reported that viral DNA extracted from the silkworm *B. mori* previously infected with a nuclear polyhedrosis virus was infectious for human amnion (FL) cells, and that such cells produced progeny virus and polyhedra. The virus produced in the human cells was morphologically and serologically similar to virus produced in insects, was infective to other test insects, and caused typical disease. These results have not yet been confirmed by other workers. Ignoffo (1968) quoted several personal communications which indicated that the iridescent virus of *T. paludosa* would not infect mammalian HeLa cells, the cytoplasmic polyhedrosis virus of *B. mori* would not infect HeLa, human amnion, or pig kidney cells, and the nuclear polyhedrosis virus of *H. zea* would not infect HeLa, African green monkey kidney, human embryonic kidney, or human diploid embryonic lung cells. Clearly, further work is required to establish whether or not vertebrate cells are susceptible to insect pathogenic viruses, and it would be useful to include cell lines derived from cold-blooded animals in such experiments. The results reported by McIntosh and Kimura (1974) demonstrate the value of such an approach.

5.2. Occurrence of Natural Antibodies to Invertebrate Viruses in Domestic and Wild Animals

In 1966, Harrap *et al.* isolated an RNA virus from the moth *Gonometa podocarpi* (Lepidoptera: Lasiocampidae). This was later characterized by Longworth *et al.* (1973*a*) and shown to have similar physical and chemical properties to the picornavirus–enterovirus group of mammalian viruses. This led Longworth *et al.* (1973*b*) to test the *Gonometa* virus against foot and mouth disease virus (FMDV) antisera. There were no cross-reactions when hyperimmune guinea pig sera were used, but marked reactions were observed when convalescent sera from pigs previously infected with FMDV were included. It was significant that antisera against *Gonometa* virus did not react with FMDV antigens. These workers concluded that the convalescent pigs had antibodies to an agent probably unrelated to FMDV. Later tests showed that similar, naturally occurring antibodies could be found in the sera of cattle, horses, sheep, dogs, and three species of deer. No cross-reacting antibodies were found in the sera of rabbits and guinea pigs. An intriguing aspect of this work was that the reacting antibodies in these domestic and wild animals were of the IgM type and not IgG as might have been expected. The conclusion drawn was that as the insect virus originated in East Africa (Uganda) it was very unlikely that the test animals could have been exposed to it in the United Kingdom. Therefore, Longworth *et al.* (1973*b*) suggested that the antibodies had been produced to a second, unknown virus that shared common antigens with the insect virus. Further, because the reactions involved IgM antibodies, the stimulus was probably of low order but of a persistent nature. Since then, two samples of human sera have been found to have antibodies capable of reacting with the *Gonometa* virus, but shortage of viral antigen has so far precluded the determination of the type of antibody involved (Tinsley, unpublished data). It is hoped to extend the survey with sera from wild and domestic animals in East Africa and with more human sera once further supplies of the *Gonometa* virus are available.

5.3. Insect Viruses as Pesticidal Agents

It has been said that insects are man's greatest competitor for food, but additionally they can cause him and his domestic animals considerable physical discomfort either by direct irritation or as vectors of disease-causing agents. Man's answer to the pest problem has been to attempt to reduce insect numbers, and chemical insecticides have emerged as the first line of defense. Since World War II, the activity of insect pests and disease vectors has been controlled in large measure by regular applications of persistent insecticides usually containing BHC and/or DDT. Such insecticides have been used to greatly reduce crop losses by insect attack and also to reduce the incidence of pathogenic diseases carried by blood-sucking insects. The ravages of leaf-defoliating caterpillars on cotton in Africa and the United States were reduced to a minor level by regular applications of such insecticides, while the eradication of malaria from certain tropical and subtropical countries was made possible only by sustained and large-scale application of DDT, which eliminated the specific mosquito vectors. However, in the last decade, serious and unpredictable events have occurred which have led to a reappraisal of the wisdom of large-scale applications of insecticides. The target pests both of crops and of man began to exhibit resistance to the pesticides, and recent estimates suggest that some 300 different insect species have developed such reactions. To remedy this, the dosage levels of the chemicals have been increased or newer compounds introduced. Unfortunately, resistance of a pest to one insecticide is frequently found to be nonspecific and so other insecticides are also similarly involved. This problem was aggravated by the discovery of chemical residues of a persistent nature in fish and wild birds, and it was evident that these contaminants owed their origin to insecticides which had now taken on a new but insidious role as environmental pollutants. A vicious circle had thus been created.

Quite recently, further complicating factors have emerged—world wide inflation and associated increased costs of petrochemicals. This has resulted in insecticides being more expensive and the costs of application rising dramatically. The need to provide cheaper, alternative methods of pest control has never been more urgent than today. Public anxiety over chemical pollution has naturally favored a reappraisal of the biological control of insect pests. Integrated control of pests is a system whereby cultural manipulation and the breeding of resistant crop varieties are combined with natural enemies of the insects such as parasites, predators, and pathogens. This is not a new concept as it has been practiced at one time or another since the turn of the century. Apart from several spectacular successes arising from the introduction of exotic parasites and predators, these natural systems were frequently rejected either because the level of control achieved was less than that obtained with chemicals or because the results were quite unpredictable in that adequate control was achieved in some years and not in others. The ability of pathogens to invade and kill their insect hosts was well known to Pasteur from his work with diseases of the silkworm. Many studies have been made since then to employ fungi, bacteria, protozoans, and viruses as pesticidal agents. Most authorities are now agreed that viruses offer the greatest potential in biological control systems in that highly virulent viruses can be isolated which can cause high levels of mortality in insect populations. It is very unlikely that insect viruses could ever replace chemical control in all situations, but they could exert a key role in the reduction of the incidence of major pests to an acceptable level. However, viruses are replicating systems and must be used in a responsible manner, for, once released, it could prove difficult to contain spread, particularly if they proved to cause disease in organisms other than the target pest. It is fundamental to realize that the indiscriminate release of insect viruses would be just as reprehensible as was the indiscriminate use of persistent chemical insectides.

There are at least seven groups of viruses which have been isolated from insects, and in this context it would be essential to consider only those virus types which appear to be confined to the class Insecta or, at very least, to Invertebrata. Such a group is the baculoviruses, which evidently have no chemical, physical, or biological properties in common with any known virus found in either vertebrates or plants. It was for this reason that WHO and FAO jointly recommended that only the baculovirus group should be considered as possible pesticidal agents (WHO Technical Report Series No. 531, 1973).

The consideration of a virus for biological control purposes involves four essential stages of investigation. These are

1. The isolation, purification, and characterization of the virus, leading to the provision of methods of unequivocal identification.
2. The testing under laboratory conditions of the efficacy of the purified virus as a pesticidal agent and the establishment of its host range.
3. The testing of the toxicological properties of the virus, together with any associated formulative materials, and investigation of the possibility of infection and replication in nontarget invertebrates and vertebrates.
4. Once the virus has satisfactorily passed these rigorous safety testing procedures, field trials to confirm efficacy and dosage rates. A system of monitoring the environment is implicit before the trials begin, during application, and for a considerable period thereafter.

Once the viruses have been cleared of any potential ecological hazard, field tests in the areas in which the pests occur could be undertaken. However, there is still a great deal of work to be done before insect viruses can be used on any effective scale, and it is vital to devise adequate and sensitive systems of monitoring virus infections in the areas of application. Unequivocal methods of identification are necessary to prove that the target insects died as a result of infection of the applied virus and not from a second, possibly unrelated virus that was either present in the population as a subclinical infection or resulted from a cross-transmission from another pest.

ACKNOWLEDGMENT

We would like to thank our colleagues Mr. J. S. Robertson, Dr. D. C. Kelly, and Dr. C. C. Payne for much help and invaluable discussions during the preparation of the manuscript. We are also indebted to Mrs. J. Bald for invaluable secretarial services.

One of us (K. A. H.) would like to thank Dr. R. W. Schlesinger, Dr. V. Stollar, and other members of the staff of the Department of Microbiology, Rutgers Medical School, for encouragement and help during the writing of parts of this chapter.

6. REFERENCES

Aizawa, K., 1954, Dissolving curve and the virus activity of the polyhedral bodies of *Bombyx mori* L. obtained 37 years ago, *Sanshi Kenkyu* **8**:52–54.

Aizawa, K., 1959, Mode of multiplication of silkworm polyhedrosis virus. II. Multiplication of the virus in the early period of the LD_{50} time curve, *J. Insect Pathol.* **1**:67–74.

Aizawa, K., 1962, Antiviral substance in the gut-juice of the silkworm *Bombyx mori* L., *J. Insect Pathol.* **4**:72–76.

Aizawa, K., 1967, Mode of multiplication of the nuclear polyhedrosis virus of the silkworm, *J. Sericult. Soc. Jap.* **36**:327.

Aizawa, K., and Iida, S., 1963, Nucleic acids extracted from the virus polyhedra of the silkworm, *Bombyx mori* (Linnaeus), *J. Insect Pathol.* **5**:344–348.

Akai, H., Gateff, E., Davis, L. E., and Schneiderman, H. A., 1967, Virus-like particles in normal and tumorous tissues of *Drosophila, Science* **157**:810–813.

Anderson, J. F., 1970, An iridescent virus infecting the mosquito *Aedes stimulans, J. Invertebr. Pathol.* **15**:219–224.

Anthony, D. W., Hazard, E. I., and Crosby, S. W., 1973, A virus disease in *Anopheles quadrimaculatus, J. Invertebr. Pathol.* **22**:1–5.

Arnott, H. J., and Smith, K. M., 1968a, An ultrastructural study of the development of a granulosis virus in the cells of the moth *Plodia interpunctella* (Hbn.), *J. Ultrastruct. Res.* **21**:251–268.

Arnott, H. J., and Smith, K. M., 1968b, Ultrastructure and formation of abnormal capsules in a granulosis virus of the moth *Plodia interpunctella* (Hbn.), *J. Ultrastruct. Res.* **22**:136–158.

Arnott, H. J., Smith, K. M., and Fullilove, S. L., 1968, Ultrastructure of a cytoplasmic polyhedrosis virus affecting the monarch butterfly, *Danaus plexippus*. I. Development of virus and normal polyhedra in the larva. *J. Ultrastruct. Res.* **24**:479–507.

Aruga, H., 1963, Induction of virus infections, in: *Insect Pathology,* Vol. I (E. A. Steinhaus, ed.), pp. 499–530, Academic Press, New York.

Aruga, H., and Watanabe, H., 1964, Resistance to *per os* infection with cytoplasmic polyhedrosis virus in the silkworm *Bombyx mori* L., *J. Insect Pathol.* **6**:387–394.

Asai, J., Kawamoto, F., and Kawase, S., 1972, On the structure of the cytoplasmic polyhedrosis virus of the silkworm, *Bombyx mori, J. Invertebr. Pathol.* **19**:279–280.

Bailey, L., and Gibbs, A. J., 1964, Acute infection of bees with paralysis virus, *J. Insect Pathol.* **6**:395–407.

Bailey, L., and Milne, R. G., 1969, The multiplication regions and interaction of acute and chronic bee paralysis viruses in adult honey bees, *J. Gen. Virol.* **4**:9–14.

Bailey, L., and Scott, H. A., 1973, The pathogenicity of Nodamura virus for insects. *Nature (London)* **241**:545.

Bailey, L., and Woods, R. D., 1974, Three previously undescribed viruses from the honey bee, *J. Gen. Virol.* **25**:175–186.

Bailey, L., Gibbs, A. J., and Woods, R. D., 1963, Two viruses from adult honey bees (*Apis mellifera*. Linnaeus), *Virology* **21**:390–395.

Bailey, L., Gibbs, A. J., and Woods, R. D., 1964, Sacbrood virus of the larval honey bee (*Apis mellifera*. Linnaeus), *Virology* **23**:425–429.

Bailey, L., Gibbs, A. J., and Woods, R. D., 1968, The purification and properties of chronic bee paralysis virus, *J. Gen. Virol.* **2**:250–260.

Bailey, L., Gibbs, A. J., and Woods, R. D., 1970, A simple way of purifying several insect viruses, *J. Gen. Virol.* **6**:175–177.

Bailey, L., Newman, J. F. E., and Porterfield, J. G., 1975, The multiplication of Nodamura virus in insect and mammalian cell cultures, *J. Gen. Virol.* **26**:15–20.

Balch, R. E., 1939, The outbreak of the European spruce sawfly in Canada and some important features of its bionomics, *J. Econ. Entomol.* **32**:412–418.

Balch, R. E., and Bird, F. T., 1944, A disease of the European spruce sawfly *Gilpinia hercyniae* (Htg.) and its place in natural control, *Sci. Agr.* **25**:65–80.

Bang, F. B., 1971, Transmissible disease, probably viral in origin, affecting the amebocytes of the European shore crab, *Carcinus maenas, Infect. Immun.* **3**:617–623.

Barefield, K. P., and Stairs, G. R., 1969, Infectious components of granulosis virus of the codling moth *Carpocapsa pomonella, J. Invertebr. Pathol.* **15**:401.

Barwise, A. H., and Walker, I. O., 1970, Studies on the DNA of a virus from *Galleria mellonella, FEBS Lett.* **6**:13–15.

Bellett, A. J. D., 1965a, The multiplication of *Sericesthis* iridescent virus in cell cultures from *Antherea eucalypti* Scott. II. An *in vitro* assay for the virus, *Virology* **26**:127–131.

Bellett, A. J. D., 1965b, The multiplication of *Sericesthis* iridescent virus in cell cultures from *Antherea eucalypti* Scott. III. Quantitative experiments, *Virology* **26**:132–141.

Bellett, A. J. D., 1968, The iridescent virus group, *Adv. Virus Res.* **13**:225–246.

Bellett, A. J. D., 1969, Relationships among the polyhedrosis and granulosis viruses of insects, *Virology* **37**:117–123.

Bellett, A. J. D., and Fenner, F., 1968, Studies of base-sequence homology among some cytoplasmic deoxyriboviruses of vertebrate and invertebrate animals, *J. Virol.* **2**:1374–1380.

Bellett, A. J. D., and Inman, R. B., 1967, Some properties of deoxyribonucleic acid preparations from *Chilo, Sericesthis* and *Tipula* iridescent viruses, *J. Mol. Biol.* **25**:425–432.

Bellett, A. J. D., and Mercer, E. H., 1964, The multiplication of *Sericesthis* iridescent virus in cell cultures from *Antherea eucalypti* Scott. I. Qualitative experiments. *Virology* **24**:645–653.

Bellett, A. J. D., Fenner, F., and Gibbs, A. J., 1973, The viruses, in: *Viruses and Invertebrates* (A. J. Gibbs, ed.), pp. 41–88, North-Holland, Amsterdam.

Benz, G., and Wäger, R., 1971, Autoradiographic studies on nucleic acid metabolism in granulosis-infected fat body of larvae of *Carpocapsa, J. Invertebr. Pathol.* **18**:70–80.

Bergoin, M., and Dales, S., 1971, Comparative observations on poxviruses of invertebrates and vertebrates, in: *Comparative Virology* (K. Maramorosch and E. Kurstak, eds.), pp. 169–205, Academic Press, New York.

Bergoin, M., and Devauchelle, G., 1972, Invertebrate poxviruses, in: *Moving Frontiers in Invertebrate Virology* (T. W. Tinsley and K. A. Harrap, eds.), No. 6 of *Monographs in Virology,* pp. 3–7, Karger, Basel.

Bergoin, M., Devauchelle, G., Duthoit, J.-L., and Vago, C., 1968a, Étude au microscope électronique des inclusions de la virose à fuseaux des Coléoptères, *C. R. Acad. Sci. Ser. D* **266**:2126–2128.

Bergoin, M., Devauchelle, G., and Vago, C., 1968b, Observations au microscope électronique sur le développment du virus de la 'maladie à fuseaux' du Coléoptère, *Melolontha melolontha* L., *C. R. Acad. Sci. Ser. D* **267**:382–385.

Bergoin, M., Devauchelle, G., and Vago, C., 1969, Electron microscopy study of the poxlike virus of *Melolontha melolontha* L. (Coleoptera: Scarabeidae), *Arch. Gesamte Virusforsch.* **28**:285–302.

Bergoin, M., Devauchelle, G., and Vago, C., 1971, Electron microscopy study of *Melolontha* poxvirus: The fine structure of occluded virions, *Virology* **43**:453–467.

Bergold, G. H., 1943, Über Polyederkrankheiten bei Insecten, *Biol. Zentralbl.* **63**:1–55.

Bergold, G. H., 1947, Die Isolierung des Polyeder-Virus und die Natur der Polyeder, *Z. Naturforsch* **26**:122–143.

Bergold, G. H., 1963a, The molecular structure of some insect virus inclusion bodies, *J. Ultrastruct. Res.* **8**:360–378.

Bergold, G. H., 1963b, The nature of nuclear polyhedrosis viruses, in: *Insect Pathology,* Vol. 1 (E. A. Steinhaus, ed.), pp. 413–456, Academic Press, New York.

Bergold, G. H., Aizawa, K., Smith, K., Steinhaus, E. A., and Vago, C., 1960, The present status of insect virus nomenclature and classification, *Int. Bull. Bacteriol. Nomencl. Taxon.* **10**:259–262.

Berkaloff, A., Bregliano, J.-C., and Ohanessian, A., 1965, Mise en évidence de virions dans des drosophiles inféctees par le virus héréditaire σ, *C. R. Acad. Sci. Ser. D* **260**:5956–5959.

Bernheimer, A. W., Caspari, E., and Kaiser, A. D., 1952, Studies on antibody formation in caterpillars, *J. Exp. Zool.* **119**:23–25.

Bird, F. T., 1953, The effect of metamorphosis on the multiplication of an insect virus, *Can. J. Zool.* **31**:300–303.

Bird, F. T., 1955, Viruses diseases of sawflies, *Can. Entomol.* **87**:124–127.

Bird, F. T., 1961a, The development of *Tipula* iridescent virus in the crane fly, *Tipula paludosa* Meig. and the wax moth *Galleria mellonella* L., *Can. J. Microbiol.* **7**:827–830.

Bird, F. T., 1961b, Transmission of some insect viruses with particular reference to ovarial transmission and its importance in the development of epizootics, *J. Insect Pathol.* **3**:352–380.

Bird, F. T., 1967, A virus disease of the European red mite *Panonychus ulmi* (Koch), *Can. J. Microbiol.* **13**:1131.

Bird, F. T., 1974, The development of spindle inclusions of *Choristoneura fumiferana* (Lepidoptera: Tortricidae) infected with entomopox virus, *J. Invertebr. Pathol.* **23**:325–332.

Bird, F. T., and Elgee, D. E., 1957, A virus disease and introduced parasites as factors controlling the European spruce sawfly, *Diprion hercyniae* (Htg.) in Central New Brunswick, *Can. Entomol.* **89**:371–378.

Bolle, J., 1894, Il Giallume del baco da seta: Notizia preliminaire, *Atti Mem. I.R. Soc. Agr. Gorizia* **33**:193.

Bonami, J. R., and Vago, C., 1971, Virus of a new type pathogenic to Crustacea, *Experientia* **27**,1363–1364.

Bonami, J. R., Grisel, H., Vago, C., and Duthoit, J. L., 1971, Recherches sur une maladie épizootique de l'huitre plate, *Ostrea* edulis linne, *Rev. Trav. Inst. Peches Marit.* **35**:415–418.

Bonnefoy, A. M., Kilenkine, X., and Vago, C., 1972, Virus-like particles in *Hydra vulgaris* (Pallas), *C. R. Acad. Sci. Ser. D* **275**:2163–2165.

Briggs, J. D., 1958, Humoral immunity in lepidopterous larvae, *J. Exp. Zool.* **138**:155–188.

Brown, F., and Hull, R., 1973, Comparative virology of the small RNA viruses, *J. Gen. Virol. Suppl.* **20**:43–60.

Brzostowski, H. W., and Grace, T. D. C., 1970, Observations of the infectivity of RNA from *Antherea* virus (AV), *J. Invertebr. Pathol.* **16**:277–279.

Bussereau, F., 1970a, Etude du symptôme de la sensibilité au CO_2 produit par le virus sigma chez la drosophile. I. Influence due lieu d'inoculation sur le delai d'apparition du symptôme, *Ann. Inst. Pasteur Paris* **118**:367–385.

Bussereau, F., 1970b, Étude du symptôme de la sensibilité au CO_2 produit par le virus sigma chez la drosophile. II. Evolution comparée du rendement des centres nerveux et de divers organes après inoculation dans l'abdomen et dans le thorax, *Ann. Inst. Pasteur Paris* **18**:626–645.

Carter, J. B., 1973a, The mode of transmission of *Tipula* iridescent virus. I. Source of infection, *J. Invertebr. Pathol.* **21**:123–130.

Carter, J. B., 1973b, Detection and assay of *Tipula* iridescent virus by the latex agglutination test, *J. Gen. Virol.* **21**:181–185.

Carter, J. B., 1974, *Tipula* iridescent virus infection on the developmental stages of *Tipula oleracea, J. Invertebr. Pathol.* **24**:271–281.

Chamberlain, R. W., 1968, Arboviruses, the arthropodborne animal viruses, *Curr. Top. Microbiol. Immunol.* **42**:38–58.

Chamberlain, R. W., and Sudia, W. D., 1961, Mechanism of transmission of viruses by mosquitoes, *Annu. Rev. Entomol.* **6**:371–390.

Chapman, H. C., Clark, T. B., Woodard, D. B., and Kellen, W. R., 1966, Additional mosquito hosts of the mosquito iridescent virus, *J. Invertebr. Pathol.* **8**:545–546.

Chapman, H. C., Peterson, J. J., Woodard, D. B., and Clark, T. B., 1968, New records of parasites of *Ceratopogonidae*, *Mosquito News* **28**:123–125.

Chapman, H. C., Clark, T. B., Anthony, D. W., and Glenn, F. E., 1971, An iridescent virus from the larvae of *Corethrella brakeleyi* (Diptera: Chaoboridae) in Louisiana, *J. Invertebr. Pathol.* **18**:284–286.

Clark, E. C., 1956, Survival and transmission of a virus causing polyhedrosis in *Malacosoma fragile*, *Ecology* **37**:728–732.

Clark, T. B., Kellen, W. R., and Lum, P. T. M., 1965, A mosquito iridescent virus (MIV) from *Aedes taeniorhynchus* (Wiedemann), *J. Invertebr. Pathol.* **7**:519–521.

Cline, G. B., Ryel, E., Ignoffo, C. M., Shapiro, M., and Strachle, W., 1970, Zonal purification studies on the nucleopolyhedrosis virus of the cotton bollworm *Heliothis zea* (Boddie), in: *Proc. IV Int. Coll. Insect Pathol.* pp. 363–370, College Park, Md., August 1970.

Cohen, A., 1963, Mechanisms of cell infection. I. Virus attachment and penetration, in: *Mechanisms of Virus Infection* (W. Smith, ed.), pp. 153–196, Academic Press, New York.

Cornalia, E., 1856, Monografia del bombice de gelso, *Mem. Ist. Lombardo* **6**:3–387.

Couch, J. A., 1974a, Free and occluded virus, similar to baculovirus, in hepatopancreas of the pink shrimp, *Nature (London)* **247**:229–231.

Couch, J. A., 1974b, An enzootic nuclear polyhedrosis virus of pink shrimp: Ultrastructure, prevalence and enhancement, *J. Invertebr. Pathol.* **24**:311–331.

Cunningham, J. C., and Hayashi, Y., 1970, Replication of *Chilo* iridescent virus in *Galleria mellonella*: Purification of the virus and the effect of actinomycin D and puromycin, *J. Invertebr. Pathol.* **16**:427–435.

Cunningham, J. C., and Longworth, J., 1968, The identification of some cytoplasmic polyhedrosis viruses, *J. Invertebr. Pathol.* **11**:196–202.

Cunningham, J. C., and Tinsley, T. W., 1968, A serological comparison of some iridescent non-occluded insect viruses, *J. Gen. Virol.* **3**:1–8.

Cunningham, J. C., Tinsley, T. W., and Walker, J. M., 1966, Haemagglutination with plant and insect viruses, *J. Gen. Microbiol.* **42**:397–401.

David, W. A. L., 1957, Breeding *Pieris brassicae* L. and *Apanteles glomeratus* L. as experimental insects, *Z. Pflanzenkr. Pflanzenschutz* **64**:572–577.

David, W. A. L., 1965, The granulosis virus of *Pieris brassicae* L., in relation to natural limitation and biological control, *Ann. Appl. Biol.* **56**:331–334.

David, W. A. L., 1970, Current problems in insect virology, in: *Proc. IV Int. Coll. Insect Pathol.* College Park, Md.

David, W. A. L., and Gardiner, B. O. C., 1960, A *Pieris brassicae* L. culture resistant to a granulosis, *J. Insect Pathol.* **2**:106–114.

Day, M. F., and Mercer, E. H., 1964, Properties of an iridescent virus from the beetle, *Sericesthis pruinosa*, *Aust. J. Biol. Sci.* **17**:892–902.

Devauchelle, G., and Durchon, M., 1973, Sur la présence d'un virus du type Iridovirus, dans les cellules mâles de *Nereis diversicolor* (O. F. Muller), *C. R. Acad. Sci. Ser. D* **277**:463–466.

Devauchelle, G., and Vago, C., 1971, Virus-like particles in *Sepia officinalis* (Cephalopod), *C. R. Acad. Sci. Ser. D* **272**:894–896.

Devauchelle, G., Bergoin, M., and Vago, C., 1971, Étude ultrastructurale du cycle de replication d'un entomopoxvirus dans les hémocytes de son hôte, *J. Ultrastruct. Res.* **37**:301–321.

Diamond, L. S., Mattern, C. F. T., and Bartgis, I. L., 1972, Viruses of *Entamoeba histolytica*. Identification of transmissible virus-like agents, *J. Virol.* **9**:326–341.

Doane, C. C., 1969, Transovum transmission of a nuclear polyhedrosis virus in the gypsy moth and the inducement of virus susceptibility, *J. Invertebr. Pathol.* **14**:199–210.

Dougherty, C. C., Ferral, D. J., Brody, B., and Gotthold, M. L., 1963, A growth anomaly and lysis with production of virus-like particles in an axenically reared microannelid, *Nature (London)* **198**:973–975.

Dunnebacke, T. H., and Schuster, F. L., 1971, Infectious agent from a free-living soil amoeba, *Naegleria gruberi, Science* **174**:516–518.

Egawa, K., and Summers, M. D., 1972, Solubilization of *Trichoplusia ni* granulosis virus proteinic crystal. I. Kinetics, *J. Invertebr. Pathol.* **19**:395–404.

Engström, A., and Kilkson, R., 1968, Molecular organization in the polyhedra of *Porthetria dispar* nuclear-polyhedrosis, *Exp. Cell Res.* **53**:305–310.

Entwistle, P. F., 1974, Epizootiology of a nuclear polyhedrosis virus in the European spruce sawfly, in: *Report of the National Environment Research Council, 1973–1974,* HMSO, London.

Estes, Z. E., and Faust, R. M., 1965, The nucleic acid composition of a virus affecting the citrus red mite *Panonychus citri* (McGregor), *J. Invertebr. Pathol.* **7**:259–260.

Estes, Z. E., and Faust, R. M., 1966, Silicon content of insect nuclear polyhedra from the corn ear worm, *Heliothis zea, J. Invertebr. Pathol.* **8**:145–149.

Farley, C. A., Banfield, W. G., Kasnie, G., Jr., and Foster, W. S., 1972, Oyster herpestype virus, *Science* **178**:759–760.

Faulkner, P., 1962, Isolation and analysis of ribonucleic acid from inclusion bodies of the nuclear polyhedrosis of the silkworm, *Virology* **16**:479–484.

Faulkner, P., and Henderson, J. F., 1972, Serial passage of a nuclear polyhedrosis disease virus of the cabbage looper (*Trichoplusia ni*) in a continuous tissue culture cell line, *Virology* **50**:920–924.

Faust, R. M., and Adams, J. R., 1966, The silicon content of nuclear and cytoplasmic viral inclusion bodies causing polyhedrosis in Lepidoptera, *J. Invertebr. Pathol.* **14**:186–193.

Faust, R. M., Dougherty, E. M., and Adams, J. R., 1968, Nucleic acid in the bluegreen and orange mosquito iridescent viruses (MIV) isolated from larvae of *Aedes taeniorhynchus, J. Invertebr. Pathol.* **10**:160.

Federici, B. A., and Hazard, E. I., 1975, Iridovirus and cytoplasmic polyhedrosis virus in the freshwater daphnid *Simocephalus expinosus, Nature (London)* **254**:327–328.

Federici, B. A., and Lowe, R. E., 1972, Studies on the pathology of a baculovirus in *Aedes triseriatus, J. Invertebr. Pathol.* **20**:14–21.

Federici, B. A., Hazard, E. I., and Anthony, D. W., 1973, A new cytoplasmic polyhedrosis virus from chironomids collected in Florida, *J. Invertebr. Pathol.* **22**:136–138.

Felluga, B., Jonsson, V., and Liljeros, M. R., 1971, Ultrastructure of new viruslike particles in *Drosophila, J. Invertebr. Pathol.* **17**:339–346.

Finch, J. T., Crowther, R. A., Hendry, D. A., and Struthers, J. K., 1974, The structure of *Nudaurelia capensis* β virus: The first example of a capsid with icosahedral surface symmetry T = 4, *J. Gen. Virol.* **24**:191–200.

Foor, W. E., 1972, Virus-like particles in nematode *Trichosomoides crassicauda, J. Parasitol.* **58**:1065–1070.

Fowler, M., and Robertson, J. S., 1972, Iridescent virus infection in field populations of *Wiseana cervinata* (Lepidoptera: Hepialidae) and *Witlesia* sp. (Lepidoptera: Pyralidae) in New Zealand, *J. Invertebr. Pathol.* **19:**154–155.

Fujii-Kawata, I., and Miura, K.-I., 1970, Segments of genome of viruses containing double-stranded ribonucleic acid, *J. Mol. Biol.* **51:**247–253.

Fukaya, M., and Nasu, S., 1966, A *Chilo* iridescent virus (CIV) from the rice stem borer, *Chilo suppressalis* Walker (Lepidoptera, Pyralidae), *Appl. Entomol. Zool.* **1:**69–72.

Furgala, B., and Lee, P. E., 1966, Acute bee paralysis virus, a cytoplasmic insect virus, *Virology* **29:**346–348.

Furuichi, Y., 1974, "Methylation-coupled" transcription by virus-associated transcriptase of cytoplasmic polyhedrosis virus containing double-stranded RNA, *Nucleic Acids Res.* **1:**809–822.

Furuichi, Y., and Miura, K., 1972, The 3′ termini of the genome RNA segments of silkworm cytoplasmic polyhedrosis virus, *J. Mol. Biol.* **64:**619–632.

Furuichi, Y., and Miura, K., 1975, A blocked structure at the 5′-terminus of mRNA from cytoplasmic polyhedrosis virus, *Nature (London)* **253:**374–375.

Furusawa, T., and Kawase, S., 1971, Synthesis of ribonucleic acid resistant to actinomycin D in silkworm midguts infected with the cytoplasmic polyhedrosis virus, *J. Invertebr. Pathol.* **18:**156–158.

Furusawa, T., and Kawase, S., 1973, Virus-specific RNA synthesis in the midgut of silkworm, *Bombyx mori,* infected with cytoplasmic polyhedrosis virus, *J. Invertebr. Pathol.* **22:**335–344.

Gartner, L. P., 1972, Virus-like particles in the adult *Drosophila* midgut, *J. Invertebr. Pathol.* **20:**364–366.

Gershenson, S. M., Kok, I. P., Vitas, K. I., Dobrovolskaya, G. N., and Skuratovskaya, I. N., 1963, Formation of a DNA-containing virus by host RNA in: *Proc. 5th Int. Congr. Biochem. Moscow, 1961,* Vol. 9, p. 150, Pergamon Press, Oxford.

Gibbs, A. J., Gay, F. J., and Wetherly, A. H., 1970, A possible paralysis virus of termites, *Virology* **40:**1063–1005.

Glitz, D. G., Hills, G. J., and Rivers, C. F., 1968, A comparison of the *Tipula* and *Sericesthis* iridescent viruses, *J. Gen. Virol.* **3:**209–220.

Gold, P., and Dales, S., 1968, Localization of nucleotide phosphohydrolase activity within vaccinia, *Proc. Natl. Acad. Sci. U.S.A.* **60:**845–852.

Goodwin, R. H., and Filshie, B. K., 1969, Morphology and development of an occluded virus from the black-soil scarab *Othnonius batesi, J. Invertebr. Pathol.* **13:**317–329.

Goodwin, R. H., and Filshie, B. K., 1975, Morphology and development of entomo-poxviruses from two Australian scarab beetle larvae (Coleoptera: Scarabaeidae), *J. Invertebr. Pathol.* **25:**35–46.

Götz, P., Huger, A. M., and Krieg, A., 1969, Über em insektenpathogenes Virus aus der Gruppe der Pockenvirus, *Naturwissenschaften* **56:**145.

Gouranton, J., 1972, Development of an intranuclear nonoccluded rod-shaped virus in some midgut cells of an adult insect, *Gyrinus natator* L. (Coleoptera), *J. Ultrastruct. Res.* **39:**281–294.

Grace, T. D. C., 1962, The development of a cytoplasmic polyhedrosis in insect cells grown *in vitro, Virology* **18:**33–42.

Grace, T. D. C., and Mercer, E. H., 1965, A new virus of the saturniid *Antherea eucalypti* Scott, *J. Invertebr. Pathol.* **7:**241–244.

Granados, R. R., 1973, Entry of an insect poxvirus by fusion of the virus envelope with the host cell membrane, *Virology* **52**:305–309.

Granados, R. R., and Roberts, D. W., 1970, Electron microscopy of a pox-like virus infecting an invertebrate host, *Virology* **40**:230–243.

Granados, R. R., McCarthy, W. J., and Naughton, M., 1974, Replication of a cytoplasmic polyhedrosis virus in an established cell line of *Trichoplusia ni* cells, *Virology* **59**:584–586.

Grimes, G. W., and Preer, J. R., 1971, Further observations on the correlation between kappa and phage-like particles in paramecium, *Genet. Res.* **18**:115–116.

Gudankas, R. T., and Canerday, D., 1968, The effect of heat, buffer salt and H-ion concentration, and ultra violet light on the infectivity of *Heliothis* and *Trichoplusia* nuclear polyhedrosis viruses, *J. Invertebr. Pathol.* **12**:405–411.

Hall, D. W., and Anthony, D. W., 1971, Pathology of a mosquito iridescent virus (MIV) infecting *Aedes taeniorhynchus, J. Invertebr. Pathol.* **18**:61–69.

Hall, D. W., and Hazard, E. I., 1973, A nuclear polyhedrosis virus of a caddisfly, *Neophylax* sp., *J. Invertebr. Pathol.* **21**:323–324.

Hamm, J. J., and Young, J. R., 1974, Mode of transmission of nuclear polyhedrosis virus to progeny of adult *Heliothis zea, J. Invertebr. Pathol.* **24**:70–81.

Harrap, K. A., 1969, in: "Viruses of Invertebrates," *Proc. 1st Int. Congress Virol.,* 1968, Helsinki, *Int. Virol.* **1**:281.

Harrap, K. A., 1970, Cell infection by a nuclear polyhedrosis virus, *Virology* **42**:311–318.

Harrap, K. A., 1972*a*, The structure of nuclear polyhedrosis viruses. I. The inclusion body, *Virology* **50**:114–123.

Harrap, K. A., 1972*b*, The structure of nuclear polyhedrosis viruses. II. The virus particle, *Virology* **50**:124–132.

Harrap, K. A., 1972*c*, The structure of nuclear polyhedrosis viruses. III. Virus assembly, *Virology* **50**:133–139.

Harrap, K. A., 1973, Virus infection in invertebrates, in: *Viruses and Invertebrates* (A. J. Gibbs, ed.), pp. 271–299, North-Holland, Amsterdam.

Harrap, K. A., and Juniper, B. E., 1966, The internal structure of an insect virus, *Virology* **29**:175–178.

Harrap, K. A., and Longworth, J. F., 1974, An evaluation of purification methods for baculoviruses, *J. Invertebr. Pathol.* **24**:55–62.

Harrap, K. A., and Robertson, J. S., 1968, A possible infection pathway in the development of a nuclear polyhedrosis virus, *J. Gen. Virol.* **3**:221–225.

Harrap, K. A., and Tinsley, T. W., 1971, A suggested latinized nomenclature of occluded insect viruses, *J. Invertebr. Pathol.* **17**:294–296.

Harrap, K. A., Longworth, J. F., Tinsley, T. W., and Brown, K. W., 1966, A non-inclusion virus of *Gonometa podocarpi* (Lepidoptera: Lasiocampidae), *J. Invertebr. Pathol.* **8**:270–272.

Harrison, S. C., Caspar, D. L. D., Camerini-Otero, R. D., and Franklin, R. M., 1971*a*, Lipid and protein arrangement in bacteriophage, PM2, *Nature (London)* New Biol. **229**:197–201.

Harrison, S. C., David, A., Jumblatt, J., and Darnell, J. E., 1971*b*, Lipid and protein organization in Sindbis virus, *J. Mol. Biol.* **60**:523–528.

Hasan, S., Croizier, G., Vago, C., and Duthoit, J. L., 1970. Infection à virus irisant dans une population naturelle d'*Aedes detritus* Haliday en France, *Ann. Zool. Ecol. Anim.* **2**:295–299.

Hayashi, Y., 1970*a*, Occluded and free virions in midgut cells, of *Malacosoma disstria* infected with cytoplasmic polyhedrosis (CPV), *J. Invertebr. Pathol.* **16**:442–450.

Hayashi, Y., 1970*b*, Properties of RNA from cytoplasmic polyhedrosis virus of the white-marked tussock moth, *Orgyia leucostigma, J. Invertebr. Pathol.* **16**:451–458.

Hayashi, Y., 1970*c*, RNA in midgut of tussock moth, *Orgyia leucostigma,* infected with cytoplasmic polyhedrosis virus, *Can J. Microbiol.* **16**:1101–1107.

Hayashi, Y., and Bird, F. T., 1968*a*, The use of sucrose gradients in the isolation of cytoplasmic polyhedrosis virus particles, *J. Invertebr. Pathol.* **11**:40–44.

Hayashi, Y., and Bird, F. T., 1968*b*, Properties of a cytoplasmic polyhedrosis virus from the white-marked tussock moth, *J. Invertebr. Pathol.* **12**:140.

Hayashi, Y., and Bird, F. T., 1970, The isolation of cytoplasmic polyhedrosis virus from the white-marked tussock moth, *Orgyia leucostigma* (Smith), *Can. J. Microbiol.* **16**:695–701.

Hayashi, Y., and Donaghue, T. P., 1971, Cytoplasmic polyhedrosis virus: RNA synthesized *in vivo* and *in vitro* in infected midgut, *Biochem. Biophys. Res. Commun.* **42**:214–221.

Hayashi, Y., and Durzan, D. J., 1971, Amino-acid composition of proteins extracted from virions of cytoplasmic polyhedrosis virus, *J. Invertebr. Pathol.* **18**:121–126.

Hayashi, Y., and Kawarabata, T., 1970, Effect of actinomycin and synthesis of cell RNA and replication of insect cytoplasmic polyhedrosis viruses *in vivo, J. Invertebr. Pathol.* **15**:461–462.

Hayashi, Y., and Kawase, S., 1964, Base pairing in ribonucleic acid extracted from the cytoplasmic polyhedra of the silkworm, *Virology* **23**:611–614.

Hayashi, Y., and Krywienczyk, J., 1972, Electrophoretic fractionation of a cytoplasmic polyhedrosis virus genome, *J. Invertebr. Pathol.* **19**:160–162.

Hayashi, Y., and Retnakaran, A., 1970, The site of RNA synthesis of a cytoplasmic polyhedrosis virus (CPV) in *Malacosoma disstria, J. Invertebr. Pathol.* **16**:150–151.

Hayashi, Y., Kawarabata, T., and Bird, F. T., 1970, Isolation of a cytoplasmic polyhedrosis virus of the silkworm, *Bombyx mori, J. Invertebr. Pathol.* **16**:378–384.

Henderson, J. F., Faulkner, P., and Mackinnon, E. A., 1974, Some biophysical properties of virus present in tissue cultures infected with the nuclear polyhedrosis virus of *Trichoplusia ni, J. Gen. Virol.* **22**:143–146.

Henry, J. E., and Omas, E. A., 1973, Ultrastructure of the replication of the grasshopper crystalline-array virus in *Schistocerca americana* compared with other picornaviruses, *J. Invertebr. Pathol.* **21**:273–281.

Henry, J. E., Nelson, B. P., and Jutila, J. W., 1969, Pathology and development of the grasshopper inclusion body virus in *Melanoplus sanguinipes, J. Virol.* **3**:605–610.

Himeno, M., Sakai, F., Onodera, K., Nakai, E. H., Fukada, T., and Kawada, Y., 1967, Formation of nuclear polyhedral bodies and nuclear polyhedrosis virus of silkworms in mammalian cells infected with viral DNA, *Virology* **33**:507–512.

Himeno, M., Yasuda, S., Kohsaka, T., and Onodera, K., 1968, The fine structure of a nuclear polyhedrosis virus of the silkworm, *J. Invertebr. Pathol.* **11**:516–519.

Hink, W. F., and Vail, P. V., 1973, A plaque assay for titration of alfalfa looper nuclear polyhedrosis virus in a cabbage looper (TN 368) cell line, *J. Invertebr. Pathol.* **22**:168–174.

Hoggan, M. D., 1971, Small DNA viruses, in: *Comparative Virology* (K. Maramorosch and E. Kurstak, eds.), Academic Press, New York.

Holmes, F. D., 1948, Order Virales, the filterable viruses, in: *Bergey's Manual of Determinative Bacteriology,* 6th ed., pp. 1223–1228, Williams and Wilkins, Baltimore.

Hosaka, Y., 1964, On the release of viral nucleic acid from the icosahedral capsid, *Biken J.* **7:**121–130.

Hosaka, Y., and Aizawa, K., 1964, The fine structure of the cytoplasmic polyhedrosis virus of the silkworm *Bombyx mori* Linnaeus, *J. Insect Pathol.* **6:**53–77.

Hostetter, D. L., and Biever, K. D., 1970, The recovery of virulent nuclear polyhedrosis virus of the cabbage looper, *Trichoplusia ni,* from feces of birds, *J. Invertebr. Pathol.* **15:**173–176.

Huger, A., 1966, A virus disease of the Indian rhinoceros beetle *Oryctes rhinoceros* (Linnaeus) caused by a new type of insect virus, *Rhabdionvirus oryctes* gen. n, sp. n., *J. Invertebr. Pathol.* **8:**35–51.

Huger, A. M., Krieg, A., Enschermann, P., and Götz, P., 1970, Further studies on *Polypoxvirus chironomi,* an insect virus of the pox group isolated from the midge, *Chironomus luridus, J. Invertebr. Pathol.* **15:**253–261.

Hughes, K. M., 1972, Fine structure and development of two polyhedrosis viruses, *J. Invertebr. Pathol.* **19:**198–207.

Hukuhara, T., and Hashimoto, Y., 1966, Serological studies of the cytoplasmic and nuclear polyhedrosis viruses of the silkworm, *Bombyx mori, J. Invertebr. Pathol.* **8:**234–239.

Hukuhara, T., and Hashimoto, Y., 1967, Multiplication of *Tipula* and *Chilo* iridescent viruses in cells of *Antherea eucalypti, J. Invertebr. Pathol.* **9:**278–281.

Hurlbut, H. S., 1951, The propagation of Japanese encephalitis virus in the mosquito by parenteral introduction and serial passage, *Am. J. Trop. Med.* **31:**448–451.

Hurlbut, H. S., 1953, The experimental transmission of a Coxsackie-like virus by mosquitoes, *J. Egypt. Med. Assoc.* **36:**495–498.

Hurlbut, H. S., 1956, West Nile virus infection in arthropods, *Am. J. Trop. Med. Hyg.* **5:**76–85.

Hurlbut, H. S., and Thomas, J. I., 1960, The experimental host range of the arthropod-borne animal viruses in anthropods, *Virology* **12:**381–407.

Hurlbut, H. S., and Thomas, J. I., 1969, Further studies on the arthropod host range of arboviruses, *J. Med. Entomol.* **6:**423–427.

Ignoffo, C., 1968, Specificity of insect viruses, *Bull. Entomol. Soc. Amr.* **14:**265–276.

Ignoffo, C. M., Shapiro, M., and Hink, W. F., 1971, Replication and serial passage of infectious *Heliothis* nuclear polyhedrosis virus in an established line of *Heliothis zea* cells, *J. Invertebr. Pathol.* **18:**131–134.

Jaques, R. P., 1962, The transmission of nuclear polyhedrosis virus in laboratory populations of *Trichoplusia ni* (Hübner), *J. Insect Pathol.* **4:**433–445.

Jaques, R. P., 1964, The persistence of a nuclear polyhedrosis virus in soil, *J. Insect Pathol.* **6:**251–254.

Jaques, R. P., 1969, Leaching of the nuclear polyhedrosis virus of *Trichoplusia ni* from the soil, *J. Invertebr. Pathol.* **13:**256–263.

Jousset, F. X., 1970, Un virus extrait de *Drosophila immigrans* provoquant un symptôme de sensibilité au CO_2 chez les mäles de *Drosophila melanogaster, C. R. Acad. Sci. Ser. D* **271:**1141–1144.

Juckes, I. R. M., Longworth, J. F., and Reinganum, C., 1973, A serological comparison of some non-occluded insect viruses, *J. Invertebr. Pathol.* **21**:119–120.

Jutila, J. W., Henry, J. E., Anacker, R. L., and Brown, W. R., 1970, Some properties of a crystalline-array virus (CAV) isolated from the grasshopper *Melanoplus bivattatus* (Say) (Orthoptera: Acrididae), *J. Invertebr. Pathol.* **15**:225–231.

Kalmakoff, J., and Robertson, J. S., 1970, Serological relationship of *Wiseana* iridescent virus to other iridescent viruses, *Proc. Univ. Otago Med. Sch.* **48**:16–18.

Kalmakoff, J., and Tremaine, J. H., 1968, Physico-chemical properties of *Tipula* iridescent virus, *J. Virol.* **2**:738–744.

Kalmakoff, J., Lewandowski, L. J., and Black, D. R., 1969, Comparison of the ribonucleic acid subunits of reovirus, cytoplasmic polyhedrosis virus and wound tumor virus, *J. Virol.* **4**:851–856.

Kalmakoff, J., Moore, S., and Pottinger, R. P., 1972, An iridescent virus from the grass grub *Costelytra zealandica:* Serological study, *J. Invertebr. Pathol.* **20**:70–76.

Kamon, E., and Shulov, A., 1965, Immune response of locusts to venom of the scorpion, *J. Invertebr. Pathol.* **7**:192–198.

Kates, J. R., and McAuslan, B. R., 1967, Messenger RNA synthesis by a "coated" viral genome, *Proc. Natl. Acad. Sci. U.S.A.,* **57**:314–320.

Kawanishi, C. Y., and Paschke, J. D., 1970, Density gradient centrifugation of the virions liberated from *Rachiplusia ou* nuclear polyhedra, *J. Invertebr. Pathol.* **16**:89–92.

Kawanishi, C. Y., Summers, M. D., Stoltz, D. B., and Arnott, H. J., 1972*a,* Entry of an insect virus *in vivo* by fusion of viral envelope and microvillus membrane, *J. Invertebr. Pathol.* **20**:104–108.

Kawanishi, C. Y., Egawa, K., and Summers, M. D., 1972*b,* Solubilization of *Trichoplusia ni* granulosis virus proteinic crystal. II. Ultrastructure, *J. Invertebr. Pathol.* **20**:95–100.

Kawarabata, T., 1974, Highly infectious free virions in the hemolymph of the silkworm (*Bombyx mori*) infected with a nuclear polyhedrosis virus, *J. Invertebr. Pathol.* **24**:196–200.

Kawarabata, T., and Hayashi, Y., 1972, Development of a cytoplasmic polyhedrosis virus in an insect cell line, *J. Invertebr. Pathol.* **19**:414–415.

Kawase, S., and Furusawa, T., 1971, Effect of actinomycin D on RNA synthesis in the midguts of healthy and CPV-infected silkworms, *J. Invertebr. Pathol.* **18**:33–39.

Kawase, S., and Kawamori, I., 1968, Chromatographic studies on RNA synthesis in the midgut of the silkworm, *Bombyx mori,* infected with a cytoplasmic-polyhedrosis virus, *J. Invertebr. Pathol.* **12**:395–404.

Kawase, S., and Miyajima, S., 1968, Infectious ribonucleic acid of the cytoplasmic polyhedrosis virus of the silkworm, *Bombyx mori, J. Invertebr. Pathol.* **11**:63–69.

Kawase, S., and Miyajima, S., 1969, Immunofluorescence studies on the multiplication of cytoplasmic-polyhedrosis virus of the silkworm *Bombyx mori, J. Invertebr. Pathol.* **13**:330–336.

Kawase, S., and Yamaguchi, K., 1974, A polyhedrosis virus forming polyhedra in midgut-cell nucleus of silkworm, *Bombyx mori.* II. Chemical nature of the virion, *J. Invertebr. Pathol.* **24**:106–111.

Kawase, S., Kawamoto, F., and Yamaguchi, K., 1973, Studies on the polyhedrosis virus forming polyhedra in the midgut-cell nucleus of the silkworm, *Bombyx mori.* I. Purification procedure and form of the virion, *J. Invertebr. Pathol.* **22**:266–272.

Kellen, W. R., Clark, T. B., Lindegren, J. F., and Sanders, R. D., 1966, A cytoplasmic polyhedrosis virus of *Culex tarsalis* (Diptera: Culicidae), *J. Invertebr. Pathol.* **8**:390–394.

Kelly, D. C., and Avery, R. J., 1974*a*, The DNA content of four small iridescent viruses: Genome size, redundancy and homology determined by renaturation kinetics, *Virology* **57**:425–435.

Kelly, D. C., and Avery, R. J., 1974*b*, Frog virus 3 deoxyribonucleic acid, *J. Gen. Virol.* **24**:339–348.

Kelly, D. C., and Robertson, J. S., 1973, Icosahedral cytoplasmic deoxyriboviruses, *J. Gen. Virol.* (*Suppl.*) **20**:17–41.

Kelly, D. C., and Tinsley, T. W., 1972, The proteins of iridescent virus types 2 and 6, *J. Invertebr. Pathol.* **19**:273–275.

Kelly, D. C., and Tinsley, T. W., 1973, Ribonucleic acid polymerase activity associated with particles of iridescent virus types 2 and 6, *J. Invertebr. Pathol.* **22**:199–202.

Kelly, D. C., and Tinsley, T. W., 1974*a*, Iridescent virus replication: A microscope study of *Aedes aegypti* and *Antherea eucalypti* cells in culture infected with iridescent virus types 2 and 6, *Microbios* **9**:75–93.

Kelly, D. C., and Tinsley, T. W., 1974*b*, Iridescent virus replication: Patterns of nucleic acid synthesis in insect cells infected with iridescent virus types 2 and 6, *J. Invertebr. Pathol.* **24**:169–178.

Kelly, D. C., and Vance, D. E., 1973, The lipid content of two iridescent viruses, *J. Gen. Virol.* **21**:417–423.

Khosaka, T., Himeno, M., and Onodera, K., 1971, Separation and structure of components of nuclear polyhedrosis virus of the silkworm, *J. Virol.* **7**:267–273.

Kislev, N., Edelman, M., and Harpaz, I., 1971, Nuclear polyhedrosis viral DNA: Characterization and comparison to host DNA, *J. Invertebr. Pathol.* **17**:199–202.

Knudson, D. L., and Harrap, K. A., 1976, Replication of a nuclear polyhedrosis virus in a continuous cell culture of *Spodoptera frugiperda:* Microscopy study of the sequence of events of the virus infection, *J. Virol.* **17**:254–268.

Knudson, D. L., and Tinsley, T. W., 1974, Replication of a nuclear polyhedrosis virus in a continuous cell culture of *Spodoptera frugiperda:* Purification, assay of infectivity and growth characteristics of the virus, *J. Virol.* **14**:934–944.

Kozlov, E. A., and Alexeenko, I. P., 1967, Electron microscope investigation of the structure of the nuclear polyhedrosis virus of the silkworm, *Bombyx mori, J. Insect Pathol.* **9**:413–419.

Kozlov, E. A., Levitina, T. L., Sidorova, N. M., Ridavski, Yu, L., and Serebryani, S. B., 1975*a,* Comparative chemical studies of the polyhedral proteins of the nuclear polyhedrosis viruses of *Bombyx mori* and *Galleria mellonella, J. Invertebr. Pathol.* **25**:103–107.

Kozlov, E. A., Sidorova, N. M., and Serebryani, S. B., 1975*b,* Proteolytic cleavage of polyhedral protein during dissolution of inclusion bodies of the molecular polyhedrosis viruses of *Bombyx mori* and *Galleria mellonella* under alkaline conditions, *J. Invertebr. Pathol.* **25**:97–101.

Krell, P., and Lee, P. E., 1974, Polypeptides in *Tipula* iridescent virus (TIV) and in TIV-infected hemocytes of *Galleria mellonella* (L.) larvae, *Virology* **60**:315–326.

Krieg, A., 1957, "Toleranzphänomen" and Latenzproblem, *Arch. Gesamte Virusforsch.* **7**:212–219.

Krywienczyk, J., 1963, Demonstration of nuclear polyhedrosis in *Bombyx mori* (Linnaeus) by fluorescent antibody technique, *J. Insect Pathol.* **5**:309–317.

Krywienczyk, J., Hayashi, Y. and Bird, F. T., 1969, Serological investigations of insect viruses. I. Comparison of three highly purified cytoplasmic-polyhedrosis viruses, *J. Invertebr. Pathol.* **13**:114–119.

Kurstak, E., 1972, The small DNA densonucleosis virus (DNV), *Adv. Virus Res.* **17**:207–241.

Kurstak, E. and Côté, J. R., 1969, Proposition de classification du virus de la densonucléose (VDN) basée sur l'étude de la structure moléculaire et des propriétés physicochemiques, *C. R. Acad. Sci. Ser. D* **268**:616–619.

Kurstak, E., and Stanislawski-Birencwajg, M., 1968, Localisation du matériel antigénique du virus de la densonucléose au cours de l'infection chez *Galleria mellonella* L., *Can. J. Microbiol.* **14**:1350–1352.

Kurstak, E., Belloncik, S., and Brailovsky, C., 1969, Transformation de cellules L de souris par un virus d'invertébrés: le virus de la densonucléose (VDN), *C. R. Acad. Sci. Ser. D* **268**:1716–1719.

Kurstak, E., Vernoux, J.-P., Niveleau, A., and Onji, P. A., 1971, Visualisation du DNA du virus de la densonucléose (VDN) à chaînes monocaténaires complémentaires de polarités inverses plus et moins, *C. R. Acad. Sci. Ser. D* **272**:762–765.

Langstroth, L. L., 1857, *A Practical Treatise on the Hive and Honey Bee*, 2nd ed., C. M. Saxton & Co., New York.

Lee, P. E., and Furgala, B., 1965, Electron microscopy of sacbrood virus *in situ*, *Virology* **25**:387–392.

Leutenegger, R., 1967, Early events of *Sericesthis* iridescent virus infection in hemocytes of *Galleria mellonella* (L.), *Virology* **32**:109–116.

Lewandowski, L. J., and Leppla, S. H., 1972, Comparison of the 3′ termini of discrete segments of the double-stranded ribonucleic acid genomes of cytoplasmic polyhedrosis virus, wound tumor virus, and reovirus, *J. Virol.* **10**:965–968.

Lewandowski, L. J., and Millward, S., 1971, Characterization of the genome cytoplasmic polyhedrosis virus, *J. Virol.* **7**:434–437.

Lewandowski, L. J., and Traynor, B. L., 1972, Comparison of the structure and polypeptide composition of three double-stranded ribonucleic acid-containing viruses (diplornaviruses): Cytoplasmic polyhedrosis virus, wound tumor virus, and reovirus, *J. Virol.* **10**:1053–1070.

Lewandowski, L. J., Kalmakoff, J., and Tanada, Y., 1969, Characterization of a ribonucleic acid polymerase activity associated with purified cytoplasmic polyhedrosis virus of the silkworm, *Bombyx mori, J. Virol.* **4**:857–865.

Linley, J. R., and Nielsen, H. T., 1968, Transmission of a mosquito iridescent virus in *Aedes taeniorhynchus*. II. Experiments related to transmission in nature, *J. Invertebr. Pathol.* **12**:17–24.

Lom, J., and Kozloff, E. N., 1969, Virus-like particles in ancistrocomid ciliate *Ignotocoma sabellarium* Kosloff, *Protistologica* **5**:173–192.

Longworth, J. F., 1973, Viruses and Lepidoptera, in: *Viruses and Invertebrates* (A. J. Gibbs, ed.), pp. 428–441, North-Holland, Amsterdam.

Longworth, J. F., and Cunningham, J. C., 1968, The activation of occult nuclear polyhedrosis viruses by foreign nuclear polyhedra, *J. Invertebr. Pathol.* **10**:361–367.

Longworth, J. F., and Harrap, K. A., 1968, A non-occluded virus isolated from four saturniid species, *J. Invertebr. Pathol.* **10**:139–145.

Longworth, J. F., and Spilling, C. R., 1970, A cytoplasmic polyhedrosis of the larch sawfly, *Anoplonyx destructor, J. Invertebr. Pathol.* **15**:276–280.

Longworth, J. F., Tinsley, T. W., Barwise, A. H., and Walker, I. O., 1968, Purification of a non-occluded virus of *Galleria mellonella, J. Gen. Virol.* **3**:167–174.

Longworth, J. F., Robertson, J. S., and Payne, C. C., 1972, The purification and properties of inclusion body protein of the granulosis virus of *Pieris brassicae, J. Invertebr. Pathol.* **19**:42–50.

Longworth, J. F., Payne, C. C., and Macleod, R., 1973*a*, Studies on a virus isolated from *Gonometa podocarpi* (Lepidoptera, Lasiocampidae), *J. Gen. Virol.* **18**:119–125.

Longworth, J. F., Robertson, J. S., Tinsley, T. W., Rowlands, D. J., and Brown, F., 1973*b*, Reactions between an insect picornavirus and naturally occurring IgM antibodies in several mammalian species, *Nature (London)* **242**:314–316.

MacKinnon, E. A., Henderson, J. F., Stoltz, D. B., and Faulkner, P., 1974, Morphogenesis of nuclear polyhedrosis virus under conditions of prolonged passage *in vitro, J. Ultrastruct. Res.* **49**:419–435.

Maestri, A., 1856, *Frammenti Anatomici Fisiologici e Patologici sul Baco da Seta,* Fusi, Pavia.

Martignoni, M. E., and Milstead, J. E., 1962, Transovum transmission of the nuclear polyhedrosis virus of *Colias eurytheme* Boisduval through contamination of the female genitalia, *J. Insect Pathol.* **4**:113–121.

Martignoni, M. E., and Schmid, P., 1961, Studies on the resistance to virus infections in natural populations of Lepidoptera, *J. Insect Pathol.* **3**:62–74.

Martignoni, M. E., Iwai, P. J., and Breillatt, J. P., 1971, Heterogeneous buoyant density in batches of viral nucleopolyhedra, *J. Invertebr. Pathol.* **18**:219–226.

Matta, J. F., 1970, The characterization of a mosquito iridescent virus. II. Physicochemical characterization, *J. Invertebr. Pathol.* **16**:157–164.

Matta, J. F., and Lowe, R. E., 1970, The characterization of a mosquito iridescent virus (MIV). I. Biological characteristics, infectivity and pathology, *J. Invertebr. Pathol.* **16**:38–41.

Mattern, C. F. T., Diamond, L. S., and Daniel, W. A., 1972, Viruses of *Entamoeba histolytica.* II. Morphogenesis of the polyhedral particles (ABRM$_2$ → HK-9) → HB-301 and the filamentous agent (ABRM)$_2$ → HK-9, *J. Virol.* **9**:342–358.

McCarthy, W. J., Granados, R. R., and Roberts, D. W., 1974, Isolation and characterization of entomopox virions from the virus-containing inclusions of *Amsacta moorei* (Lepidoptera: Arctiidae), *Virology* **59**:59–69.

McCarthy, W. J., Granados, R. R., Sutter, G. R., and Roberts, D. W., 1975, Characterization of entomopox virions of the army cutworm, *Euxoa auxiliaris* (Lepidoptera: Noctuidae), *J. Invertebr. Pathol.* **25**:215–220.

McIntosh, A. H., and Kimura, M., 1974, Replication of the insect *Chilo* iridescent virus (CIV) in a poikilothermic vertebrate cell line, *Intervirology* **4**:257–267.

McKinley, E. B., 1929, Filterable virus and *Rickettsia* diseases, *Philipp. J. Sci.* **39**:1–413.

McLean, D. M., 1955, Multiplication of viruses in mosquitoes following feeding and injection into the cavity, *Austr. J. Exp. Biol. Med. Sci.* **33**:53–66.

Merrill, M. H., and ten Broeck, C., 1935, The transmission of equine encephalomyelitis virus by *Aedes aegypti, J. Exp. Med.* **62**:687–695.

Meynadier, G., 1966, Étude d'une maladie à virus chez l'Orthoptère *Gryllus bimaculatus* Geer, *C. R. Acad. Sci. Ser. D* **263**:742–744.

Meynadier, G., Vago, C., Plantevin, G., and Atger, P., 1964, Viruse d'un type inhabituel chex le lépidoptère *Galleria mellonella* L., *Rev. Zool. Agri. Appl.* **63**:207–208.

Meynadier, G., Fosset, J., Vago, C., Duthoit, J. L., and Bres, N., 1968, Une virose à inclusions ovoides chez un lépidoptère, *Ann. Epiphytol.* **19**:703–705.

Miller, J. H., and Schwartzwelder, J. C., 1960, Virus-like particles in an *Entamoeba histolytica* trophozoite, *J. Parasitol.* **46**:523–524.

Miura, K., Fujii, I., Sakaki, T., Fuke, M., and Kawase, S., 1968, Double-stranded ribonucleic acid from cytoplasmic polyhedrosis virus of the silkworm, *J. Virol.* **2**:1211–1222.

Miura, K., Fujii-Kawata, I., Iwata, H., and Kawase, S., 1969, Electron microscopic observations of a cytoplasmic polyhedrosis virus from the silkworm, *J. Invertebr. Pathol.* **14**:262–265.

Miura, K., Watanabe, K., and Sugiura, M., 1974, 5′-Terminal nucleotide sequences of the double stranded RNA of silkworm cytoplasmic polyhedrosis virus, *J. Mol. Biol.* **86**:31–48.

Miyajima, S., and Kawase, S., 1968, Changes in virus-infectivity titer in the hemolymph and midgut during the course of a cytoplasmic polyhedrosis in the silkworm, *J. Invertebr. Pathol.* **12**:329–334.

Miyajima, S., and Kawase, S., 1969, Haemagglutination with cytoplasmic polyhedrosis virus of the silkworm, *Bombyx mori, Virology,* **39**:347–348.

Miyajima, S., Kimura, I., and Kawase, S., 1969, Purification of a cytoplasmic polyhedrosis virus of the silkworm, *Bombyx mori, J. Invertebr. Pathol.* **13**:296–302.

Monsarrat, P., Meynadier, G., Croizier, G., and Vago, C., 1973a, Recherches cytopathologiques sur une maladie virale du Coléoptère *Oryctes rhinoceros* L., *C. R. Acad. Sci. Ser. D* **276**:2077–2080.

Monsarrat, P., Veyrunes, J., Meynadier, G., Croizier, G., and Vago, C., 1973b, Purification et étude structurale du virus du Coléoptère *Oryctes rhinoceros* L., *C. R. Acad. Sci. Ser. D* **277**:1413–1415.

Morel, G., 1975, Un virus cytoplasmique chez le scorpion *Buthus occitanus* Amoreux, *C. R. Acad. Sci. Ser. D* **280**:2893–2894.

Morgante, J. S., Da Cunha, A. B., Pavan, C., Biesele, J. J., Riess, R. W., and Garrido, M. C., 1974, Development of a nuclear polyhedrosis in cells of *Rhynchosciara angelae* (Diptera: Sciaridae) and patterns of DNA synthesis in infected cells, *J. Invertebr. Pathol.* **24**:93–105.

Morris, O. N., 1968a, Metabolic changes in diseased insects. I. Autoradiographic studies in DNA synthesis on normal and polyhedrosis-virus-infected Lepidoptera, *J. Invertebr. Pathol.* **10**:28–38.

Morris, O. N., 1968b, Metabolic changes in diseased insects. II. Radioautographic studies on DNA and RNA synthesis in nuclear-polyhedrosis and cytoplasmic polyhedrosis virus infections, *J. Invertebr. Pathol.* **11**:476–486.

Morris, O. N., 1970, Metabolic changes in diseased insects. III. Nucleic acid metabolism in Lepidoptera infected by densonucleosis and *Tipula* iridescent viruses, *J. Invertebr. Pathol.* **16**:180–186.

Morris, O. N., 1971, Metabolic changes in diseased insects. IV. Radioautographic studies on protein changes in nuclear polyhedrosis, densonucleosis and *Tipula* iridescent virus infections, *J. Invertebr. Pathol.* **18**:191–206.

Munyon, W., Paoletti, E., and Grace, J. T., Jr., 1967, RNA polymerase activity in purified infectious vaccinia virus, *Proc. Natl. Acad. Sci. U.S.A.* **58**:2280–2287.

Munyon, W., Paoletti, E., Ospina, J., and Grace, J. T., Jr., 1968, Nucleotide phosphohydrolase in purified vaccinia virus, *J. Virol.* **2**:167–172.

Murphy, F. A., Scherer, W. F., Harrison, A. K., Dunne, H. W., and Gary, G. W.,

1970, Characterization of Nodamura virus, an arthropod transmissible picornavirus, *Virology* **40**:1008–1021.

Newman, J. F. E., and Brown, F., 1973, Evidence for a divided genome in Nodamura virus, an arthropod-borne picornavirus, *J. Gen. Virol.* **21**:371–384.

Newman, J. F. E., Brown, F., Bailey, L., and Gibbs, A. J., 1973, Some physico-chemical properties of two honey-bee picornaviruses, *J. Gen. Virol.* **19**:405–409.

Nielson, M. M., and Elgee, D. E., 1960, The effect of storage on the virulence of a polyhedrosis virus, *J. Insect Pathol.* **2**:165–171.

Nielson, M. M., and Elgee, D. E., 1968, The method and role of vertical transmission of a nucleo-polyhedrosis virus in the European spruce sawfly, *Diprion hercyniae, J. Invertebr. Pathol.* **12**:132–139.

Nishimura, A., and Hosaka, Y., 1969, Electron microscope study on RNA of cytoplasmic polyhedrosis virus of the silkworm, *Virology* **38**:550–557.

Nordin, G. L., and Maddox, J. V., 1971, Observations on the nature of the carbonate dissolution process on inclusion bodies of a nuclear polyhedrosis virus of *Pseudaletia unipuncta, J. Invertebr. Pathol.* **18**:316–321.

Nysten, P. H., 1808, *Recherches sur les Maladies des Vers à Soie,* Impr. Impériale, Paris.

Ohanessian, A., 1971, Sigma virus multiplication in *Drosophila* cell lines of different genotypes, *Curr. Top. Microbiol. Immunol.* **55**:230–233.

Ohanessian, A., and Echalier, G., 1967, Multiplication of *Drosophila* hereditary virus (σ virus) in *Drosophila* embryonic cells cultivated *in vitro, Nature (London)* **213**:1049–1050.

Onodera, K., Komano, T., Himeno, M., and Sakai, F., 1965, The nucleic acid of nuclear polyhedrosis virus of the silkworm, *J. Mol. Biol.* **13**:532–539.

Ossowski, L. L. J., 1958, Occurrence of strains of the nuclear polyhedrosis virus of the Wattle bagworm, *Nature (London)* **181**:648.

Ossowski, L. L. J., 1960, Variation in virulence of a Wattle bagworm virus, *J. Insect Pathol.* **2**:35–43.

Payne, C. C., 1972, Actinomycin D and the replication of an invertebrate RNA virus, *Monogr. Virol.* **6**:11–15.

Payne, C. C., 1974, The isolation and characterization of a virus from *Oryctes rhinoceros, J. Gen. Virol.* **25**:105–116.

Payne, C. C., and Kalmakoff, J., 1973, The synthesis of virus-specific single-stranded RNA in larvae of *Bombyx mori* infected with a cytoplasmic polyhedrosis virus, *Intervirology* **1**:34–40.

Payne, C. C., and Kalmakoff, J., 1974, Biochemical properties of polyhedra and virus particles of the cytoplasmic polyhedrosis virus of *Bombyx mori, Intervirology* **4**:354–364.

Payne, C. C., and Tinsley, T. W., 1974, The structural proteins and RNA components of a cytoplasmic polyhedrosis virus from *Nymphalis io* (Lepidoptera: Nymphalidae), *J. Gen. Virol.* **25**:291–302.

Peers, R. R., 1972, Bunyamwera virus replication in mosquitoes, *Can. J. Microbiol.* **18**:741–745.

Phillips, J. H., 1960, Immunological processes and recognition of foreigness in the invertebrates, in *Phylogeny of Immunity* (R. T. Smith, R. A. Meiseler, and R. A. Good, eds.), pp. 133–140, University of Florida Press, Gainesville, Fla.

Philpott, D. E., Weibel, J., Atlan, H., and Miquel, J., 1969, Virus-like particles in the fat body, oenocytes and central nervous tissue of *Drosophila melanogaster* imagoes, *J. Invertebr. Pathol.* **14**:31–38.

Plus, N., and Duthoit, J. L., 1969, Un nouveau virus de *Drosophila melanogaster*, le virus P, *C. R. Acad. Sci. Ser. D* **268**:2313–2315.

Plus, N., Jousset, F. X., David, J., and Croizier, G., 1972, Detection of small icosahedral viruses in apparently normal populations of *Drosophila melanogaster* and *Drosophila immigrans, Monogr. Virol.* **6**:24–26.

Pogo, B. G. T., and Dales, S., 1969, Two deoxyribonuclease activities within purified vaccinia virus, *Proc. Natl. Acad. Sci. U.S.A.* **63**:820–827.

Pogo, B. G. T., Dales, S., Bergoin, M., and Roberts, D. W., 1971, Enzymes associated with an insect poxvirus, *Virology* **43**:306–309.

Polson, A., Stannard, L., and Tripconey, D., 1970, The use of haemocyanin to determine the molecular weight of *Nudaurelia cytherea capensis* β virus by direct particle counting, *Virology* **41**:680–687.

Preer, J. R., and Jurand, A., 1968, The relation between virus-like particles and R bodies of *Paramecium aurelia, Genet. Res.* **12**:331–340.

Preer, J. R., and Preer, L. B., 1967, Virus-like bodies in killer *Paramecia, Proc. Natl. Acad. Sci. U.S.A.* **58**:1774–1781.

Printz, P., 1967, Mise en évidence d'un variant du virus de la stomatite vésiculaire (souche Indiana) conférant une sensibilité retardée au gaz carbonique chez *Drosophila melanogaster, C. R. Acad. Sci. Ser. D* **265**:169–172.

Printz, P., 1973, Relationship of sigma virus to vesicular stomatitis virus, *Adv. Virus Res.* **18**:143–157.

von Prowazek, S., 1907, Chlamydozoa. II. Gelbsucht der Seidenraupen, *Arch. Protistenkd.* **10**:348–364.

Putman, W. L., 1970, Occurrence and transmission of a virus disease of European red mite *Panonychus ulmi, Can. Entomol.* **102**:305–321.

Raghow, R., and Grace, T. D. C., 1974, Studies on a nuclear polyhedrosis virus in *Bombyx mori* cells *in vitro*. I. Multiplication kinetics and ultrastructural studies, *J. Ultrastruct. Res.* **47**:384–399.

Ramoska, W. A., and Hink, W. F., 1974, Electron microscope examination of two plaque variants from a nuclear polyhedrosis virus of the alfalfa looper, *Autographa californica, J. Invertebr. Pathol.* **23**:197–201.

Reed, D. K., and Hall, I. M., 1972, Electron microscopy of a rod-shaped non-inclusion virus infecting the citrus red mite, *J. Invertebr. Pathol.* **20**:272–278.

Reed, D. K., Hall, I. M., Rich, J. E., and Shaw, J. G., 1972, Birefringent crystal formation in citrus red mites associated with non-inclusion virus disease, *J. Invertebr. Pathol.* **20**:170–175.

Rehaček, J., 1965, Development of animal viruses and Rickettsiae in ticks and mites, *Annu. Rev. Entomol.* **10**:1–24.

Reinganum, C., O'Loughlin, G. T., and Hogan, T. W., 1970, A nonoccluded virus of the field crickets *Teleogryllus oceanicus* and *T. commodus* (Orthoptera: Gryllidae), *J. Invertebr. Pathol.* **16**:214–220.

Reuter, M., 1975, Virus like particles in *Gyratrix hermaphroditis* (Turbellaria: Rhadocoela), *J. Invertebr. Pathol.* **25**:79–95.

Revet, B., and Monsarrat, P., 1974, L'acide nucléique du virus du Coléoptère *Oryctes rhinoceros* L.: Un ADN superhélicoidal de haut poids moleculaire, *C. R. Acad. Sci. Ser. D* **278**:331–334.

Richards, W. C., and Hayashi, Y., 1971, Effect of some organic solvents on the cytoplasmic polyhedrosis virus of the forest tent caterpillar, *Malacosoma disstria, J.
Invertebr. Pathol.* **17**:42–47.

Richards, W. C., and Hayashi, Y., 1972, Preferential separation of cytoplasmic polyhedrosis virus (CPV) RNAs from infected midgut cells, *J. Invertebr. Pathol.* **20:**
200–207.

Richardson, J., Sylvester, E. S., Reeves, W. C., and Hardy, J. L., 1974, Evidence of
two inapparent nonoccluded viral infections of *Culex tarsalis, J. Invertebr. Pathol.*
23:213–224.

Rivers, C. F., 1959, Virus resistance in larvae of *Pieris brassicae* L., in: *Trans. 1st Int.
Conf. Insect Pathol. Biol. Control,* Prague, 1958, pp. 205–210.

Rivers, C. F., and Longworth, J. F., 1972, A nonoccluded virus of *Junonia coenia*
(Nymphalidae: Lepidoptera), *J. Invertebr. Pathol.* **20:**369–370.

Roberts, D. W., and McCarthy, W. S., 1972, Partial characterization of virions and
proteins of an entomopoxvirus from lepidopterous insects, *Monogr. Virol.* **6:**7–9.

Robertson, J. S., Harrap, K. A., and Longworth, J. F., 1974, Baculovirus morphogenesis: The acquisition of the virus envelope, *J. Invertebr. Pathol.* **23:**248–251.

Rottman, F., Shatkin, A. J., and Perry, R. P., 1974, Sequences containing methylated
nucleotides at the 5′ termini of messenger RNAs: Possible implications for processing, *Cell* **3:**197–199.

Rungger, D., Rastelle, M., Braendle, E., and Malsberger, R. G., 1971, A virus-like
particle associated with lesions in the muscles of *Octopus vulgaris, J. Invertebr.
Pathol.* **17:**72–79.

Salt, G., 1970, The cellular defence reactions of insects, in: *Cambridge Monographs in
Experimental Biology,* No. 16, Cambridge University Press, Cambridge.

Scherer, W. F., and Hurlbut, H. S., 1967, Nodamura virus from Japan: A new unusual
arbovirus resistant to diethyl ether and chloroform, *Am. J. Epidemiol.* **86:**271–285.

Schuster, F. L., 1969, Intranuclear virus-like bodies in the amoeboflagellate *Naegleria
gruberi, J. Protozool.* **16:**724–727.

Schuster, F. L., and Dunnebacke, T. H., 1971, Formation of bodies associated with
virus-like particles in the amoeboflagellate *Naegleria gruberi, J. Ultrastruct. Res.*
36:659–668.

Scott, H. A., and Young, S. Y., 1973, Antigens associated with a nuclear polyhedrosis
virus from cabbage looper larvae, *J. Invertebr. Pathol.* **21:**315–317.

Scott, H. A., Young, S. Y., and McMasters, J. A. 1971, Isolation and some properties
of components of nuclear polyhedra from the cabbage looper, *Trichoplusia ni, J.
Invertebr. Pathol.* **18:**177–182.

Shapiro, M., and Ignoffo, C. M., 1971a, Protein and free amino acid changes in the
hemolymph of *Heliothis zea* larvae during nucleopolyhedrosis, *J. Invertebr. Pathol.*
17:327–332.

Shapiro, M., and Ignoffo, C. M., 1971b, Amino-acid and nucleic acid analyses of
inclusion bodies of the cotton bollworm *Heliothis zea,* nucleo-polyhedrosis virus, *J.
Invertebr. Pathol.* **18:**154–155.

Shigematsu, H., and Suzuki, S., 1971, Relationship of crystallization to the nature of
polyhedron protein with reference to a nuclear-polyhedrosis virus of the silkworm,
Bombyx mori, J. Invertebr. Pathol. **17:**375–382.

Shimotohno, K., and Miura, K., 1973, Single-stranded RNA synthesis *in vitro* by the
RNA polymerase associated with cytoplasmic polyhedrosis virus containing double-
stranded RNA, *J. Biochem. (Tokyo)* **74:**117–125.

Shimotohno, K., and Miura, K., 1974, 5′-Terminal structure of messenger RNA transcribed by the RNA polymerase of silk worm cytoplasmic polyhedrosis virus containing double-stranded RNA, *J. Mol. Biol.* **86**:21–30.

Sidor, C., 1959, Susceptibility of larvae of the large white butterfly *Pieris brassicae* L. to two virus diseases, *Ann. Appl. Biol.* **47**:109–113.

Sinha, R. C., 1967, Response of wound-tumor virus infection in insects to vector age and temperature, *Virology* **31**:746–748.

Smirnoff, W. A., 1962, Transovum transmission of a virus of *Neodiprion swainei* Middleton (Hymenoptera-Tenthredinidae), *J. Insect Pathol.* **4**:192–200.

Smirnoff, W. A., 1965, Observations on the effect of virus infections on insect behaviour, *J. Invertebr. Pathol.* **7**:387–388.

Smith, K. M., 1967, *Insect Virology,* Academic Press, New York.

Smith, K. M., 1971, The viruses causing the polyhedroses and granuloses of insects, in: *Comparative Virology* (K. Maramorosch and E. Kurstak, eds.), pp. 479–507, Academic Press, New York.

Smith, K. M., and Hills, G. T., 1962, Replication and ultrastructure of insect viruses, *Proc. IX Int. Congr. Entomol.* Vienna, 1960, **2**:823.

Smith, K. M., Hills, G. J., Munger, F., and Gilmore, J. E., 1959, A suspected virus disease of the citrus red mite *Panonychus citri* (McG.), *Nature (London)* **184**:70.

Smith, K. M., Hills, C. J., and Rivers, C. F., 1961, Studies on the cross inoculation of the *Tipula* iridescent virus, *Virology* **13**:233–241.

Sohi, S. S., and Cunningham, J. C., 1972a, Replication of a nuclear polyhedrosis virus in serially transferred insect hemocyte cultures, *J. Invertebr. Pathol.* **19**:51–61.

Sohi, S. S., and Cunningham, J. C., 1972b, Replication of a nuclear polyhedrosis virus in serially transferred hemocyte cultures, *Monogr. Virol.* **6**:35–42.

Sohi, S. S., Bird, F. T., and Hayashi, Y., 1970, Development of *Malacosoma disstria* cytoplasmic polyhedrosis virus in *Bombyx mori* ovarian and tracheal tissue cultures, in: *Proc. IV Int. Colloq. Inst. Pathol.,* College Park, Md., pp. 340–351.

Stairs, G. R., 1964, Selection of a strain of insect granulosis virus producing only cubic inclusion bodies, *Virology* **24**:520–521.

Stairs, G. R., 1968, Inclusion-type insect viruses, *Curr. Top. Microbiol. Immunol.* **42**:1–23.

Steinhaus, E. A., 1949, *Principles of Insect Pathology,* McGraw Hill, New York.

Steinhaus, E. A., 1959, Possible virus disease in European red mite, *J. Insect Pathol.* **1**:435–437.

Steinhaus, E. A., 1960, The duration of viability and infectivity of certain insect pathogens, *J. Insect Pathol.* **2**:225–229.

Steinhaus, E. A., and Leutenegger, R., 1963, Icosahedral virus from a scarab (*Sericesthis*), *J. Insect Pathol.* **5**:226–270.

Stephens, J. M., 1959, Immune responses of some insects to some bacterial antigens, *Can. J. Microbiol.* **5**:203–228.

Stephens Chadwick, J., 1967a, Serological responses of insects, *Fed. Proc.* **26**:1675–1679.

Stephens Chadwick, J., 1967b, Some aspects of immune responses in insects, in: *Differentiation and Defence Mechanisms in Lower Organisms: In Vitro,* Vol. 3 (M. M. Segel, ed.), pp. 120–128, Symposium, Tissue Culture Association, Williams and Wilkins, Baltimore.

Stoltz, D. B., 1971, The structure of icosahedral cytoplasmic deoxyriboviruses, *J. Ultrastruct. Res.* **37**:219–239.

Stoltz, D. B., 1973, The structure of icosahedral cytoplasmic deoxyriboviruses. II. An alternative model, *J. Ultrastruct. Res.* **43**:58–74.

Stoltz, D. B., and Hilsenhoff, W. L., 1969, Electron-microscope observations on the maturation of a cytoplasmic polyhedrosis virus, *J. Invertebr. Pathol.* **14**:39–48.

Stoltz, D. B., and Summers, M. D., 1971, Pathway of infection of mosquito iridescent virus. I. Preliminary observations on the fate of ingested virus, *J. Virol.* **8**:900–909.

Stoltz, D. B., and Summers, M. D., 1972, Observations on the morphogenesis and structure of a hemocytic poxvirus in the midge *Chironomus attenuatus*, *J. Ultrastruct. Res.* **40**:581–598.

Stoltz, D. B., Hilsenhoff, W. L., and Stich, H. F., 1968, A virus disease in *Chironomus plumosus*, *J. Invertebr. Pathol.* **12**:118–128.

Stoltz, D. B., Pavan, C., and Da Cunha, A. B., 1973, Nuclear polyhedrosis virus: A possible example of *de novo* intranuclear membrane morphogenesis, *J. Gen. Virol.* **19**:145–150.

Storer, G. B., Shepherd, M. G., and Kalmakoff, J., 1973a, Enzyme activities associated with cytoplasmic polyhedrosis virus from *B. mori*. I. Nucleotide phosphohydrolase and nuclease activities, *Intervirology* **2**:87–94.

Storer, G. B., Shepherd, M. G., and Kalmakoff, J., 1973b, Enzyme activities associated with cytoplasmic polyhedrosis virus from *Bombyx mori*. II. Comparative studies, *Intervirology* **2**:193–199.

Storey, H. H., 1932, The inheritance by an insect vector of the ability to transmit a plant virus, *Proc. R. Soc. London Ser. B.* **112**:46–60.

Storey, H. H., 1939, Transmission of plant viruses by insects, *Bot. Rev.* **5**:240–272.

Struthers, J. K., and Hendry, D. A., 1974, Studies of the protein and nucleic acid components of *Nudaurelia capensis β* virus, *J. Gen. Virol.* **22**:355–362.

Summers, M. D., 1969, Apparent *in vivo* pathway of granulosis virus invasion and infection, *J. Virol.* **4**:188–190.

Summers, M. D., 1971, Electron microscope observations on granulosis virus entry, uncoating and replication processes during infection of the midgut cells of *Trichoplusia ni*, *J. Ultrastruct. Res.* **35**:606–625.

Summers, M. D., and Anderson, D. L., 1972a, Granulosis virus deoxyribonucleic acid: A closed double-stranded molecule, *J. Virol.* **9**:710–713.

Summers, M. D., and Anderson, D. L., 1972b, Characterization of deoxyribonucleic acid isolated from the granulosis virus of the cabbage looper *Trichoplusia ni* and the fall armyworm *Spodoptera frugiperda*, *Virology* **50**:459–471.

Summers, M. D., and Anderson, D. L., 1973, Characterization of nuclear polyhedrosis virus DNAs, *J. Virol.* **12**:1336–1346.

Summers, M. D., and Arnott, H. J., 1969, Ultrastructural studies on inclusion formation and virus occlusion in nuclear polyhedrosis and granulosis virus-infected cells of *Trichoplusia ni* (Hübner), *J. Ultrastruct. Res.* **28**:462–480.

Summers, M. D., and Egawa, K., 1973, Physical and chemical properties of *Trichoplusia ni* granulosis virus granulin, *J. Virol.* **12**:1092–1103.

Summers, M. D., and Paschke, J. D., 1970, Alkali-liberated granulosis virus of *Trichoplusia ni*. I. Density gradient purification of virus components and some of their *in vitro* chemical and physical properties, *J. Invertebr. Pathol.* **16**:227–240.

Tanada, Y., and Leutenegger, R., 1970, Multiplication of a granulosis virus in larval midgut cells of *Trichoplusia ni* and possible pathways of invasion into the hemocoel, *J. Ultrastruct. Res.* **30**:589–600.

Tandler, B., 1972, Virus-like particles in testis of *Drosophila virilis, J. Invertebr. Pathol.* **20**:214–215.

Teninges, D., 1968, Mise en évidence de virions sigma dans les cellules de la lignée germinale mâle de drosophile stabilisées, *Arch. Gesamte Virusforsch.* **23**:378–387.

Teninges, D., and Plus, N., 1972, P virus of *Drosophila melanogaster*, as a new picornavirus, *J. Gen. Virol.* **16**:103–109.

Terzakis, J. A., 1969, A protozoan virus, *Mil. Med.* **134**:916–921.

Thomas, R. S., 1961, The chemical composition and particle weight of *Tipula* iridescent virus, *Virology* **14**:240–252.

Tinsley, T. W., 1975, Factors affecting virus infection of insect gut tissue, in: *Invertebrate Immunity: Mechanisms of Invertebrate Vector–Parasite Relations* (K. Maramorosch and R. E. Shope, eds.), pp. 55–63, Academic Press, New York.

Tinsley, T. W., and Entwistle, P. F., 1974, The use of pathogens in the control of insect pests, in: *Biology in Pest and Disease Control*, 13th Symposium of the British Ecological Society, Oxford, 1972 (D. Price Jones and M. E. Solomon, eds.), Blackwell, Oxford.

Tinsley, T. W., and Kelly, D. C., 1970, An interim nomenclature system for the iridescent group of insect viruses, *J. Invertebr. Pathol.* **16**:470–472.

Tinsley, T. W., and Longworth, J. F., 1973, Parvoviruses, *J. Gen. Virol.* (*Suppl.*) **20**:71–75.

Tinsley, T. W., and Melnick, J. L., 1974, Potential ecological hazards of pesticidal viruses, *Intervirology* **2**:206–208.

Tinsley, T. W., Robertson, J. S., Rivers, C. F., and Service, M. W., 1971, An iridescent virus of *Aedes cantans* in Great Britain, *J. Invertebr. Pathol.* **18**:427–428.

Tripconey, D., 1970, Studies on a nonoccluded virus of the pine tree emperor moth, *J. Invertebr. Pathol.* **15**:268–275.

Truffaut, N., Berger, G., Niveleau, A., May, P., Bergoin, M., and Vago, C., 1967, Recherches sur l'acide nucléique du virus de la densonucléose du lépidoptère *Galleria mellonella* L., *Arch. Gesamte Virusforsch.* **21**:469–474.

Vago, C., 1963, A new type of insect virus, *J. Insect Pathol.* **5**:275–276.

Vago, C., 1966, Virus disease in Crustacea (*Macropipius depurator* L.), *Nature* (*London*) **209**:1290

Vago, C., and Bergoin, M., 1963, Développment des virus à corps d'inclusion du lépidoptère *Lymantria dispar* en cultures cellulaires, *Entomophaga* **8**:253–261.

Vago, C., and Bergoin, M., 1968, Viruses of invertebrates, *Adv. Virus Res.* **13**:247–303.

Vago, C., Amargier, A., Hurpin, B., Meynadier, G., and Duthoit, J. L., 1968*a*, Virose à fuseaux d'un scarabéide d'Amérique du Sud, *Entomophaga* **13**:373–375.

Vago, C., Monsarrat, P., Duthoit, J. L., Amargier, A., Meynadier, G., and van Waerebeke, D, 1968*b*, Nouvelle virose à fuseaux observée chez un Lucanide (Coleoptera) de Madagascar, *C. R. Acad. Sci. Ser. D* **266**:1621–1623.

Vago, C., Rioux, J.-A., Duthoit, J.-L., and Dedet, J.-P., 1969*a*, Infection spontanée à virus irisant dans une population d'*Aedes detritus* (Hal. 1933) des environs de Tunis, *Ann. Parasitol. Hum. Comp.* **44**:667–676.

Vago, C., Robert, P., Amargier, A., and Duthoit, J.-L., 1969*b*, Nouvelle virose à sphéroides et à fuseaux observée chez le coleoptère *Phyllopertha corticola* L., *Mikroscopie* **25**:378–386.

Vago, C., Aizawa, K., Ignoffo, C. M., Martignoni, M. E., Tarasevitch, L., and Tinsley, T. W., 1974, Present status of the nomenclature and classification of invertebrate viruses, *J. Invertebr. Pathol.* **23**:133–134.

Vail, P. V., Jay, D. L., and Hink, W. F., 1973, Replication and infectivity of the nuclear polyhedrosis virus of the alfalfa looper, *Autographa californica,* produced in cells grown *in vitro, J. Invertebr. Pathol.* **22**:231–237.

Varma, M. G. R., 1972, Invertebrate host specificity in arthropod borne animal viruses, in: *Moving Frontiers in Invertebrate Virology* (T. W. Tinsley and K. A. Harrap, eds.), No. 6 of *Monographs in Virology,* Karger, Basel.

Vaughn, J. L., and Stanley, M. S., 1970, A micromethod for the assay of insect viruses in primary cultures of insect tissue, *J. Invertebr. Pathol.* **16**:357–362.

Wäger, R., and Benz, G., 1971, Histochemical studies on nucleic acid metabolism in granulosis-infected *Carpocapsa pomonella, J. Invertebr. Pathol.* **17**:107–115.

Wagner, G. W., Pashke, J. D., Campbell, W. R., and Webb, S. R., 1973, Biochemical and biophysical properties of two strains of mosquito iridescent virus, *Virology* **52**:72–80.

Wagner, G. W., Paschke, J. D., Campbell, W. R., and Webb, S. R., 1974a, Proteins of two strains of mosquito iridescent virus, *Intervirology* **3**:97–105.

Wagner, G. W., Webb, S. R., Paschke, J. D., and Campbell, W. R., 1974b, A picornavirus isolated from *Aedes taeniorhyncus* and its interaction with mosquito iridescent virus, *J. Invertebr. Pathol.* **24**:380–382.

Watanabe, H., 1967, Site of viral RNA synthesis within the midgut cells of the silkworm, *Bombyx mori,* infected with cytoplasmic-polyhedrosis virus, *J. Invertebr. Pathol.* **9**:480–487.

Watanabe, H., 1972, An electron microscope radioautography of DNA synthesis in the fat cell of silkworm, *Bombyx mori* infected with a nuclear polyhedrosis virus, *J. Invertebr. Pathol.* **20**:223–225.

Weiser, J., 1965, A new virus of mosquito larvae, *Bull. WHO* **33**:586–588.

Weiser, J., 1968, Iridescent virus from the blackfly *Simulium ornatum* Meigen in Czechoslovakia, *J. Invertebr. Pathol.* **12**:36–39.

Weiser, J., 1969, A pox-like virus in the midge *Camptochironomus tentans, Acta Virol.* (*Prague*) **13**:549–553.

Wellington, E. F., 1954, The amino-acid composition of some insect viruses and their characteristic inclusion-body proteins, *Biochem. J.* **57**:334–338.

White, G. F., 1917, Sacbrood, *USDA Bull.,* No. 431.

Whitfield, S. G., Murphy, F. A., and Sudia, D. W., 1973, St. Louis encephalitis virus: An ultrastructural study of infection in a mosquito vector, *Virology* **56**:70–87.

Wildy, P., 1971, Classification and nomenclature of viruses, in: *Monographs in Virology,* Vol. 5, Karger, Basel.

Williams, R. C., and Smith, K. M., 1958, The polyhedral form of *Tipula* iridescent virus, *Biochem. Biophys. Acta* **28**:464–469.

Willison, J. H. M., and Cocking, E. C., 1972, Frozen-fractured viruses: A study of virus structure using freeze-etching, *J. Microsc.* (*Oxford*) **95**:397–411.

WHO Technical Report No. 531, 1973, The use of viruses for the control of insect pests and disease vectors: Report of a joint FAO/WHO meeting on insect viruses, *WHO Tech. Report Ser.,* No. 531, Geneva.

Wrigley, N. G., 1969, An electron microscope study of the structure of *Sericesthis* iridescent virus, *J. Gen. Virol.* **5**:123–134.

Wrigley, N. G., 1970, An electron microscope study of the structure of *Tipula* iridescent virus, *J. Gen. Virol* **6**:169–173.

Xeros, N., 1954, A second virus disease of the leather jacket, *Tipula paludosa, Nature* (*London*) **174**:562–563.

Xeros, N., 1956, The virogenic stroma in nuclear and cytoplasmic polyhedroses, *Nature (London)* **178**:412–413.

Young, S. Y., and Johnson, D. R., 1972, Nuclear polyhedrosis virus-specific soluble antigens in fatbody of *Trichoplusia ni* larvae, *J. Invertebr. Pathol.* **20**:114–117.

Young, S. Y., and Lovell, J. S., 1971, Hemolymph proteins of *Trichoplusia ni* during the course of a nuclear polyhedrosis virus infection, *J. Invertebr. Pathol.* **17**:410–418.

Young, S. Y., and Lovell, J. S., 1973, Virion proteins of the *Trichoplusia ni* nuclear polyhedrosis virus, *J. Invertebr. Pathol.* **22**:471–472.

Younghusband, H. B., and Lee, P. E., 1969, Virus-cell studies of *Tipula* iridescent virus in *Galleria mellonella* (L.), *Virology* **38**:247–254.

Younghusband, H. B., and Lee, P. E., 1970, The cytochemistry and autoradiography of *Tipula* iridescent virus in *Galleria mellonella, Virology* **40**:757–760.

Yule, B. G., and Lee, P. E., 1973, A cytological and immunological study of *Tipula* iridescent virus-infected *Galleria mellonella* larval hemocytes, *Virology* **51**:409–423.

Zherebtsova, E. N., Strokovskaya, L. I., and Gudz-Gorban, A. P., 1972, Subviral infectivity in nuclear polyhedrosis of the great wax moth (*Galleria mellonella* L.), *Acta Virol.* **16**:427–431.

CHAPTER 2

Viruses in Fungi

K. N. Saksena and P. A. Lemke

Mellon Institute
Carnegie-Mellon University
Pittsburgh, Pennsylvania 15213

1. INTRODUCTION

Fungal virology represents a relatively new and interdisciplinary area of science. Viruses and viruslike structures have been recognized in a variety of fungi. The vast majority of these viruses have been shown to possess double-stranded RNA (dsRNA) genomes and to constitute latent or nonsymptomatic infections.

The viruses that occur naturally and multiply in fungi are called mycoviruses. Viruses other than mycoviruses are often associated with fungi or with fungal cell surfaces. For example, some fungi serve as vectors for plant viruses, and still others are reported to serve as surrogate hosts in experiments dealing with nonfungal viruses. Thus a distinction is made here between *bona fide* mycoviruses and the alien viruses associated with fungi. While all virus–fungus relationships may be of concern to virologists, emphasis is given herein to mycoviruses. Readers interested generally in this subject are referred to other reviews (Bozarth, 1972; Hollings and Stone, 1971; Lemke and Nash, 1974; Lemke, 1976; Saksena, 1977; Spire, 1971; Yamashita, 1974).

2. HISTORY

2.1. Early Observations

The viral nature of abnormalities or unusual biological activities associated with fungi was often speculated about in early observations.

One such observation dates back to 1936 and involves an anomalous lysis in a yeast culture (Wiebols and Wieringa, 1936). Later, an apparently infectious disorder of yeast was again described and the presumed virus was even assigned the name "zymophage" (Lindegren and Bang, 1961). Ultrastructural examination of affected yeast cells revealed pleomorphic, vesicular structures approximately 0.1 μm in diameter (Hirano *et al.*, 1962), but these structures were unlike any known virus and their identification to date remains uncertain.

Disorders of two mushrooms, *Agaricus bisporus* (Sinden and Hauser, 1950) and *Laccaria laccata* (Blattný and Pilát, 1957), were described but no experimental evidence for virus accompanied these reports. Other fungal abnormalities have been observed and studied as elements of cytoplasmic inheritance. Examples include the "stunted" colonies of *Helminthosporium victoriae* (Lindberg, 1959, 1960), the phenomenon of "vegetative death" in *Aspergillus glaucus* (Jinks, 1959), the "senescence" syndrome in *Podospora anserina* (Marcou, 1961), and the abnormal characteristics for growth in *Pestalozzia annulata* (Chevaugeon and Digbeu, 1960). In all cases, cytoplasmic determinants for abnormal growth or development were transmissible to healthy strains following hyphal anastomosis (plasmogamy). These early studies often led to inferences as to the viral nature of the cytoplasmic determinant, but these speculations have remained unconfirmed.

2.2. Discovery of Mycoviruses

The viruses of fungi were discovered through developments in two independent areas of research involving three species of fungi. Observations on the transmissible disease of the cultivated mushroom, *A. bisporus* (Sinden and Hauser, 1950, 1957), paralleled in time the demonstration that culture filtrates from *Penicillium funiculosum* (Shope, 1948, 1953) and *P. stoloniferum* (Powell *et al.*, 1952) possessed antiviral activity. Although the causal agent for mushroom disease was unknown, induction of disease by hyphal anastomosis (Gandy, 1960) heightened speculation as to its viral nature. The first experimental evidence of mycoviruses came with the demonstration that virus particles were indeed associated with diseased mushrooms, and at least three morphologically distinct viruses were identified (Gandy and Hollings, 1962; Hollings, 1962). Virus particles containing dsRNA were subsequently shown to be present in *Penicillium* molds (Ellis and Kleinschmidt, 1967; Kleinschmidt and Ellis, 1968; Banks *et al.*, 1968), and the antiviral activity associated with these molds was shown to be attributable to interferon induction by mycoviral dsRNA (Lampson *et al.*, 1967; Banks *et al.*, 1968).

3. PHYSICOCHEMICAL PROPERTIES

Viruses* and viruslike particles (VLP) have now been observed in numerous species of fungi (Table 1), but only few of these particles have been isolated and studied in detail. Purification of these particles is often quite difficult and may be complicated by the multicomponent nature of the virus or by the presence of more than one virus in the fungal host. Regardless, purification is impractical without efficient methods for breaking fungal cells, and yields even then are generally low since effective methods for cell disruption may damage virus particles.

Mycoviruses which have been characterized biophysically and biochemically are discussed here. It is evident that the mycoviruses studied thus far share a number of characteristics. The majority are isometric particles, 25–50 nm in diameter, and exhibit centrifugal and density heterogeneity. They are frequently multicomponent systems; the dsRNA is segmented and individual segments are separately encapsidated. Despite their often high concentration in cells, most mycoviruses remain latent and do not cause any significant or overt damage to their hosts.

3.1. *Agaricus bisporus* Viruses

An infectious, degenerative disease of the cultivated mushroom *Agaricus bisporus* (Sinden and Hauser, 1950) is caused by a complex of viruses (Gandy and Hollings, 1962; Hollings, 1962). These viruses are transmitted through hyphal anastomosis (Gandy, 1960) and carried by spores (Schisler *et al.*, 1967). Artificial injection of purified virus into mushroom primordia has been reported to incite disease symptoms (Dieleman-van Zaayen and Temmink, 1968; Hollings, 1962).

---

* Particles of nucleoprotein composition and viruslike morphology are herein called "viruses," but this term is used with some reservation, since purified particles have not been rigorously proven to be infectious. The more extensively characterized particles, however, have been shown to contain double-stranded RNA genomes and polymerase activity (see text for details). Such features are consistent with the viral nature of particles, but none of the particles isolated from fungi has thus far satisfied all criteria expected of a virus. Particles that morphologically resemble viruses but have not been isolated or characterized are designated as VLP in the text.

TABLE 1
Occurrence of Viruses and Viruslike Particles in Fungi[a]

Host[b]	Particle morphology and size (nm)	Key reference
	BASIDIOMYCETES	
Agaricus bisporus	Isometric, 25	Hollings (1962)
	Isometric, 29	Hollings (1962)
	Isometric, 34	Dieleman-van Zaayen and Temmink (1968)
	Isometric, 50	Hollings and Stone (1971)
	Rod, 19 × 50	Hollings (1962)
	Rod, 17 × 350	Dieleman-van Zaayen (1967)
Boletus edulis	Isometric, 28	Huttinga *et al.* (1975)
	Isometric, 32	Huttinga *et al.* (1975)
	Isometric, 42	Huttinga *et al.* (1975)
	Rod, 13 × 500	Huttinga *et al.* (1975)
Coprinus lagopus	Spherical, 130	Shahriari *et al.* (1973)
Corticium rofsii	Isometric, 28	Yamashita *et al.* (1975)
	Isometric, 43	Yamashita *et al.* (1975)
Laccaria amethystina	Isometric, 28	Blattný and Králík (1968)
Laccaria laccata	Isometric, 28	Blattný and Králík (1968)
Lentinus edodes	Isometric, 25	Ushiyama and Nakai (1975)
	Isometric, 30	Ushiyama and Nakai (1975)
	Isometric, 39	Ushiyama and Nakai (1975)
	Rod, 15 × 700 ~900	Ushiyama and Hashioka (1973)
Puccinia alli	Isometric, 40	Yamashita *et al.* (1975)
Puccinia graminis	Isometric, 38	Rawlinson and Maclean (1973)
Puccinia horiana	Isometric, 40	Yamashita *et al.* (1975)
Puccinia malvacearum	Isometric, 34	Lecoq *et al.* (1974)
Puccinia miscanthi	Isometric, 40	Yamashita *et al.* (1975)
Puccinia recondita	Isometric, 40	Yamashita *et al.* (1975)
Puccinia striiformis	Isometric, 34	Lecoq *et al.* (1974)
Puccinia suaveolens	Isometric, 40	Yamashita *et al.* (1975)
Puccinia triticina	Isometric, 40	Yamashita *et al.* (1975)
Schizophyllum commune	Spherical, 130	Koltin *et al.* (1973)
Tilletiopsis sp.	Isometric, 40	Bozarth (1972)
Uromyces alcoperuri	Isometric, 40	Yamashita *et al.* (1975)
Uromyces durus	Isometric, 40	Yamashita *et al.* (1975)
Ustilago maydis	Isometric, 41	Wood and Bozarth (1972)
	PHYCOMYCETES	
Allomyces arbuscula	Spherical, 40	Khandjian *et al.* (1975)
Aphelidium sp.	Iridescent type, ~200	Schnepf *et al.* (1970)
Pythium sylvaticum	Tobacco mosaic virus[c]	Brants (1971)
Thraustochytrium sp.	Herpes type, 110	Kazama and Schornstein (1972)
	ASCOMYCETES	
Diplocarpon rosae	Isometric, 32–34	Bozarth *et al.* (1972a)
Erisiphe graminis	Isometric, 40	Yamashita *et al.* (1975)
	Tobacco mosaic virus	Nienhaus (1971)

TABLE 1. (Continued)

Host[b]	Particle morphology and size (nm)	Key reference
Gaeumannomyces graminis	Isometric, 27	Rawlinson et al. (1973)
	Isometric, 29	Lapierre et al. (1970)
	Isometric, 35	Rawlinson et al. (1973)
Microsphaera mougeotii	Isometric, 32	Yamashita et al. (1975)
	Rod, 19 × 48	Yamashita et al. (1975)
Neurospora crassa	Isometric, 20	Tuveson et al. (1975)
	Isometric, 30	Tuveson and Sargent (1976)
	Isometric, 60 ~ 130	Tuveson and Peterson (1972)
	Isometric, 80	Lechner et al. (1972)
	Pleomorphic, 200 ~ 400	Küntzel et al. (1973)
Peziza ostrachoderma	Isometric, 25	Spire (1971)
	Rod, 17 × 350	Dieleman-van Zaayen (1967)
Saccaromyces carlsbergensis	Bacteriophagelike head 70 ~ 80 × 70	Volkoff and Walters (1970)
Saccaromyces cerevisiae	Isometric, 30	Border (1972)
	Isometric, 38–40	Border (1972)
	Influenza virus[c]	Lavroushin and Treagan (1972)
	Newcastle disease virus[c]	Lavroushin and Treagan (1972)
	Polyoma virus[c]	Kovács et al. (1969)
	Tobacco mosaic virus[c]	Coutts et al. (1972)
Saccharomyces ludwigii	Isometric, 60	Kozlova (1973)
Sphaerotheca lanestris	Tobacco mosaic virus	Nienhaus (1971)

FUNGI IMPERFECTII

Host[b]	Particle morphology and size (nm)	Key reference
Alternaria tenius	Isometric, 30–40	Isaac and Gupta (1964)
Aspergillus flavus	Isometric, 30	Mackenzie and Adler (1972)
Aspergillus foetidus	Isometric, 40–42	Banks et al. (1970)
Aspergillus glaucus	Isometric, 25	Hollings and Stone (1971)
Aspergillus niger	Isometric, 40–42	Banks et al. (1970
Candida albicans	Polyoma virus[c]	Kovács et al. (1969)
Candida tropicalis	Isometric, 150	Nesterova et al. (1973)
Candida utilis	Isometric, 50	Kozlova (1973)
Cephalosporium acremonium	Isometric, 30	Day and Ellis (1971)
Cochiobolus miyabeanus	Isometric, 30	Yamashita et al. (1975)
Colletotrichum atramentarium	Isometric, 26	Yamashita et al. (1975)
Colletotrichum lindemuthianum	Isometric, 30	Rawlinson (1973)
Fusarium moniliforme	Isometric, 40	Bozarth (1972)
Fusarium roseum	Isometric, 25	Chosson et al. (1973)
	Isometric, 40	Chosson et al. (1973)
Gonatobotrys sp.	Isometric, 30	Spire et al. (1972b)
Helminthosporium carbonum	Isometric, 39	Dunkle (1974b)
Helminthosporium maydis	Isometric, 40	Bozarth et al. (1972b)
Helminthosporium oryzae	Isometric, 25	Spire et al. (1972a)

(Continued)

TABLE 1. (Continued)

Host[b]	Particle morphology and size (nm)	Key reference
Helminthosporium	Isometric, 18	Yamashita *et al.* (1975)
sacchari	Isometric, 30	Yamashita *et al.* (1975)
	Isometric, 45	Yamashita *et al.* (1975)
Helminthosporium	Isometric, 40	Bozarth (1972)
victoriae		
Mycogone perniciosa	Isometric, 42	Lapierre *et al.* (1972)
	Rod, 18 × 120	Lapierre *et al.* (1972)
Penicillium	Isometric, 40	Wood *et al.* (1971)
brevicompactum	Bacteriophagelike head 45; 52	Velikodvorskaya *et al.* (1972)
Penicillium chrysogenum	Isometric, 23	Volkoff *et al.* (1972)
	Isometric, 35	Banks *et al.* (1969*b*)
	Isometric, 40	Bozarth and Wood (1971)
Penicillium citrinum	Isometric, 40–50	Borré *et al.* (1971)
Penicillium claviforme	Isometric, 25–30	Lai and Zachariah (1975)
	Isometric, 50–70	Lai and Zachariah (1975)
Penicillium cyaneofulvum	Isometric, 32	Banks *et al.* (1969*a*)
Penicillium funiculosum	Isometric, 25–30	Banks *et al.* (1968)
Penicillium multicolor	Isometric, 32–34	Mackenzie and Adler (1972)
Penicilium notatum	Isometric, 25	Hollings and Stone (1971)
Penicillium stoloniferum	Isometric, 25–30	Ellis and Kleinschmidt (1967)
	Isometric, 32–34	Banks *et al.* (1968)
Penicillium variable	Isometric, 40–50	Borré *et al.* (1971)
Periconia circinata	Isometric, 32	Dunkle (1974*a*)
Piricularia oryzae	Isometric, 25	Férault *et al.* (1971)
	Isometric, 36	Yamashita *et al.* (1971)
Rhodotorula glutinis	Bacteriophagelike, 30 × 20; 60 × 50	Kozlova (1973)
Sclerotium cepivorum	Isometric, 30	Lapierre and Faivre-Amiot (1970)
Stemphylium botryosum	Isometric, 25	Hollings and Stone (1971)
Thielaviopsis basicola	Isometric, 40	Yamashita *et al.* (1975)
Verticilium fungicola	Isometric, 48	Lapierre *et al.* (1973)
	Rod, 17 × 35	Lapierre *et al.* (1973)

[a] The majority of reports cited here are based exclusively on electron microscopic observation of viruslike particles in tissue extracts or in thin sections of intact cells. Viruslike particles have been reported in a number of other fungi (Bozarth, 1972), but without details of morphology or size, and the names of these fungi are excluded from this table.

[b] In addition to the fungi listed here, serological evidence for dsRNA has been reported for *Colletotrichum falcatum*, *C. graminicola*, *Endothia parasitica*, *Helminthosporium turcicum*, *Penicillium cyclopium*, and *Pythium butleri* (Moffitt and Lister, 1974; 1975).

[c] Alien virus reported to be experimentally introduced in the fungal host.

Although as many as six different viruses have been reported to occur in this fungus (Hollings and Stone, 1971), generally three particle types occur (Fig. 1), often in combination, and are most common (Dieleman-van Zaayen, 1972*b*; Hollings, 1962; Saksena, 1975). These represent two spherical particles, 25–27 nm (AbV-25) and 34–36 nm

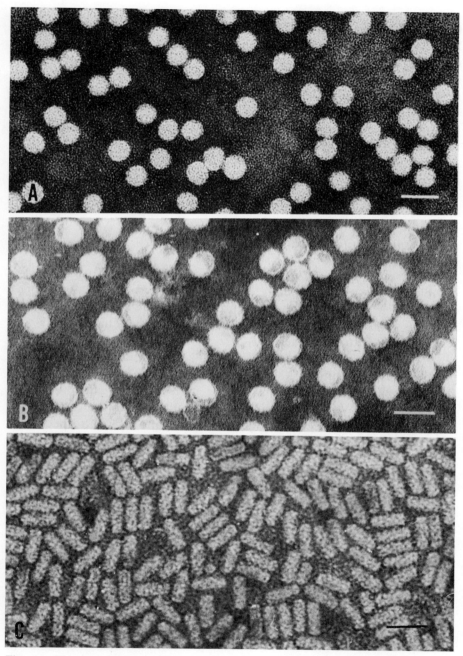

Fig. 1. Electron micrographs of viruses isolated from the cultivated mushroom, *Agaricus bisporus*. Bar equals 50 nm. A: Spherical particles (AbV-25) averaging 26 nm in diameter. B: Spherical particles (AbV-34) averaging 35 nm in diameter. C: Bacilliform particles (AbV-19/50) averaging 20 × 51 nm.

(AbV-34) in diameter, and bacilliform particles, 19 × 50 nm (AbV-19/ 50). Lack of critical separation of these viruses has hampered detailed characterization. AbV-25 and AbV-34, however, are serologically distinct and have single capsid proteins of 24,000 and 63,000 daltons, respectively (Barton, 1975). Both viruses contain dsRNA (Barton, 1975; Molin and Lapierre, 1973). AbV-25 and AbV-34, respectively, contain four and two species of dsRNA (Barton, 1975), which may represent the six species of dsRNA (0.67, 1.6, 1.70, 1.76, 1.89, and 2.17×10^6 daltons) identified among nucleic acids from virus-infected mushrooms (Marino et al., 1976). AbV-19/50 contains ssRNA of 0.9×10^6 daltons (Molin and Lapierre, 1973).

3.2. *Aspergillus foetidus* Viruses

Viruses in *Aspergillus foetidus* (Banks et al., 1970) represent a complex of two serologically distinct but morphologically similar isometric particles. These viruses are distinguishable as fast (AfV-F) and slow (AfV-S) by their relative electrophoretic mobility (Buck and Ratti, 1975a; Ratti and Buck, 1972).

The two viruses differ in their amino acid composition and have no polypeptides in common. AfV-F has one major ($\phi3$) and two minor ($\phi2$ and $\phi1$) polypeptides, with respective molecular weights of 87,000, 100,000, and 125,000. AfV-S has one major ($\sigma1$) and one minor ($\sigma2$) polypeptide, with molecular weights of 83,000 and 78,000, respectively. The molar ratio of $\phi3:\phi2:\phi1$ polypeptides from AfV-F is about 120:1:1 and remains essentially constant, while the ratio of $\sigma1:\sigma2$ polypeptides from AfV-S is variable. There is no phenotypic and genotypic mixing of these two viruses. Viruses identical to AfV-F and AfV-S have been found in another *Aspergillus* species, *A. niger* (Buck et al., 1973a).

Both AfV-F and AfV-S are multicomponent in that genomic segments of dsRNA are separately encapsidated. Particle heterogeneity results from differing amounts of nucleic acids. AfV-F is composed of five particle classes, FO, F1, F2, F3, and F4, and each class contains a single dsRNA component, respectively, 1.24, 1.44, 1.70, 1.87, and 2.31 $\times 10^6$ daltons. AfV-S has four major particle classes, S1, S2, S3, and S4. The S1 and S2 classes can be resolved further into subclasses S1a, S1b, S2a, and S2b. Analysis of various AfV-S particles reveals differences in sedimentation values and corresponding differences in RNA content. Three species of dsRNA, 0.1, 2.24, and 2.76×10^6 daltons, are involved. The subclasses S1a and S2a each contain one molecule of dsRNA, respectively, 2.24 and 2.76 $\times$ 10^6 daltons, whereas S1b and S2b contain, respectively, 2.24 and 2.76 $\times$ 10^6 dalton dsRNA, as well as an additional small dsRNA molecule of 0.1 $\times$ 10^6 daltons. S4 particles contain two molecules of dsRNA, both 2.24 $\times$ 10^6 daltons, and S3 particles have an equivalent of 1.5 molecules of this dsRNA. S3 and S4 particles are relatively more sensitive to host proteinases than are S1 and S2 (Buck and Ratti, 1975a,b).

Proteinaceous VLPs have been isolated from an aflatoxin-producing strain of *Aspergillus flavus* (Wood *et al.*, 1974). These particles are isometric, are 27–30 nm in diameter, have a buoyant density of 1.28 g/ml in cesium chloride, have a sedimentation value of 49 S, and contain no nucleic acid. Apparently no correlation exists between the presence or absence of the VLP and aflatoxin production among strains of *A. flavus*.

3.3. *Penicillium chrysogenum* Virus

Virus particles in *Penicillium chrysogenum* (PcV) are isometric, are 35–40 nm in diameter, and have been well characterized (Banks *et al.*, 1969b; Bozarth and Wood, 1971; Buck *et al.*, 1971; Lemke and Ness, 1970; Nash *et al.*, 1973; Wood and Bozarth, 1973). PcV is carried in spores (Yamashita *et al.*, 1973) and can be transmitted by hyphal anastomosis (Lemke *et al.*, 1976). Purified preparations have been reported to infect host protoplasts (Pallett, 1972).

PcV has a sedimentation coefficient of 150 S, a buoyant density of 1.27 g/ml in potassium tartrate (Buck *et al.*, 1971) or 1.35 g/ml in cesium chloride (Bozarth and Wood, 1971), and a diffusion coefficient of 1.03 $\times$ 10^{-7} cm^2/sec. (Wood and Bozarth, 1972). PcV has a single capsid protein with a molecular weight of 125,000 (Yamashita *et al.*, 1975).

Virus particles contain dsRNA which is separable into three molecular weight species of 1.89, 1.99, and 2.18 $\times$ 10^6 (Buck *et al.*, 1971; Wood and Bozarth, 1972; Nash *et al.*, 1973). The PcV dsRNA has a sedimentation coefficient of 13 S, a density of 1.60 g/ml in cesium sulfate, and a thermal melting point (T_m) of 100°C in standard saline citrate (SSC) or 88–92°C in 0.1 $\times$ SSC, and it contains equimolar amounts of G:C and A:U. A minor ssRNA component (ρ = 1.69 g/ml) has been reported, but its significance has not been ascertained. PcV virions have a particle weight of 13 $\times$ 10^6 and contain about 15% RNA. These data suggest that each PcV particle contains a single molecule of dsRNA. Electron microscopy of partially degraded PcV particles also indicates that individual particles contain one dsRNA molecule (Wood and Bozarth, 1973). Molecules of PcV dsRNA have an average contour length of 0.86 μm (Nash *et al.*, 1973), which corresponds to about 2.0×10^6 daltons. Studies on secondary structure of PcV dsRNA, including circular dichroism, thermal denaturation, and intercalation by ethidium bromide, further confirm its double helical nature (Cox *et al.*, 1970; Douthart *et al.*, 1973; Nash *et al.*, 1973).

A virus isolated from *P. brevicompactum* is serologically related to PcV and has similar biophysical properties (Wood *et al.*, 1971). Ultrastructural analysis of *P. brevicompactum* virus suggests that virions have icosahedral symmetry with a triangulation (T) number equal to 3, and are composed of 180 structural units that are grouped into 12 pentamers and 20 hexamers (Grigoriev *et al.*, 1974).

3.4. *Penicillium stoloniferum* Viruses

The virus particles in *Penicillium stoloniferum* (Ellis and Kleinschmidt, 1967; Banks *et al.*, 1968) are now recognized as a complex of two serologically distinct but morphologically similar viruses (Bozarth *et al.*, 1971; Buck and Kempson-Jones, 1970, 1973).

These isometric particles, 30–34 nm in diameter, are separable by electrophoresis (Bozarth *et al.*, 1971) or by ion-exchange chromatography (Buck and Kempson-Jones, 1970), and on the basis of their relative mobilities they have been designated as fast (PsV-F) and slow (PsV-S). PsV-F and PsV-S each have two capsid polypeptides but none in common. The major (S2) and the minor (S1) polypeptides of PsV-S have molecular weights of 42,500 and 55,500, respectively, while in PsV-F the major (F2) and the minor (F1) polypeptides have molecular weights of 47,000 and 59,000, respectively. These molecular weight values were obtained by polyacrylamide gel electrophoresis and confirmed by amino acid analysis. Based on the estimated weight of the PsV-S particle (5×10^6), the capsid should contain 120 molecules of polypeptide S2 and one molecule of polypeptide S1. This is consistent with an icosahedral structure having 60 structural units, each corresponding to two polypeptide chains (Buck and Kempson-Jones, 1974).

Both PsV-F and PsV-S are heterogeneous for centrifugal (66–113 S) and density (1.299–1.376 g/ml in cesium chloride) components.

They contain segmented RNA genomes encapsulated into separate particles (Buck and Kempson-Jones, 1973; Buck, 1975). In both viruses, the particles contain either dsRNA, ssRNA, or both types of nucleic acid, but individual particle fractions of PsV-F have not been analyzed. The physicochemical properties of PsV-S fractions and analysis of RNA derived therefrom clearly demonstrate that the centrifugal heterogeneity of this virus is due to different amounts of nucleic acid within particles. PsV-S involves four major particle classes, E particles (no nucleic acid), M particles (ssRNA), L particles (dsRNA), and H particles (both dsRNA and ssRNA). The M, L, and H particles are further resolved into subclasses. The M1 and M2 subclasses contain one molecule of dsRNA, respectively, 0.47 and 0.56 $\times$ 10^6 daltons; the L1 and L2 subclasses each contain one molecule of dsRNA, respectively, 0.94 and 1.11 $\times$ 10^6 daltons; and the H1 and H2 subclasses each contain one molecule of dsRNA (molecular size same as for L1 and L2 RNA) as well as some ssRNA. Amino acid composition, polypeptide ratios, and electrophoretic mobilities of M, L, and H particles are similar.

PsV-S and PsV-F can occur together and apparently replicate without any observable genotypic or phenotypic mixing (Bozarth *et al.*, 1971). Immunofluorescent studies show that both viruses occur in cytoplasm (Adler and Mackenzie, 1972). Both PsV-S and PsV-F may be transmitted to virus-free strains of *P. stoloniferum* by hyphal anastomosis (Lhoas, 1971*a*). In an attempt to infect protoplasts with extracellular virus, only PsV-S could be detected, an indication that this virus can replicate independently (Lhoas, 1971*b*).

3.5. Other Viruses

Several other mycoviruses have been characterized, but not extensively. These, too, have general features characteristic of dsRNA mycoviruses.

3.5.1. *Gaeumannomyces graminis* Virus

Virus particles have been detected in *Gaeumannomyces* (*Ophiobolus*) *graminis,* the fungus associated with take-all disease in cereals (Frick and Lister, 1975; Lapierre *et al.,* 1970; Lemaire *et al.,* 1970; Rawlinson *et al.,* 1973). Isometric particles, 27 and 35 nm in diameter, often occur in combination. Preparations of 35-nm particles sediment into two major and one minor components, and the 27-nm particles sediment into two components with a density range of 1.29–1.37 g/ml.

3.5.2. *Helminthosporium maydis* Virus

A pathogen of corn, *Helminthosporium maydis,* contains isometric virus particles, 48 nm in diameter, which sediment as three components (152 S, 212 S, and 283 S) and contain, respectively, 0%, 17%, and 32% nucleic acid (Bozarth, 1976). The 283 S component contains dsRNA with a density of 1.6062 g/ml and a molecular weight of 6.3×10^6.

3.5.3. *Helminthosporium victoriae* Virus

Another phytopathogenic fungus, *Helminthosporium victoriae,* responsible for blight of oats, contains virus particles (Sanderlin and Ghabrial, 1975). The particles sediment as two centrifugal components at 170 S and 190 S. They can be resolved into two electrophoretic components, and the two component proteins have molecular weights of 83,000 and 88,000. The virus particles contain dsRNA which sediments as a single component (17.3 S) on linear log sucrose gradients.

3.5.4. *Periconia circinata* Virus

Certain strains of *Periconia circinata,* a pathogen of sorghum, contain isometric virus particles, 32 nm in diameter, with major sedimentation components of 66 S, 140 S, and 150 S and two minor components of 36 S and 105 S (Dunkle, 1974*a,b*). The dsRNA isolated either from virus or from virus-infected mycelium has an asymmetrical sedimentation profile with a peak value at 11.5 S and a prominent shoulder at 13.5 S, indicating multiple species. This dsRNA is resolved by polyacrylamide gel electrophoresis into six molecular weight forms $(0.43, 0.48, 1.10, 1.25, 1.40, \text{ and } 1.75 \times 10^6)$ which may be encapsidated separately.

3.5.5. *Thielaviopsis basicola* Viruses

A complex of virus particles has been isolated from another pathogenic fungus, *Thielaviopsis basicola* (Bozarth and Goenaga, 1976).

These particles are 40 nm in diameter and sediment as eight centrifugal components. Immunoelectrophoresis yields four serologically distinct components. At least five species of dsRNA (2.7, 3.8, 3.9, 4.2, and 4.5×10^6 daltons) are represented among these particles.

4. BIOLOGICAL ASPECTS

4.1. Infectivity and Transmission

An expected criterion for any virus is infectivity. The mycoviruses have not yet proven to be infectious as purified particles. Fungal viruses, as revealed by electron microscopy, are often present at high titers but do not lyse the host cell. Natural transmission of fungal viruses apparently is accomplished by hyphal anastomosis (plasmogamy) between cells and by heterokaryosis—a genetically controlled phenomenon that involves fusion between compatible cell lines (Gandy, 1960; Hollings and Stone, 1971; Lemke and Nash, 1974; Lhoas, 1971a,b). Adaptation of fungal viruses to cell-mediated transmission would effectively delimit the spread of viruses within a species. Viruses carried in spores from infected strains could ultimately travel to and enter uninfected strains following spore germination and intraspecific hyphal fusions. Attempts to artificially introduce purified virus either by microinjection or in protoplast cultures have been rather inefficient and the results are erratic.

4.1.1. Natural Transmission

4.1.1a. Hyphal Anastomosis and Heterokaryosis

Although certain disease symptoms in fungi were earlier shown to be inducible through hyphal anastomosis, the possible viral nature of such diseases could only be suggested by these studies (Gandy, 1960; Caten, 1972; Jinks, 1959; Lindberg, 1959, 1960, 1971a,b; Marcou, 1961). Following the discovery of mycoviruses in *A. bisporus* (Gandy and Hollings, 1962; Hollings, 1962), virus transmission through anastomosis has been confirmed in several fungal species. Using genetically marked strains of *P. stoloniferum* and *A. niger,* viruses were transmitted by heterokaryosis from infected to normal strains (Lhoas, 1971a). Other instances of successful virus transmission through heterokaryosis include studies with *G. graminis* (Lemaire *et al.,* 1971;

Rawlinson, *et al.,* 1973), *P. chrysogenum* (Lemke *et al.,* 1976), *P. claviforme* (Metitiri and Zachariah, 1972), and *Ustilago maydis* (Day and Anagnostakis, 1972; Day *et al.,* 1972; Koltin and Day, 1976a,b; Wood and Bozarth, 1972).

4.1.1b. Spores

Transmission of mushroom viruses through infected spores is important epidemiologically. As few as ten infected spores per 5 ft^2 area can establish the disease in mycelia of *A. bisporus*. Virus-infected spores of this species can remain viable and infectious even after 5 years of storage at 4°C (Dieleman-van Zaayen, 1974). The presence of viruses in spores of other fungi (Hooper *et al.,* 1972; Yamashita *et al.,* 1973) suggests that spores are an important factor in dissemination and maintenance of viruses in a fungal population.

4.1.1c. Vectors

Phorid flies, *Megaselia halterata,* reared aseptically and fed on purified virus of *A. bisporus*, are, at best, a very inefficient vector for mushroom viruses. Such flies and other insects could aid in transfer of diseased spores or mycelial fragments in a mushroom farm (Hollings and Stone, 1971). No vector has yet been described for efficient or direct transfer of mycoviruses.

4.1.2. Artificial Transmission

Attempts to mechanically infect fungal cells by purified viruses have met with only limited success. Purified virus injected into mushroom initials of *A. bisporus* resulted in some diseased mushrooms subsequently produced surrounding the area of injection. Virus particles were detected in these mushrooms but not in control (noninjected) mushrooms (Dieleman-van Zaayen, 1972a,b; Hollings, 1962). However, artificial transmission of mycoviruses in general has proven to be both difficult and inefficient. The occurrence of virus particles in apparently normal fungal cultures complicates interpretation of data where only a low level of infectivity is recognized. Reports of virus particles in symptomless mushrooms (Passmore and Frost, 1974) underscore the need for critical reevaluation of data on virus infectivity by mechanical means. Unfortunately, lack of an effective biological assay for fungal viruses remains a major handicap in evaluation of infectivity.

Several obvious experiments have been conducted despite the lack of a rapid assay for infectivity, and these have generally yielded negative results. No infection is achieved if mycelial cultures are shaken or abraded in the presence of purified virus. Failure to directly infect fungi by mechanical means has prompted the use of isolated protoplasts for viral uptake (Lhoas, 1971b; Pallett, 1972). Apparently, yeast cells are, under certain conditions, susceptible to infection by purified viruses. For example, viruses from *A. niger* and *P. stoloniferum* have been transmitted to *Saccharomyces cerevisiae* by incubating compatible haploid yeast cells under conditions suitable for conjugation and zygote formation (Border, 1972; Lhoas, 1972). Similarly, transmission of a virus from *Candida tropicalis* to *S. cerevisiae* was achieved following conjugation of *S. cerevisiae* cells (Nesterova *et al.*, 1975). This method of virus transfer into yeast cells may be a nonrestrictive mechanism for uptake of viruses by yeasts under natural conditions. A protoplast fusion technique has been utilized for transmission of *Piricularia oryzae* viruses from infected to healthy strains (Boissonet-Menes and Lecoq, 1976). This method may be particularly useful in fungi where anastomosis is difficult or occurs at low frequency.

4.2. Virulence and Pathogenicity

Among the viruses of fungi, those viruses associated with the cultivated mushroom, *A. bisporus*, represent the best example of virulence and pathogenicity. Severity of this disease is correlated with the concentration of the viruses (Hollings, 1962; Dieleman-van Zaayen and Temmink, 1968; Dieleman-van Zaayen, 1972b). Disease symptoms, although somewhat variable, include deformed mushrooms and degeneration of mycelium in compost (Sinden and Hauser, 1950; Last *et al.*, 1974). Mycelia isolated from diseased mushrooms grow more slowly, and attempts to cure strains of these viruses by heat treatment have been only partially successful (Gandy, 1960; Hollings, 1962, Dieleman-van Zaayen, 1972b; Nair, 1973; Rasmussen *et al.*, 1972). Viruses found in two other mushroom species, *Boletus edulis* (Huttinga *et al.*, 1975) and *Lentinus edodes* (Ushiyama and Hashioka, 1973; Ushiyama and Nakai, 1975), have not been shown to be pathogenic.

Many plant pathogenic fungi harbor viruses and, as a consequence, their pathogenicity may be influenced. Preliminary studies with

G. graminis indicated that virus infection may impair pathogenicity of this fungus (Lapierre, 1970; Lemaire *et al.,* 1970, 1971; Rawlinson *et al.,* 1973). Virus-infected strains showed slow growth and poor sporulation, and were often weak pathogens. Infection of normal strains diminished phytopathogenicity, suggesting use of these viruses in biological control. Unfortunately, further studies have indicated inconsistency in the correlation of virus infection with altered morphology and decreased pathogenicity (Lapierre *et al.,* 1972; Rawlinson, 1975). Similar inconsistency between virus infection and reduced pathogenicity has been obtained in studying the viruses of *H. victoriae* (Bozarth *et al.,* 1972*b*), *P. oryzae* (Férault *et al.,* 1971; Yamashita *et al.,* 1971) and *U. maydis* (Wood and Bozarth, 1973). Apparently, the viruses of plant pathogenic fungi do little harm to their hosts, and the possibility of using these viruses in biological control of pathogenic fungi seems remote. Nevertheless, the killer system of *U. maydis* (Puhalla, 1968) would probably be the best system for investigation in this regard, since killer proteins produced by *U. maydis* are toxic to other species of smut fungi pathogenic to cereals (Koltin and Day, 1975).

4.3.   Fungal Metabolites

Although synthesis of host RNA, DNA, or protein was essentially comparable in uninfected and virus-infected strains of *P. stoloniferum* (Still *et al.,* 1975), the possible influence of mycoviruses on host metabolism, particularly on the production of secondary metabolites, has been suggested (Lemke and Nash, 1974). Unfortunately, no positive correlation seems to exist between the presence of mycoviruses and the synthesis of certain antibiotics. For example, all strains of *P. chrysogenum,* regardless of virus concentration, are apparently able to produce penicillin, and a strain of *P. notatum,* apparently virus free, can produce penicillin (Lemke and Nash, 1974; Volkoff *et al.,* 1972). In *P. stoloniferum,* the presence of viruses has been correlated with high galactosamine content of cell wall (Buck *et al.,* 1969), and in *S. cerevisiae* and *U. maydis* the presence of specific molecular weight forms of dsRNA has been correlated with toxin production (Berry and Bevan, 1972; Bevan *et al.,* 1973*a,b*; Day and Anagnostakis, 1973; Koltin and Day, 1975; Vodkin *et al.,* 1974).

Effects of antimetabolites on mycovirus replication have been observed, but their mode of action in this regard is unclear. Both cycloheximide (Lemke *et al.,* 1973) and mycophenolic acid (Borré *et al.,* 1971; Detroy *et al.,* 1973) reportedly interfere with mycovirus replication. Cycloheximide (10 μg/ml) reduces PsV-F replication by 95% relative to untreated controls without adversely affecting fungal growth (Detroy and Still, 1975). Evidently, cycloheximide acts as a selective inhibitor of PsV-F replication. The observation that some strains of *P. stoloniferum* which produced mycophenolic acid were also virus free and that a virus-infected strain did not produce this compound suggested that mycophenolic acid production might somehow inhibit viral replication in this fungal system (Detroy *et al.,* 1973). While exogenously supplied mycophenolic acid (300 μg/ml) was shown to reduce PsV-F replication by 25%, it also reduced growth by 36%. Therefore, no definitive evidence for a specific antimycoviral effect of this compound has been obtained.

Among other fungal metabolites tested, patulin and gliotoxin (50–100 μg/ml) appear to inhibit virus replication and have little effect on fungal growth, RNA, or protein synthesis (Detroy and Still, 1975, 1976).

4.4. Interferon Induction

An explanation for host-mediated interference of viral synthesis followed from the discovery of interferon (Isaacs and Lindenmann, 1957). It was initially assumed that only viruses or viral components stimulated interferon production in cells. Later, reports that heterologous and synthetic nucleic acids were able to produce interferon (Rotem *et al.,* 1963) prompted reevaluation of materials that earlier exhibited antiviral activity. The antiviral activity of a culture filtrate from *P. stoloniferum* (Kleinschmidt and Probst, 1962; Powell *et al.,* 1952) was accordingly reexamined as an inducer of interferon (Kleinschmidt and Murphy, 1965; Kleinschmidt *et al.,* 1964). The activity proved to be associated with viruslike particles (Kleinschmidt and Ellis, 1968; Ellis and Kleinschmidt, 1967) and was later identified as dsRNA (Banks *et al.,* 1968; Kleinschmidt *et al.,* 1968). A similar antiviral activity from cultures of *P. funiculosum* (Shope, 1948, 1953) had earlier been recognized to involve a ribonucleoprotein (Lewis *et al.,* 1959). Further characterization of this active material showed that it also contained dsRNA (Lampson *et al.,* 1967). Isometric virus particles containing dsRNA were subsequently isolated from *P. funiculosum* (Banks *et al.,* 1968).

Virus particles capable of inducing interferon in animals were soon identified in several species of *Penicillium* (Banks *et al.,* 1968, 1969a,b, 1970; Cooke and Stevenson, 1965a,b; Ellis and Kleinschmidt, 1967; Kleinschmidt and Ellis, 1968; Lemke and Ness, 1970; Planterose *et al.,* 1970) as well as from other genera of fungi (Banks *et al.,* 1970; Chandra *et al.,* 1971; Maheshwari and Gupta, 1973a,b). Both intact virus particles and derived dsRNA are active inducers of interferon.

The therapeutic and prophylactic potential of dsRNAs has been extensively investigated. Evidently, free viral dsRNA is more active as inducer of interferon than intact particles with equivalent levels of dsRNA (Nemes *et al.,* 1969). Intravenous injection of viral dsRNA results in earlier and higher serum interferon levels in mice than that obtained by intact *P. chrysogenum* virus (Buck *et al.,* 1971). The slower stimulation of interferon production by virus particle is probably due to a time lag for release of dsRNA. The lower yield of interferon induced by virus than by an equivalent amount of viral dsRNA is surprising and has not yet been adequately explained. Similar results were obtained following intraperitoneal injection of mice with dsRNA from *P. stoloniferum* virus (Planterose *et al.,* 1970). Furthermore, virus particles differ from derived dsRNA in their potential to induce interferon in cell cultures.

In addition to conferring cellular immunity against several viruses by inducing interferon (Kleinschmidt, 1972), dsRNA has been shown to inhibit cellular protein synthesis (Robertson and Mathews, 1973) and specifically to bind the initiation factor (IF-3) necessary for recycling of ribosomes (Kaempfer and Kaufman, 1973).

4.5. Killer Systems

4.5.1. *Saccharomyces cerevisiae*

The killer system in yeast, *Saccharomyces cerevisiae,* was initially recognized as a genetic phenomenon (Makowar and Bevan, 1963). Three phenotypes, i.e., killer, sensitive, and neutral, were described. Killer strains secrete a protein toxic to sensitive strains but are themselves immune to toxin. Neutral strains do not produce toxin but are immune (Bussey, 1972; Woods and Bevan, 1968). Preliminary studies on the structure and mode of action of toxin suggest that the yeast killer protein is unstable, is sensitive to protease and detergents, and apparently adsorbs to yeast cell walls (Bussey, 1972; Woods and Bevan, 1968).

Analysis of the inheritance of killer and immune functions has established that the killer phenotype is determined by an extrachromosomal factor that is dependent on nuclear genes for its maintenance. Several nuclear genes associated with killer and immune functions or their suppression have been recognized (Bevan and Somers, 1969; Bevan *et al.*, 1973*a,b*; Fink and Styles, 1972; Somers and Bevan, 1969; Naumov, 1974; Vodkin *et al.*, 1974; Wickner, 1974*a,b*, 1976; Wickner and Leibowitz, 1976*a,b*; Leibowitz and Wickner, 1976). At least ten chromosomal genes are essential for maintenance or replication of the killer factor. These are *M* (Somers and Bevan, 1969), *PETS* (Fink and Styles, 1972), and *MAK 1* through *MAK 8* (for maintenance of killer) (Wickner, 1974*a*; Wickner and Leibowitz, 1976*a*). In addition, three other genes are necessary for expression of killer function (Wickner, 1974*b*). Two of these, *KEX 1* and *KEX 2* (for killer expression), are required for secretion of toxin but not for resistance to toxin or for maintenance of the killer function. *KEX 2* is also required for normal mating and meiotic sporulation (Leibowitz and Wickner, 1976). A third, locus *REX 1* (for resistance expression), confers resistance to killing.

The normal killer strain can be "cured" of killing potential by growth in the presence of cycloheximide (Fink and Styles, 1972) or at high temperatures (Wickner, 1974*a*). A conversion to nonkiller or sensitive phenotype results with concurrent loss of the nonchromosomal killer determinant.

A complex of dsRNA species is known to be associated with this cytoplasmically inherited killer system (Berry and Bevan, 1972; Herring *et al.*, 1973; Vodkin *et al.*, 1974). Killer strains contain two species of dsRNA, referred to here as L-dsRNA and M-dsRNA, respectively, 2.5 and 1.4 $\times$ 10^6 daltons. Sensitive strains either possess L-dsRNA only or lack dsRNA. A correlation between the presence of M-dsRNA and the expression of the killer character has been established (Adler *et al.*, 1976; Bevan *et al.*, 1973*a,b*; Herring and Bevan, 1974; Mitchell *et al.*, 1973; Vodkin and Fink, 1973). The strains altered for the killer or resistance functions, including those deficient in both functions, reveal both qualitative and quantitative differences in their dsRNA profiles. Superkillers, for example, contain an excess of M-dsRNA. The M-dsRNA is routinely associated with killer and immune phenotypes and is never found in the absence of L-dsRNA (Bevan *et al.*, 1973*a,b*; Herring and Bevan, 1974; Vodkin *et al.*, 1974). Another RNA species, XL-dsRNA (3.8 $\times$ 10^6 daltons) has been detected in killer and immune phenotypes (Wickner and Leibowitz, 1976*a*). The XL-dsRNA may represent a replicative complex of the more commonly recognized L-dsRNA and M-dsRNA. Certain other dsRNA species, e.g., S-dsRNA (5×10^5 daltons) in suppressive nonkillers and I-dsRNA (2.5×10^6 daltons) in immune-minus strains, have also been observed (Vodkin et al., 1974), but these, like M-dsRNA, are found only when L-dsRNA is present.

Isometric particles containing dsRNA occur in yeast (Adler et al., 1976; Buck et al., 1971; Herring and Bevan, 1974). Apparently, the transmission of these particles and the inheritance of killer factor are related. Although particles may be recognized in all phenotypes (Adler et al., 1976), the particles differ in RNA content. Initially, virus particles with only L-dsRNA were reported (Buck et al., 1971), but subsequently a particle fraction containing two morphologically identical particles, both 39 nm in diameter, was shown to contain L-dsRNA and M-dsRNA (Herring and Bevan, 1974). These two dsRNA species are probably separately encapsidated. Since the replication of M-dsRNA is seemingly controlled by nuclear genes, the M-dsRNA virus in undoubtedly closely integrated with the host cell (Herring and Bevan, 1974). Since M-dsRNA virus is found only in the presence of L-dsRNA virus, the former may be defective and dependent on the latter. Other small and apparently extraparticulate dsRNAs such as S-dsRNA and I-dsRNA are found in certain mutants, but their significance and origin are not fully understood (Vodkin et al., 1974). Low molecular weight species of dsRNA (1.19 and 1.29×10^6) have been found associated with partially purified virus particles (Adler et al., 1976). Such RNA may be extraparticulate or may be enveloped in labile particles that degrade during processing.

Killer strains are common among laboratory strains of S. cerevisiae (Fink and Styles, 1972) and apparently also occur in commercial varieties of yeast (Maule and Thomas, 1973; Naumov, 1974; Philliskirk and Young, 1975). Since the killing occurs at low pH and not at the pH of standard media, the killer property is often not observed.

The killer factor from S. cerevisiae kills not only sensitive S. cerevisiae but also the pathogenic yeast Torulopsis glaberata, evidently by a common mechanism involving membrane damage (Bussey and Skipper, 1976). Other yeast strains, e.g., Crytococcus neoformis, Candida albicans, and Schizosaccharomyces pombe, are not affected, indicating some specificity of toxin action and limited potential as an antifungal antibiotic. Killer strains have been found in other yeast genera, Candida, Debaryomyces, Hansenula, Kluyveromyces, Pichia, and Torulopsis (Philliskirk and Young, 1975).

4.5.2. Ustilago maydis

A killer system similar to that of *S. cerevisiae* exists in *Ustilago maydis* (Puhalla, 1968). At least three killer specificities (P1, P4, and P6) with corresponding immunities are recognized (Koltin and Day, 1976a,b). A correlation exists between the presence of isometric virus particles, 41 nm in diameter, and dsRNA components in the killer and immune strains (Wood and Bozarth, 1973; Koltin and Day, 1976a,b). As many as nine species of dsRNA have been identified from the various killer and immune strains. Certain species of dsRNA are common to killer strains with different specificities but there are differences recognized in dsRNA profiles of specific killers. Two dsRNA components (2.6 and 2.9 $\times$ 10^6 daltons) are shared by all three killers. Immune strains possess additional forms of dsRNA and may lack one or both of the two dsRNA components identified in killers. Sensitive strains either lack dsRNA or possess cytoplasmic genes which functionally suppress the killer or immune response. Heterokaryon transfer of virus particles effectively transmits killer and immune factors. In crosses between killer and sensitive strains, the progeny are preferentially killer in phenotype and possess the dsRNA profile of the killer parent. Similarly, immune strains derived from crosses of immune with sensitive strains carry dsRNA components found in the immune parent (Day and Anagnostakis, 1973; Koltin and Day, 1976a,b; Wood and Bozarth, 1973).

Although nuclear genes are known to maintain the expression of the cytoplasmic genes in the yeast killer system (Bevan *et al.,* 1969; Nesterova and Zeknov, 1973), there is as yet no evidence that nuclear genes are involved in the maintenance of the killer determinant in *U. maydis* (Day and Anagnostakis, 1973; Koltin and Day, 1976a,b). A nuclear gene with two alleles influences the expression of immunity in *U. maydis*. The s^+ allele of this chromosomal gene confers immunity, while sensitive nonkillers carry the alternate or s^- allele. To date, no nuclear genes for maintenance or suppression of killer function have been found (Koltin and Day, 1976a,b). A determinant for the killer suppression has been identified in *U. maydis,* but this determinant is cytoplasmically transmitted and is a species of dsRNA. Unlike the yeast killer system, the information for killer and immunity may reside in different dsRNA molecules, but the suppressive action for killer in *U. maydis* operates on the immunity function as well (Koltin and Day, 1976b). If the dsRNA profiles of the various killer specificities reflect differences in one viral genotype, then specific suppression may result from a recognition mechanism permitting selective replication or interference of dsRNA segments (Koltin and Day, 1976b). While the precise functional significance of various dsRNA components remains unresolved at this time, the killer and immune responses are apparently associated with qualitative differences in dsRNAs.

The potential use of killer toxin in the control of smut fungi has been considered, and a large number of fungi have been screened for sensitivity to *U. maydis* killer toxins (Koltin and Day, 1975). Except for certain taxonomically related grass smuts, other fungi and bacteria are resistant.

4.6. Conditional Lysis

Production of lytic plaques of mycoviral origin was first reported in *Penicillium citrinum* and *P. variable* (Borré et al., 1971). Patches of white asporogenic mycelium often develop in *Penicillium* strains grown on standard laboratory media, but the expression of lytic plaques is dependent on growth on a special unbuffered, lactose-containing medium. On this medium, certain *P. chrysogenum* cultures produce asporogenic patches and exhibit an asynchronous lysis. This phenomenon is apparently correlated with the presence of the dsRNA virus and has been observed only in certain mutant strains of *P. chrysogenum* (Lemke et al., 1973, 1976). Auxotrophic mutants as well as strains grown in the presence of antimetabolites or subjected to heat show a reduction in viral titer and relatively fewer plaques. One heat-treated strain, Δ, was apparently cured of virus and produced neither asporogenic patches nor lytic plaques.

In *P. chrysogenum,* virus transmission and lytic plaque formation have been studied further through heterokaryon tests (Lemke et al., 1976). The wild-type strain (NRRL 1951), although infected with virus, does not form plaques. However, in a mutant strain, E-15, plaque formation occurs, and its frequency is related to viral titer. Analysis of crosses involving appropriately marked strains (spore color and nutritional requirements) of NRRL 1951 × E-15 indicated that the progeny, like parents, contain virus, but the segregation pattern for plaque formation among progeny suggested that a nuclear allele (s^-) was responsible for plaque formation. In another cross involving the E-15 parent and the Δ strain, all progeny contained virus and formed plaques when grown on lactose medium. These results support the contention that lysis is due to viral infection but is conditional on nutritional source and at least one nuclear allele. Presumably, strain E-15 carries that allele (s^-). By implication, the wild type (NRRL 1951) carries the dominant allele (s^+) for resistance to plaque formation, and indeed a diploid formed by E-15 (virus infected) and NRRL 1951 did not form plaques.

Such conditional and asynchronous lysis, although somewhat reminiscent of bacterial lysis and plaque formation by bacteriophages, is clearly an exceptional phenomenon in *P. chrysogenum* and is unsuitable as a biological assay for unit infectivity by virus.

5. ULTRASTRUCTURE AND INTRACELLULAR FEATURES

5.1. Morphological Diversity

A comprehensive list of fungal viruses and viruslike particles of different morphological types found among fungi is presented in Table 1. Most fungal viruses are isometric, but particles of other morphological types occur as well (Saksena, 1978). The uncommon particle types include bacilliform particles (Hollings, 1962), rigid rods (Dieleman-van Zaayen *et al.*, 1970; Ushiyama and Hashioka, 1973), flexuous rods (Huttinga *et al.*, 1975), herpes-type particles (Kazama and Schornstein, 1972; 1973), and bacteriophagelike particles (see Chapter 4) (Tikchonenko *et al.*, 1974; Velikodvorskaya *et al.*, 1972; Volkoff and Walters, 1970).

5.2. *Agaricus bisporus* Viruses

Detailed ultrastructural studies on the intracellular distribution of fungal viruses have been conducted in only a few cases. *Agaricus bisporus* viruses seem to accumulate in vegetative mycelium as well as in fruiting bodies and spores (Albouy, 1972; Dieleman-van Zaayen, 1972a,b; Dieleman-van Zaayen and Igesz, 1969). The AbV-34 particles are often scattered in the cytoplasm. Thin sections reveal that AbV-34 particles are present in and around the septal pores, suggesting cell-to-cell migration. In spores, virus particles occur in cytoplasm or in vacuoles. In mushroom tissue, virus particles appear in vacuoles or as crystalline arrays of linear or tubular arrangement, the latter resembling a "corncob" (Fig. 2). Scattered AbV-25 particles are often difficult to distinguish from ribosomes, but careful examination shows that these virus particles are either dispersed or aggregated in cyto-

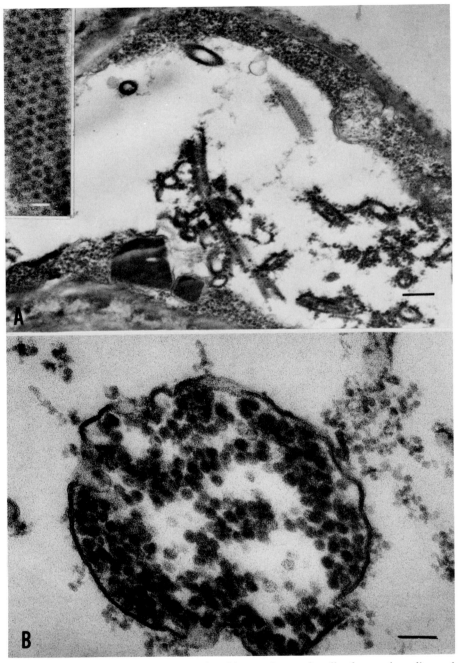

Fig. 2. Electron micrographs of ultrathin sections of cells from virus-diseased mushroom, *Agaricus bisporus*. A: Various aggregates of AbV-34 particles. Bar equals 500 nm. Inset shows an enlargement of an aggregate in "corncob" configuration. Bar equals 50 nm. B: Accumulation of AbV-34 particles within a vesicle. Bar equals 100 nm.

plasm and, again, are often found in vacuoles. The AbV-19/50 particles (bacilliform) are seen only occasionally in thin sections.

5.3. *Penicillium* Viruses

In *Penicillium chrysogenum* (Crosse and Mason, 1974; Yamashita *et al.*, 1973; Volkoff *et al.*, 1972), isometric virus particles occur singly or in aggregates in the cytoplasm or in vacuoles, and, occasionally, crystalline arrangements of particles are observed. In older cells, an excessive amount of virus accompanies some degeneration of the cytoplasm and cellular organelles. In *P. stoloniferum* (Hooper *et al.*, 1972), virus particles occur in abundance in mycelium or in spores, and the intracellular appearance of these particles is quite similar to that seen in *P. chrysogenum*. Evidently, virus multiplies well in the intact host cell, although some cellular aberration is seen in older cells.

5.4. *Thraustochytrium* Virus

The ultrastructure of a herpes-type virus described from a species of *Thraustochytrium* has been studied in detail (Kazama and Schornstein, 1972, 1973). The replicative cycle of this virus is essentially similar to that of known herpes viruses. After assembly in the nucleus, the nucleocapsid is temporarily enveloped by two nuclear membranes during passage into the cytoplasm. Here, the particles lose the nuclear membranes and the nucleocapsids subsequently acquire a fibrillar coat. The final envelopment of the particles occurs by budding into various cytoplasmic organelles or envelopment by cytoplasmic vesicles, as has been observed with other herpes-type viruses.

6. REPLICATION

6.1. RNA Polymerase

An RNA-dependent RNA polymerase activity has been described for virus particles derived from at least three fungi, *A. foetidus* (Ratti and Buck, 1975; Buck and Ratti, 1975*a*), *P. chrysogenum* (Nash *et al.*, 1973), and *P. stoloniferum* (Alaoui *et al.*, 1974; Buck, 1975; Chater and Morgan, 1974; Lapierre *et al.*, 1971). The *in vitro* reactions use dsRNA as a template, and activity is magnesium dependent and inhibited by ethidium bromide but not by actinomycin D. The product of polymerase activity has not been characterized in all cases. However, the product of PsV-S polymerase activity is reported to be a dsRNA that remains within the virus particles (Buck, 1975; Chater and Morgan, 1974). Polymerase activity of PsV-S is restricted to H particles (i.e., particles which contain both ss- and dsRNA), and the resultant product of this activity is either one new molecule of dsRNA or two complementary single strands of RNA (Buck, 1975). On the other hand, the major reaction product for AfV-S polymerase seems to be ssRNA (Ratti and Buck, 1975). Preparations of AfV-S, however, are heterogeneous for RNA content and include particles which apparently contain dsRNA products as well as ssRNA molecules, the latter possibly intermediates of virus replication.

6.2. A Model

Studies on viral replication have shown that the RNA polymerase of PsV-S is capable of copying both strands of dsRNA and therefore acts essentially as a replicase. Accordingly, a model for the replication of dsRNA mycoviruses has been proposed (Buck and Ratti, 1975b). Simply stated, the model involves a particle-associated duplication of dsRNA. In the model, the virion-associated RNA polymerase utilizes dsRNA as a template and can bind to the 3′ end of either strand of dsRNA with equal probability. The reaction yields a molecule of ssRNA and this product apparently remains within the particle. Such particles would then contain both dsRNA and ssRNA and indeed have been observed (S3 particles of AfV-S and H particles of PsV-S). The complementary ssRNA molecules from different particles could, if released, anneal to form extraparticulate dsRNA molecules, but this is probably not in the normal sequence of events. Subsequent to the synthesis of one strand of ssRNA, the polymerase can apparently switch positions from the 5′ end of one template strand to the 3′ end of a complementary strand. This results in further ssRNA synthesis and ultimately gives rise to particles with an equivalent of two molecules of dsRNA per particle (S4 particles of AfV-S and P particles of PsV-S. *In vivo* proteolytic uncoating of these product type (S4 and P) particles would release dsRNA template along with associated RNA polymerase. The polymerase could then act as a transcriptase and produce copies of viral messenger RNA which eventually are translated into virus-specific protein. Finally, the two dsRNA molecules released from a product particle are separately encapsidated, resulting in two virus particles each containing one dsRNA molecule. Thus, for every particle originally present, a duplicate particle is ultimately produced.

This cycle of replication is especially suited for the mycoviruses as contrasted to that of other dsRNA viruses (Wood, 1973), because they remain latent and replicate indefinitely without lysis of the host cell. Thus far, experimental data are consistent with such a model. Perhaps it should be emphasized, however, that the model is based on a limited sample of viruses so far characterized, and there is no *a priori* reason to rule out alternative mechanisms for replication of other dsRNA mycoviruses. Furthermore, mycoviruses containing ssRNA or DNA should eventually reveal other interesting models.

7. ALIEN VIRUSES ASSOCIATED WITH FUNGI

7.1. Animal Viruses

Propagation of mammalian viruses in axenic cultures of the yeasts *S. cerevisiae* and *C. albicans* have been reported (Kovács, 1967; Kovács and Bucz, 1967; Kovács *et al.,* 1966; Lavroushin and Treagen, 1972). It was further reported that DNA from polyoma virus was infectious in three yeast strains (Kovács, 1970; Kovács *et al.,* 1969). The yield of mammalian viruses in these experiments was enhanced considerably by addition of dimethylsulfoxide to the yeast cultures. The visualization of fluorochrome-labeled virus in yeasts provided direct evidence for adsorption, penetration, and multiplication of mammalian viruses in these protists.

Infection was accompanied by characteristic morphological and biological changes in the yeast cells, and a parallel between increased viral titer and cell mortality was observed. The induction of these cytopathic effects by a virus suggests the suitability of the system as a model for the study of the mechanism of viral infection in a relatively simple eukaryotic cell line (Kovács, 1967).

The synthesis of virus-specific RNAs in isolated yeast (*S. cerevisiae*) mitochondria inoculated with RNA from Venezuelian equinine encephalomyelitis virus has also been reported (Ershov *et al.,* 1974). Apparently, the mitochondrial translation apparatus for protein synthesis is nonspecific and can accommodate different classes of messenger RNAs.

7.2. Bacteriophages

Observations that extracts from certain species of *Penicillium* can lyse bacteria led to the discovery of another group of viruses, the *Penicillium*-derived bacterial viruses (PBV) (Bobkova *et al.*, 1975; Chaplygina *et al.*, 1975; Lebed *et al.*, 1975; Tikchonenko *et al.*, 1974; Velikodvorskaya *et al.*, 1972, 1974). The precise role of these viruses with the host fungus is still uncertain, but it has been implied that PBV have a temperate lysogenic relationship with the fungus (Tikchonenko *et al.*, 1974). (See Chapter 4 for a more detailed discussion.)

Like the dsRNA mycoviruses, the viruses of the PBV series can be isolated from cultures of *Penicillium brevicompactum* and *P. stoloni-ferum* as well as from certain penicillin-producing strains of *P. chryso-genum*. Attempts to infect healthy cultures of *Penicillium* with purified PBVs have thus far failed, but these viruses infect and can be propagated in strains of the bacterium *Escherichia coli*.

To date, five types of PBV have been recognized: PBV-1, PBV-2, and PBV-3 from *P. brevicompactum;* PBV-4 from *P. stoloniferum;* and PBV-5 from *P. chrysogenum*. These viruses can be distinguished by their host specificity to strains of *E. coli* and by differences in physical properties. PBV-1, PBV-2, and PBV-3 differ in density—respectively, 1.38, 1.40, and 1.42 g/ml—and these densities are very different from those typical of dsRNA mycoviruses. PBV-1 and PBV-3 have icosahe-dral heads and long noncontractile tails, and resemble λ phage. PBV-2 has an icosahedral head with a short tail and resembles T5 phage (Velikodvorskaya *et al.*, 1972). The nucleic acid of PBV-1 and PBV-2 is infectious and can lyse several strains of *E. coli*. PBV-5 replicates preferentially in *E. coli* strain C. An apparent correlation between PBV-5 titer in the mycelium and the ability of the fungus for penicillin production has been reported (Velikodvorskaya *et al.*, 1974).

7.3. Plant Viruses

Several phytopathogenic viruses are now known to be transmitted by fungi (Grogan and Campbell, 1966). The fungus in such instances merely serves as a vector and apparently does not support virus multi-plication. More recently, it has been proposed that certain plant viruses may actually use fungi as an alternate host, but host relationships of these viruses have not been investigated extensively (Nienhaus, 1971; Nienhaus and Mack, 1974; Yarwood and Hecht-Poinar, 1973).

Based on tests for infectivity in higher plants, tobacco mosaic virus (TMV) or TMV-like viruses have been found associated with a variety of fungi. These fungi include certain rusts and powdery mildews (Nienhaus, 1971; Yarwood and Hecht-Poinar, 1973) as well as some nonpathogenic fungi (Coutts *et al.*, 1972).

Artificial infection of *Pythium* species with TMV has been reported (Brants, 1969, 1971; Wieringa-Brants, 1973). Apparently, the virus persists and presumably multiplies in this mycelium. However, only certain species of *Pythium* seem to support virus multiplication. For example, a significant increase in TMV occurs in *P. arrhenomonas,* but a rapid decrease in virus titer is observed in *P. debaryanum* and *P. ultimum.* Similarly, tobacco necrosis virus increases only in *P. arrhenomonas* (Nienhaus and Mack, 1974).

The transmigration of TMV from tobacco callus cultures through hyphae of *P. debaryanum* has also been shown (Subbarayudu *et al.,* 1973). It is not clear if the virus in this case was merely transported or if it actually multiplied in the mycelium. Regardless, transportation of viruses by fungal hyphae may be of some significance in plant disease complexes where fungi and alien viruses are closely associated.

TMV has been reported to infect yeast protoplasts but with no dramatic increase in virus titer following the infection (Coutts *et al,* 1972).

8. CONCLUDING REMARKS

After a belated discovery, viruses in fungi have been found to occur commonly. The majority of mycoviruses are multicomponent systems, contain dsRNA, and are remarkably similar in general features. They are often latent, and evidence to date suggests that mycoviruses are suited for transmission by plasmogamy. The difficulty in demonstrating infectivity by cell-free virus preparations may reflect inefficiency of current techniques for handling mycoviruses. Reliable and rapid assays for infectivity are needed, particularly in assessment of biological significance of these viruses. Despite technical difficulties, considerable progress has already been made in understanding virus–fungus interactions, especially in the killer systems of two fungi. Indeed, mycoviruses offer investigators a challenge as well as an opportunity to study virus-related phenomena in host cell systems that are both eukaryotic and microbial.

ACKNOWLEDGMENTS

Preparation of this chapter was supported by the Butler County Mushroom Farm, Inc., Worthington, Pennsylvania, through a fellowship to the Mellon Institute.

9. REFERENCES

Adler, J. P., and Mackenzie, D. W., 1972, Intrahyphal localization of *Penicillium stoloniferum* viruses by fluorescent antibody, *Abstr. Annu. Meet. Am. Soc. Microbiol.* **1972**:68

Adler, J., Wood, H. A., and Bozarth, R. F., 1976, Virus-like particles in killer, neutral, and sensitive strains of *Saccharomyces cerevisiae, J. Virol.* **17**:472–476.

Alaoui, R., Jacquemi, J. M., and Boccardo, G., 1974, Ribonucleic acid polymerase activity associated with virus-like particles isolated from *Penicillium stoloniferum,* ATCC 14586, *Acta Virol.* **18**:193–202.

Albouy, J., 1972, Etude ultramicroscopique du complexe viral de la "goutte seche" de carpophores d'*Agaricus bisporus, Ann. Phytopathol.* **4**:39–44.

Banks, G. T., Buck, K. W., Chain, E. B., Himmelweit, F., Marks, J. E., Tyler, J. M., Hollings, M., Last, F. T., and Stone, O. M. 1968, Viruses in fungi and interferon stimulation, *Nature* (*London*) **218**:542–545.

Banks, G. T., Buck, K. W., Chain, E. G., Darbyshire, J. E., and Himmelweit, F., 1969*a*, *Pennicillium cyaneo-fulvum* and interferon stimulation, *Nature* (*London*) **223**:155–158.

Banks, G. T., Buck, K. W., Chain, E. B., Darbyshire, J. E., and Himmelweit, F., 1969*b*, Virus-like particles in penicillin producing strains of *Penicillium chrysogenum, Nature* (*London*) **222**:89–90.

Banks, G. T., Buck, K. W., Chain, E. B., Darbyshire, J. E., Himmelweit, F., Ratti, G., Sharpe, T. J., and Planterose, D. N., 1970, Antiviral activity of double stranded RNA from a virus isolated from *Aspergillus foetidus, Nature* (*London*) **227**:505–507.

Barton, R. J., 1975, Purification and some properties of two viruses infecting *Agaricus bisporus:* The cultivated mushroom, *3rd Int. Congr. Virol.* (*Madrid*) *Abstr.,* p. 147.

Berry, E. A., and Bevan, E. A., 1972, A new species of double stranded RNA from yeast, *Nature* (*London*) **239**:279–280.

Bevan, E. A., and Somers, J. M., 1969, Somatic segregation of the killer (K) and neutral (N) cytoplasmic genetic determinants in yeast, *Genet. Res.* **14**:71–76.

Bevan, E. A., Somers, J., and Theivandirajah, K., 1969, Genes controlling the expression of the killer character in yeast (*Saccharomyces cerevisiae*), *11th Int. Bot. Congr.* (*Seattle*) *Abstr.,* p. 14.

Bevan, E. A., Herring, A. J., and Mitchell, D. J., 1973*a*, Preliminary characterization of two species of dsRNA in yeast and their relationship to killer character, *Nature* (*London*) **245**:81–86.

Bevan, E. A., Mitchell, D. J., and Herring, A., 1973*b*, The association between the cytoplasmically inherited "killer" character and double-stranded RNA in *Saccharomyces cerevisiae, Genetics* **74**:s23.

Blattný, C., and Králík, O., 1968, A virus disease of *Laccaria laccata* (Scop. ex Fr.) Cooke and some other fungi, *Česká Mykol.* **22**:161–166.

Blattný, C., and Pilát, A., 1957, Možnost existence viros u vyšších hub.—Die Möglichkeit der Existenz von Virosen bei den Hutpilzen, *Česká Mykol.* **11**:205–211.

Bobkova, A. F., Velikodvorskaya, G. A., Tikchonenko, T. I., Lebed, E. S., Egorov, A. A., and Bartoshevich, Y. E., 1975, Viruses infective for bacteria in strains of *Penicillium chrysogenum* with various antibiotic activity, *Antibiotiki* **20**:600–606.

Boissonet-Menes, M., and Lecoq, H., 1976, Transmission de virus per fusion de protoplastes chez *Pyricularia oryzae* Briosi and Cav, *Physiol. Veg.* **14**:251–257.

Border, D. J., 1972, Electron microscopy of cells of *Saccharomyces cerevisiae* infected with double stranded RNA viruses from *Aspergillus niger* and *Penicillium stoloniferum, Nature (London) New Biol.* **236**:87–88.

Borré, E., Morgantini, L. E., Ortali, V., and Tonolo, A., 1971, Production of lytic plaques of viral origin in *Penicillium, Nature (London)* **229**:568–569.

Bozarth, R. F., 1972, Mycoviruses: A new dimension in microbiology, in: *Environmental Health Perspectives,* pp. 23–39, U.S. Department of Health, Education, and Welfare, Washington, D.C.

Bozarth, R. F., 1976, Properties of the dsRNA of a virus-like particle from *Helminthosporium maydis, Beltsville Symp. Virol. Agr. Abstr.,* p. 31.

Bozarth, R. F., and Goenaga, A., 1976, A complex of virus-like particles from *Thielaviopsis basicola, Proc. Am. Phytopathol. Soc.* **3:** in press.

Bozarth, R. F., and Wood, H. A., 1971, Purification and properties of the virus-like particles of *Penicillium chrysogenum, Phytopathology* **61**:886.

Bozarth, R. F., Wood, H. A., and Mandelbrot, A., 1971, The *Penicillium stoloniferum* virus complex: Two similar double-stranded RNA viruslike particles in a single cell, *Virology* **45**:516–523.

Bozarth, R. F., Wood, H. A., and Goenaga, A., 1972a, Virus-like particles from a culture of *Diplocarpon rosae, Phytopathology* **62**:493.

Bozarth, R. F., Wood, H. A., and Nelson, R. R., 1972b, Virus-like particles in virulent strains of *Helminthosporium maydis, Phytopathology* **62**:748.

Brants, D. H., 1969, Tobacco mosaic virus in *Pythium* spec, *Neth. J. Plant Pathol.* **75**:296–299.

Brants, D. H., 1971, Infection of *Pythium sylvaticum in vitro* with tobacco mosaic virus, *Neth. J. Plant Pathol.* **77**:175–177.

Buck, K. W., 1975, Replication of double-stranded-RNA in particles of *Penicillium stoloniferum* virus S, *Nucl. Acids Res.* **2**:1889–1902.

Buck, K. W., and Kempson-Jones, G. F., 1970, Three types of virus particle in *Penicillium stoloniferum, Nature (London)* **225**:945–946.

Buck, K. W., and Kempson-Jones, G. F., 1973, Biophysical properties of *Penicillium stoloniferum* virus S, *J. Gen. Virol.* **18**:223–235.

Buck, K. W., and Kempson-Jones, G. F., 1974, Capsid polypeptides of 2 viruses isolated from *Penicillium stoloniferum, J. Gen. Virol.* **22**:441–445.

Buck, K. W., and G. Ratti, G., 1975a, Biophysical and biochemical properties of 2 viruses isolated from *Aspergillus foetidus, J. Gen. Virol.* **27**:211–224.

Buck, K. W., and Ratti, G., 1975b, A model for the replication of double-stranded ribonucleic acid mycoviruses, *Biochem. Soc. Trans.* **3**:542–544.

Buck, K. W., Chain, E. B., and Darbyshire, J. E., 1969, High cell wall galactosamine content and virus particles in *Penicillium stoloniferum, Nature (London)* **223**:1273.

Buck, K. W., Chain, E. B., and Himmelweit, F., 1971, Comparison of interferon induction in mice by purified *Penicillium chrysogenum* virus and derived double-stranded RNA, *J. Gen. Virol.* **12**:131–139.

Buck, K. W., Girvan, R. F., and Ratti, G., 1973*a*, Two serologically distinct double-stranded ribonucleic acid viruses isolated from *Aspergillus niger, Biochem. Soc. Trans.* **1**:1138–1140.

Buck, K. W., Lhoas, P., Border, D. J., and Street, B. K., 1973*b*, Virus particles in yeast, *Biochem. Soc. Trans.* **1**:1141–1142.

Bussey, H., 1972, Effects of yeast killer factor on sensitive cells, *Nature (London) New Biol.* **235**:73–77.

Bussey, H., and Skipper, N., 1975, Membrane-mediated killing of *Saccharomyces cerevisiae* by glycoproteins from *Torulopsis glabrata, Can. J. Genet. Cytol.* **17**:457.

Bussey, H., and Skipper, N., 1976, Killing of *Torulopsis glabrata* by *Saccharomyces cerevisiae* killer factor, *Antimicrob. Agents Chemother.* **9**:352–354.

Caten, C. E., 1972, Vegetative incompatibility and cytoplasmic infection in fungi, *J. Gen. Microbiol.* **72**:221–229.

Chandra, K., Gupta, B. M., and Maheshwari, R. K., 1971, A fungal filtrate with pronounced inhibitory action against Semliki Forest virus (SFV) in mice, *Curr. Sci.* **40**:571–572.

Chaplygina, N. M., Velikodvorskaya, G. A., and Tikchonenko, T. I., 1975, Comparative characteristics of PBV-1 and PBV-3 viruses isolated from *Penicillium brevicompactum, Vopr. Virusol.* **1975**:476–480.

Chater, K. F., and Morgan, D. H., 1974, Ribonucleic acid synthesis by isolated viruses of *Penicillium stoloniferum, J. Gen. Virol.* **23**:307–317.

Chevaugeon, J., and Digbeu, S., 1960, Un second facteur cytoplasmique infectant chez *Pestalozzia annulata, C. R. Acad. Sci. (Paris) Ser. D* **250**:2247–2249.

Chosson, J. F., Lapierre, H., Kusiak, C., and Molin, G., 1973, Presence de particules de type viral chez les champignons du genre *Fusarium, Ann. Phytopathol.* **5**:324.

Cooke, P. M., and Stevenson, J. W., 1965*a*, An antiviral substance from *Penicillium cyaneo-fulvum* Biourge. I. Production and partial modification, *Can. J. Microbiol.* **11**:913–914.

Cooke, P. M., and Stevenson, J. W., 1965*b*, An antiviral substance from *Penicillium cyaneo-fulvum* Biourge. II. Biological activity against RNA containing viruses, *Can. J. Microbiol.* **11**:921–928.

Coutts, R. H. A., Cocking, E. C., and Kassanis, B., 1972. Infection of protoplasts from yeast with tobacco mosaic virus, *Nature (London)* **240**:466–467.

Cox, R. A., Kanagalingam, K., and Sutherland, E. S., 1970, Double-helical character of ribonucleic acid from virus-like particles found in *Penicillium chrysogenum, Biochem. J.* **120**:549–558.

Crosse, R., and Mason, P. J., 1974, Virus-like particles in *Penicillium chrysogenum, Trans. Br. Mycol. Soc.* **62**:603–610.

Day, L. E., and Ellis, L. F., 1971, Virus-like particles in *Cephalosporium acremonium, Appl. Microbiol.* **22**:919–920.

Day, P. R., and Anagnostakis, S. L., 1972, Heterokaryon transfer of the "killer" factor in *Ustilago maydis, Phytopathology* **62**:494.

Day, P. R., and Anagnostakis, S. L., 1973, The killer system in *Ustilago maydis:* Heterokaryon transfer and loss of determinants, *Phytopathology* **63**:1017–1018.

Day, P. R., Anagnostakis, S. L., Wood, H. A., and Bozarth, R. F., 1972, Positive correlation between extracellular inheritance and mycovirus transmission in *Ustilago maydis, Phytopathology* **62**:753.

Detroy, R. W., and Still, P. E., 1975, Fungal metabolites and viral replication in *Penicillium stoloniferum, Dev. Ind. Microbiol.* **16**:145–151.

Detroy, R. W., and Still, P. E., 1976, Patulin inhibition of mycovirus replication in *Penicillium stoloniferum, J. Gen. Microbiol.* **92**:167–174.

Detroy, R. W., Freer, S. N., and Fennel, D. I., 1973, Relationship between the biosynthesis of virus-like particles and mycophenolic acid in *Penicillium brevi-compactum, Can. J. Microbiol.* **19**:1459–1462.

Dieleman-van Zaayen, A., 1967, Virus-like particles in a weed mould growing on mushroom trays, *Nature (London)* **216**:595–596.

Dieleman-van Zaayen, A., 1972a, Intracellular appearance of mushroom virus in fruiting bodies and basidiospores of *Agaricus bisporus, Virology* **47**:94–104.

Dieleman-van Zaayen, A., 1972b, *Mushroom Virus Disease in the Netherlands: Symptoms, Etiology, Electron Microscopy, Spread and Control,* 130 pp., Centre for Agricultural Publishing and Documentation, Wageningen, Netherlands.

Dieleman-van Zaayen, A., 1974, Epidemiology and control of virus disease in the cultivated mushroom, *Champignon* **153**:13–26.

Dieleman-van Zaayen, A., and Igesz, O., 1969, Intracellular appearance of mushroom virus, *Virology* **39**:147–152.

Dieleman-van Zaayen, A., and Temmink, J. H. M., 1968, A virus disease of cultivated mushrooms in the Netherlands, *Neth. J. Plant Pathol.* **74**:48–51.

Dieleman-van Zaayen, A., Igesz, O., and Finch, J. T., 1970, Intracellular appearance and some morphological features of virus-like particles in an ascomycete fungus, *Virology* **42**:534–537.

Douthart, R. J., Burnett, J. P., Beasley, F. W., and Frank, B. H., 1973, Binding of ethidium bromide to double-stranded ribonucleic acid, *Biochemistry* **12**:214–220.

Dunkle, L. D., 1974a, Double-stranded RNA mycovirus in *Periconia circinata, Physiol. Plant Pathol.* **4**:107–116.

Dunkle, L. D., 1974b, The relation of virus-like particles to toxin producing fungi in corn and sorghum, *Proc. 28th Annu. Corn. Sorghum Res. Conf.,* pp. 72–81.

Ehrenfeld, E., and Hunt, T., 1971, Double stranded poliovirus RNA inhibits initiation of protein synthesis of reticolocyte lysate, *Proc. Natl. Acad. Sci. U.S.A.* **68**:1075–1078.

Ellis, L. F., and Kleinschmidt, W. J., 1967, Virus-like particles of a fraction of statolon, a mould product, *Nature (London)* **215**:694–650.

Ershov, F. I., Gaitskhoki, V. S., Kiselev, O. I., Golubkov, V. I., Zaitseva, O. V., Men'shikh, L. K., Naifakh, S. A., and Zhdanov, V. M., 1974, The replication of infectious RNA of Venezuelan equine encephalomyelitis virus in isolated yeast mitochondria, *Biokhimiya* **39**:835–839.

Férault, A. C., Spire, D., Rapilly, F., Bertrandy, J., Skajennikoff, M., and Bernaux, P., 1971, Observation de particules virales dans des souches de *Piricularia oryzae*, Briosi et Cav., *Ann. Phytopathol.* **3**:267–269.

Fink, G. R., and Styles, C. A., 1972, Curing of a killer factor in *Saccharomyces cerevisiae, Proc. Natl. Acad. Sci. U.S.A.* **69**:2846–2849.

Frick, L. J., and Lister, R. M., 1975, Purification and some properties of VLP's from an Indiana isolate of *Gaeumannomyces graminis, Proc. Am. Phytopathol. Soc.* **2**:84.

Gandy, D. G., 1960, "Watery stipe" of cultivated mushrooms, *Nature (London)* **185**:482–483.

Gandy, D. G., and Hollings, M., 1962, Die-back of mushrooms: A disease associated with a virus, *Rep. Glasshouse Crops Res. Inst.* **1961**:103–108.

Grigoriev, V. B., Chaplygina, N. M., Velikodvorskaya, G. A., and Klimenko, S. M., 1974, Shell structure of virion of mycovirus isolated from fungi of genus *Penicillium, Dokl. Acad. Nauk USSR* **218**:461–463.

Grogan, R. G., and Campbell, R. N., 1966, Fungi as vectors and hosts of viruses, *Annu. Rev. Phytopathol.* **4**:29–51.

Herring, A. J., and Bevan, E. A., 1974, Virus-like particles associated with double-stranded-RNA species found in killer and sensitive strain of yeast *Saccharomyces cerevisiae, J. Gen. Virol.* **22**:387–394.

Herring, A. J., Bevan, E. A., and Mitchell, D. J., 1973, Characterization of yeast double-stranded RNA associated with killer character in yeast, *Heredity* **31**:134.

Herring, A. J., Mitchell, D. J., Rogers, D., and Bevan, E. A., 1975, *ds* RNA virus of yeast and its relationship to the killer character, *3rd Int. Congr. Virol.* (*Madrid*) *Abstr.,* p. 148.

Hirano, T., Lindegren, C. C., and Bang, Y. N., 1962, Electron microscopy of virus-infected yeast cells, *J. Bacteriol.* **83**:1363–1364.

Hollings, M., 1962, Viruses associated with a die-back disease of cultivated mushroom, *Nature* (*London*) **196**:962–965.

Hollings, M., and Stone, O. M., 1971, Viruses that infect fungi, *Annu. Rev. Phytopathol.* **9**:93–118.

Hooper, G. R., Wood, H. A., Myers, R., and Bozarth, R. F., 1972, Virus-like particles in *Penicillium brevi-compactum* and *P. stoloniferum* hyphae and spores, *Phytopathology* **62**:823–825.

Huttinga, H., Wichers, H. J., and Dieleman-van Zaayen, A., 1975, Filamentous and polyhedral virus-like particles in *Boletus edulis, Neth. J. Plant Pathol.* **81**:102–106.

Isaac, P. K., and Gupta, S. K., 1964, A virus-like infection of *Alternaria tenius, 10th Int. Bot. Congr.* (*Edinburgh*) *Abstr.,* pp. 390–391.

Isaacs, A., and Lindenmann, J., 1957, Virus interference. 1. The interferon, *Proc. R. Soc.* (*London*) *Ser. B* **147**:258.

Jinks, J. L., 1959, Lethal suppressive cytoplasms in aged clones of *Aspergillus glaucus, J. Gen. Microbiol.* **21**:397–409.

Kaempfer, R., and Kaufman, J., 1973, Inhibition of cellular protein synthesis by double-stranded RNA: Inactivation of an initiation factor, *Proc. Natl. Acad. Sci. U.S.A.* **70**:1222–1226.

Kandel, J. S., Grill, L., Adler, J. P., Trimmell, D., and Coats, H., 1976, Detection of double-stranded RNA in yeast by inhibition of protein synthesis, *Annu. Meet. Am. Soc. Microbiol. Abstr.,* p. 181.

Kazama, F. Y., and Schornstein, K. L., 1972, Herpes-type virus particles associated with a fungus, *Science* **177**:696–697.

Kazama, F. Y., and Schornstein, K. L., 1973, Ultrastructure of a fungus herpes-type virus, *Virology* **52**:478–487.

Khandjian, E. W., Roos, U.-P., and Turian, G., 1975, Mycovirus d'*Allomyces, Pathol. Microbiol.* **42**:250–251.

Kleinschmidt, W. J., 1972, Biochemistry of interferon and its inducers, *Annu. Rev. Biochem.* **41**:517–542.

Kleinschmidt, W. J., and Ellis, L. F., 1968, Statolon as an inducer of interferon, in: *Ciba Foundation Symposium on Interferon, 1967* (G. E. Wolstenholme and M. O'Conner, eds.), pp. 39–46, J. and A. Churchill, London.

Kleinschmidt, W. J., and Murphy, E. B., 1965, Investigations on interferon induced by statolon, *Virology* **27**:484–489.

Kleinschmidt, W. J., and Probst, G. W., 1962, The nature of statolon, an antiviral agent, *Antibiot. Chemother.* **12**:289–309.

Kleinschmidt, W. J., Cline, J. C., and Murphy, E. G., 1964, Interferon production induced by statolon, *Proc. Natl. Acad. Sci. U.S.A.* **52**:741–744.

Kleinschmidt, W. J., Ellis, L. F., van Frank, R. M., and Murphy, E. B., 1968, Interferon stimulation by a double stranded RNA of a mycophage in statolon preparations, *Nature (London)* **220**:167–168.

Koltin, Y., and Day, P. R., 1975, Specificity of *Ustilago maydis* killer proteins, *Appl. Microbiol.* **30**:694–696.

Koltin, Y., and Day, P. R., 1976a, Inheritance of killer phenotypes and double-stranded RNA in *Ustilago maydis, Proc. Natl. Acad. Sci. U.S.A.* **73**:594–598.

Koltin, Y., and Day, P. R., 1976b, Suppression of the killer phenotype in *Ustilago maydis, Genetics* **82**:629–637.

Koltin, Y., Berick, R., Stamberg, J., and Ben-Shaul, Y., 1973, Virus-like particles and cytoplasmic inheritance of plaques in a higher fungus, *Nature (London) New Biol.* **241**:108–109.

Kovács, E., 1967, Change in population densities, viability, or multiplication of yeasts and *Tetrahymena* infected experimentally with encephalomyocarditis virus, *J. Cell Biol.* **35**:73A–74A.

Kovács, E., 1970, Activation of virus production by DMSO in *C. albicans* experimentally infected with polyoma-DNA, *Experientia* **26**:1396–1397.

Kovács, E., and Bucz, B., 1967, Propagation of mammalian viruses in protista. II. Isolation of complete virus from yeast and *Tetrahymena* experimentally infected with picorna viral particles or their infectious RNA, *Life Sci.* **6**:347–358.

Kovács, E., Bucz, B., and Kolompár, G., 1966, Propagation of mammalian viruses in protista. I. Visualization of fluorochrome labelled EMC virus in yeast and *Tetrahymena, Life Sci.* **5**:2117–2126.

Kovács, E., Bucz, B., and Kolompár, G., 1969, Propagation of mammalian viruses in protista. IV. Experimental infection of *C. albicans* and *S. cerevisiae* with polyoma virus, *Proc. Soc. Exp. Biol. Med.* **132**:971–977.

Kozlova, T. M., 1973, Virus-like particles in yeast cells, *Microbiologiya* **42**:745–747.

Küntzel, H., Barath, Z., Ali, I., Kind, J., and Althaus, N.-H., 1973, Virus-like particles in an extranuclear mutant of *Neurospora crassa, Proc. Natl. Acad. Sci. U.S.A.* **70**:574–578.

Lai, H. C., and Zachariah, K., 1975, Detection of virus-like particles in coremia of *Penicillium claviforme, Can. J. Genet. Cytol.* **17**:525–533.

Lampson, G. P., Tytell, A. A., Field, A. K., Nemes, M. M., and Hilleman, M. R., 1967, Inducers of interferon and host resistance. I. Double-stranded RNA from extracts of *Penicillium funiculosum, Proc. Natl. Acad. Sci. U.S.A.* **58**:782–789.

Lapierre, H., and Faivre-Amiot, A., 1970, Presence de particules virales chez différentes souches de *Sclerotium cepivorum, 7th Congr. Int. Prot. Plantes (Paris)*, pp. 542–543.

Lapierre, H., Lemaire, J.-M., Jouan, B., and Molin, G., 1970, Mise en evidence de particules virales associees à une perte de pathogénicité chez le Piétin-echaudage des céréales, *Ophiobolus graminis* Sacc, *C. R. Acad. Sci. (Paris) Ser. D* **271**:1833–1836.

Lapierre, H., Astier-Manifacier, S., and Cornuet, P., 1971, Activité RNA polymérase associée aux preparations purifées de virus du *Penicillium stoloniferum, C. R. Acad. Sci. (Paris) Ser. D* **273**:992–994.

Lapierre, H., Faivre-Amiot, A., Kusiak, C., and Molin, G., 1972, Párticules de type viral associèes au *Mycogone perniciosa magnus,* agent d'une des môles du champignon de couche, *C. R. Acad. Sci. (Paris) Ser. D* **274**:1867–1870.

Lapierre, H., Faivre-Amiot, A., and Molin, G., 1973, Isolement de particules de type viral associees au *Verticillium fungicola:* Agent d'une mole du champignon de couche, *Ann. Phytopathol.* **5**:323.

Last, F. T., Hollings, M., and Stone, O. M., 1974, Effects of cultural conditions on mycelial growth of healthy and virus-infected cultivated mushroom, *Agaricus bisporus, Ann. Appl. Biol.* **76**:99–111.

Lavroushin, A., and Treagen, L., 1972, Experimental infection of *Saccharomyces cerevisiae* with mammalian viruses, *Wasmann J. Biol.* **30**:97–92.

Lebed, E. S., Bartoshevich, Y. E., Egorov, A. A., Bobkova, A. F., Velikodvorskaya, G. A., and Tikchonenko, T. I., 1975, Induction of virus PBV 5 in mycelium of *Penicillium chrysogenum, Antibiotiki* **20**:606–610.

Lechner, J. F., Scott, A., Aaslestad, H. G., Fuscaldo, A. A., and Fuscaldo, K. E., 1972, Analysis of a virus-like particle from *N. crassa, Abstr. Annu. Meet. Am. Soc. Microbiol.* **1972**:188.

Lecoq, H., Spire, D., Rapilly, F., and Bertrandy, J., 1974, Mise en evidence de particules de type viral chez les *Puccinia, C. R. Acad. Sci. (Paris) Ser. D* **27**:1599–1602.

Leibowitz, M. J., and Wickner, R. B., 1975, "Killer" character: Relation to the sexual cycle of *Saccharomyces cerevisiae, Abstr. Annu. Meet. Am. Soc. Microbiol.* **1975**:115.

Leibowitz, M. J., and Wickner, R. B., 1976, A chromosomal gene required for killer plasmid expression, mating and spore maturation in *Saccharomyces cerevisiae, Proc. Natl. Acad. Sci. U.S.A.* **73**:2061–2065.

Lemaire, J.-M., Lapierre, H., Jouan, B., and Bertrand, G., 1970, Découverte de particules virales chez certaines souches d'*Ophiobolus graminis* agent du Piétin-echaudage des céréales, conséquences agronomiques previsibles, *C. R. Acad. Agr. (Fr.)* **56**:134–138.

Lemaire, J.-M., Jouan, B., Perraton, B., and Sailly, M., 1971, Perspectives de lutte biologique contre les parasites des céréales d'origine tellurique en particulier *Ophiobolus graminis* Sacc., *Sci. Agron. Rennes* **1971**:1–8.

Lemke, P. A., 1976, Viruses of eucaryotic microorganisms, *Annu. Rev. Microbiol.* **30**:105–145.

Lemke, P. A., and Nash, C. H., 1974, Fungal viruses, *Bacteriol. Rev.* **38**:29–56.

Lemke, P. A., and Ness, T. M., 1970, Isolation and characterization of a double-stranded ribonucleic acid from *Penicillium chrysogenum, J. Virol.* **6**:813–891.

Lemke, P. A., Nash, C. A., and Pieper, S. W., 1973, Lytic plaque formation and variation in virus titre among strains of *Penicillium chrysogenum, J. Gen. Microbiol.* **76**:265–275.

Lemke, P. A., Saksena, K. N., and Nash, C. H., 1976, Viruses of industrial fungi, in: *Genetics of Industrial Microorganisms* (K. D. Macdonald, ed.), pp. 323–355, Academic Press, New York.

Lewis, U. J., Rickes, E. L., McClelland, L., and Brink, N. G., 1959, Purification and characterization of the antiviral agent helenine, *J. Am. Chem. Soc.* **81**:4115.

Lhoas, P., 1971a, Infection of protoplasts from *Penicillium stoloniferum* with double-stranded RNA viruses, *J. Gen. Virol.* **13**:365–367.

Lhoas, P., 1971b, Transmission of double-stranded RNA viruses to a strain of *Penicillium stoloniferum* through heterokaryosis, *Nature (London)* **230**:248–249.

Lhoas, P., 1972, Mating pairs of *Saccharomyces cerevisiae* infected with double stranded RNA viruses from *Aspergillus niger, Nature (London) New Biol.* **236**:86–87.

Lindberg, G. D., 1959, A transmissible disease of *Helminthosporium victoriae, Phytopathology* **49**:29–32.

Lindberg, G. D., 1960, Reduction in pathogenicity and toxin production in diseased *Helminthosporium victoriae, Phytopathology* **50**:457.

Lindberg, G. D., 1971a, Disease-induced toxin production in *Helminthosporium oryzae, Phytopathology* **61**:420.

Lindberg, G. D., 1971b, Influence of transmissible diseases on toxin production in *Helminthosporium maydis, Phytopathology* **61**:900–901.

Lindegren, C. C., and Bang, Y. N., 1961, The zymophage, *Antonie van Leeuwenhoek, J. Microbiol. Serol.* **27**:1–18.

Mackenzie, D. E., and Adler, J. P., 1972, Virus-like particles in toxigenic *Aspergilli, Abstr. Annu. Meet. Am. Soc. Microbiol.* **1972**:68.

Maheshwari, R. K., and Gupta, B. M., 1973a, A new antiviral agent designated 6-MFA from *Aspergillus flavus.* II. General physicochemical characteristics of 6-MFA, *J. Antibiot.* **26**:328–334.

Maheshwari, R. K., and Gupta, B. M., 1973b, Antiviral agents from fungi effective against Semliki Forest Virus (SFV) in mice, *Indian J. Med. Res.* **61**:1292–1298.

Makowar, M., and Bevan, E. A., 1963, The inheritance of a killer character in yeast, *Saccharomyces cerevisiae, Proc. 11th Int. Congr. Genet.* **1**:202.

Marcou, D., 1961, Notion de longevité et nature cytoplasmique du determinant de la senescence chez quelques champignons, *Ann. Sci. Nat. Bot. Biol. Veg.* **2**:653–764.

Marino, R., Saksena, K. N., Schuler, M., Mayfield, J. E., and Lemke, P. A., 1976, Double-stranded RNA from *Agaricus bisporus, Appl. Environ. Microbiol.* **32**:433–438.

Maule, A. P., and Thomas, P. D., 1973, Strains of yeast lethal to brewery yeasts, *J. Inst. Brew. (London)* **79**:137–141.

Metitiri, P. O., and Zachariah, K., 1972, Virus-like particles and inclusion bodies in penicillus cells of a mutant of *Penicillium, J. Ultrastruc. Res.* **40**:272–283.

Mitchell, D. J., Bevan, E. A., and Herring, A. J., 1973, Correlation between dsRNA in yeast and killer character, *Heredity* **31**:133–134.

Moffitt, E. M., and Lister, R. M., 1974, Detection of mycoviruses using antiserum specific for double-stranded RNA, *Virology* **52**:301–304.

Moffitt, E. M., and Lister, R. M., 1975, Application of a serological test for detecting double-stranded RNA mycoviruses, *Phytopathology* **65**:851–859.

Molin, G., and Lapierre, H., 1973, L'acide nucleique des virus de champignons: Cas des virus de l'*Agaricus bisporus, Ann. Phytopathol.* **5**:233–240.

Nair, N. G., 1973, Heat therapy of virus-infected cultures of the cultivated mushroom *Agaricus bisporus, Aust. J. Agr. Res.* **24**:533–541.

Nash, C. H., Douthart, R. J., Ellis, L. F., Van Frank, R. M., Burnett, J. P., and Lemke, P. A., 1973, On the mycophage of *Penicillium chrysogenum, Can. J. Microbiol.* **19**:97–103.

Naumov, G. I., 1974, Comparative genetics of yeast. XIV. Analysis of wine strains of *Saccharomyces* neutral to killer strain of type k2, *Genetika* **10**:130–136.

Nemes, M. M., Tytell, A. A., Lampson, G. P., Field, A. K., and Hilleman, M. R., 1969, Inducers of interferon and host resistance. VII. Antiviral efficacy of double-stranded RNA of natural origin, *Proc. Soc. Exp. Biol. Med.* **132**:784–789.

Nesterova, G. F., and Zekhnov, A. M., 1973, Genetic control of killer character in *Saccharomyces, Genetika* **9**:171–172.

Nesterova, G. F., Kyarner, Y., and Soom, Y. O., 1973, Virus-like particles in *Candida tropicalis, Mikrobiologiya* **42**:162–165.

Nesterova, G. F., Kärner, J., and Kozlova, T. M., 1974a, Possible intercellular transfer of virus-like particles from *Candida tropicalis* and *Saccharomyces cerevisiae, Mikrobiologiya* **43**:930–932.

Nesterova, G. F., Kärner, J., and Kozlova, T. M., 1974b, On the possibility of transfer of virus-like particles from the cells of *Candida tropicalis* into the cells of *Saccharomyces cerevisiae, Mikrobiologiya* **43**:1095–1097.

Nesterova, G. F., Zekhnov, A. M., and Ingevech, S. G., 1975, Dominant nonsense suppressors which inhibit killer activity in yeast *Saccharomyces cerevisiae, Genetika* **11**:96–103.

Nienhaus, E., 1971, Tobacco mosaic virus strains extracted from conidia of powdery mildews, *Virology* **46**:504–505.

Nienhaus, F., and Mack, C., 1974, Infection of *Pythium arrhenomanes in vitro* with tobacco mosaic virus and tobacco necrosis virus, *Z. Pflanzenkr. Pflanzenschutz* **81**:728–731.

Pallett, I., 1972, Production and regeneration of protoplasts from various fungi and their infection with fungal viruses. *3rd Int. Symp. Yeast Protoplasts* (*Salamanca*) *Abstr.,* p. 78.

Passmore, E. L., and Frost, R. R., 1974, The detection of virus-like particles in mushrooms and mushroom spawns. *Phytopathol. Z.* **80**:85–87.

Philliskirk, G., and Young, T. W., 1975, Occurrence of "killer" character in yeasts of various genera, *Antonie van Leeuwenhoek, J. Microbiol. Serol.* **41**:147–151.

Planterose, D. M., Birch, P. J., Pilch, D. J. F., and Sharpe, T. J., 1970, Antiviral activity of double stranded RNA and virus-like particles from *Penicillium stoloniferum, Nature* (*London*) **227**:504–505.

Powell, H. M., Culbertson, C. G., McGuire, J. M., Hoehn, M. M., and Baker, L. A., 1952, A filtrate with chemoprophylactic and chemotherapeutic action against MM and Semliki Forest viruses in mice, *Antibiot. Chemother.* **2**:432–434.

Puhalla, J. E., 1968, Compatibility reactions on solid medium and interstrain inhibition in *Ustilago maydis, Genetics* **60**:461–475.

Rasmussen, C. R., Mitchell, R. E., and Slack, C. I., 1972, "Heat-treatment" of cultures from apparently healthy and virus-infected mushrooms and the subsequent effect on cropping yields, *Mushroom Sci.* **8**:239–251.

Ratti, G., and Buck, K. W., 1972, Virus particles in *Aspergillus foetidus:* A multicomponent system, *J. Gen. Virol.* **14**:165–175.

Ratti, G., and Buck, K. W., 1975, RNA-polymerase activity in double-stranded ribonucleic acid virus particles from *Aspergillus foetidus, Biochem. Biophys. Res. Commun.* **66**:706–711.

Rawlinson, C. J., 1973, Virus-like particles in plant pathogenic fungi, *2nd Int. Congr. Plant Pathol.* (*Minneapolis*) *Abstr.,* No. 0911.

Rawlinson, C. J., 1975, Role of fungal viruses in the pathogenecity of fungi, *3rd Int. Congr. Virol. (Madrid) Abstr.,* p. 147.

Rawlinson, C. J., and Maclean, D. J., 1973, Virus-like particles in axenic cultures of *Puccinia graminis tritici, Trans. Br. Mycol. Soc.* **61**:590–593.

Rawlinson, C. J., Hornby, D., Pearson, V., and Carpenter, J. M., 1973, Virus-like particles in the take-all fungus, *Gaeumannomyces graminis, Ann. Appl. Biol.* **74**:209.

Robertson, H. D., and Mathews, M. B., 1973, Double-stranded RNA as an inhibitor of protein synthesis and as a substrate for a nuclease in extracts of Krebs II ascites cell, *Proc. Natl. Acad. Sci. U.S.A.* **70**:225–229.

Rotem, Z., Cox, R. A., and Isaacs, A., 1963, Inhibition of virus multiplication by foreign nucleic acid, *Nature (London)* **197**:564–566.

Rytel, M. W., Shope, R. E., and Kilbourne, E. D., 1966, An antiviral substance from *Penicillium funiculosum*. V. Induction of interferon by helenine, *J. Exp. Med.* **123**:577–584.

Saksena, K. N., 1975, Isolation and large-scale purification of mushroom viruses, *Dev. Ind. Microbiol.* **16**:134–144.

Saksena, K. N., 1978, Mycoviruses, in: *Atlas of Insect and Plant Viruses: Including Mycoplasmas and Viroids* (K. Maramorosch, ed.), pp. 421–430, Academic Press, New York.

Sanderlin, R. S., and Ghabrial, S. A., 1975, Virus-like particles containing double stranded RNA in normal and diseased *Helminthosporium victoriae, Proc. Am. Phytopathol. Soc.* **2**:140.

Schisler, L. C., Sinden, J. W., and Sigel, E. M., 1967, Etiology, symptomatology, and epidemiology of a virus disease of cultivated mushrooms, *Phytopathology* **57**:519–526.

Schnepf, E., Soeder, C. J., and Hegewald, E., 1970, Polyhedral virus-like particles lysing the aquatic phycomycete *Aphelidium* sp., a parasite of the green alga, *Scenedesmus armatus, Virology* **42**:482–487.

Shahriari, H., Kirkham, J. B., and Casselton, L. A., 1973, Virus-like particles in the fungus *Coprinus lagopus, Heredity* **31**:428.

Shope, R. E., 1948, The therapeutic activity of a substance from *Penicillium funiculosum* Thom against swine influenza virus infection of mice, *Am. J. Bot.* **35**:803.

Shope, R. E., 1953, An antiviral substance from *Penicillium funiculosum*. I. Effect upon infection in mice with swine influenza virus and Columbia SK encephalomyelitis virus, *J. Exp. Med.* **97**:601–625.

Sinden, J. W., and Hauser, E., 1950, Report on two new mushroom diseases, *Mushroom Sci.* **1**:96–100.

Sinden, J. W., and Hauser, E., 1957, It is "La France," *Mushroom Growers Assoc. Bull.* **95**:407–409.

Somers, J., and Bevan, E. A., 1969, The inheritance of the killer character in yeast, *Genet. Res.* **13**:71–83.

Spire, D., 1971. Virus des champignons, *Physiol. Veg.* **9**:555–567.

Spire, D., Ferault, A. C., and Bertrandy, J., 1972a, Observation de particles de type viral dans une souche d'*Helminthosporium oryzae,* Br. de H., *Ann. Phytopathol.* **4**:359–360.

Spire, D., Ferault, A. C., Bertrandy, J., Rapily, F., and Skajennikoff, M., 1972b, Particules de type viral dans un champignon hyperparasite: *Gonatobotrys, Ann. Phytopathol.* **4**:419.

Still, P. E., Detroy, R. W., and Hesseltine, C. W., 1975, *Penicillium stoloniferum* virus: Altered replication in ultraviolet-derived mutants, *J. Gen. Virol.* **27**:275–281.

Subbarayudu, S., Padma, R., Gupta, M. D., and Raychaudhuri, S. P., 1973, Translocation of tobacco mosaic virus from tobacco callus cultures through hyphae of *Pythium debaryanum* Hesse., *Curr. Sci.* **42**:325–326.

Tikchonenko, T. I., Velikodvorskaya, G. A., Bobkova, A. F., Bartoshevich, Y. E., Lebed, E. P., Chaplygina, N. M., and Maksimova, T. S., 1974, New fungal viruses capable of reproducing in bacteria, *Nature (London)* **249**:454–456.

Tuveson, R. W., and Peterson, J. F., 1972, Virus-like particles in certain slow-growing strains of *Neurospora crassa, Virology* **47**:527–531.

Tuveson, R. W., and Sargent, M. L., 1976, Characterization of virus-like particles from slow-growing strains of *Neurospora crassa, Genetics* **83**:s77.

Tuveson, R. W., Sargent, M. L., and Bozarth, R. F., 1975, Purification of a small virus-like particle from strains of *Neurospora crassa, Annu. Meet. Am. Soc. Microbiol. Abstr.,* p. 216.

Tyurina, L. V., and Buryan, N. I., 1975, Phenotypes (killer, neutral, sensitive) of yeasts of genus *Saccharomyces* in viticulture and methods of their determination, *Mikrobiologiya* **44**:316–320.

Ushiyama, R., 1975, Virus-like particles in shiitake mushroom, *Lentinus edodes.* (Berk.). Sing., *Proc. 1st Intersectional Congr. Int. Assoc. Microbiol. Soc. (Tokyo)* **3**:402–406.

Ushiyama, R., and Hashioka, Y., 1973, Viruses associated with hymenomycetes. I. Filamentous virus-like particles in the cells of a fruit body of shiitake, *Lentinus edodes* (Berk.) Sing., *Rep. Tottori Mycol. Inst.* **10**:797–805.

Ushiyama, R., and Nakai, Y., 1975, Viruses associated with hymenomycetes. II. Presence of polyhedral virus-like particles in shiitake mushrooms, *Lentinus edodes* (Berk.) Sing., *Rep. Tottori Mycol. Inst.* **12**:53–60.

Velikodvorskaya, G. A., Bobkova, A. F., Maksimova, T. S., Klimenko, S. M., and Tikchonenko, T. I., 1972, New viruses isolated from the culture of a fungus of *Penicillium* genus, *Bull. Eksp. Biol. Med.* **73**:90–93.

Velikodvorskaya, G. A., Chaplygina, N. M., Petrovsky, G. V., Grigoriev, V. B., Maksimova, T. S., and Tikchonenko, T. I., 1974, Physicochemical properties of mycovirus isolated from *Penicillium brevi-compactum, Vopr. Virusol.* **1974**:442–445.

Vodkin, M. H., and Fink, G. R., 1973, A nucleic acid associated with a killer strain of yeast, *Proc. Nat. Acad. Sci. U.S.A.* **70**:1069–1072.

Vodkin, M., Katterman, F., and Fink, G. R., 1974, Yeast killer mutants with altered double-stranded ribonucleic acid, *J. Bacteriol.* **117**:681–686.

Volkoff, O., and Walters, T., 1970, Virus-like particles in abnormal cells of *Saccharomyces carlsbergensis, Can. J. Genet. Cytol.* **12**:621–626.

Volkoff, O., Walters, T., and Dejardin, R. A., 1972, An examination of *Penicillium chrysogenum* type virus particles, *Can. J. Microbiol.* **18**:1352–1353.

Wickner, R. B., 1974*a*, Killer character of *Saccharomyces cerevisiae:* Curing by growth at elevated-temperature, *J. Bacteriol.* **117**:1356–1357.

Wickner, R. B., 1974*b*, Chromosomal and nonchromosomal mutations affecting the "killer character" of *Saccharomyces cerevisiae, Genetics* **76**:423–432.

Wickner, R. B., 1976, Mutants of the killer plasmid of *Saccharomyces cerevisiae* dependent on chromosomal diploidy for expression and maintenance, *Genetics* **82**:273–285.

Wickner, R. B., and Leibowitz, M. J., 1976a, Chromosomal genes essential for replication of double stranded RNA plasmid of *Saccharomyces cerevisiae:* The killer character of yeast, *J. Mol. Biol.* **105**:427–443.

Wickner, R. B., and Leibowitz, M. J., 1976b, Two chromosomal genes required for killer expression in killer strains of *Saccharomyces cerevisiae, Genetics* **82**:429–442.

Wiebols, G. L. W., and Wieringa, K. T., 1936, Bacteriophagie een algemeen voorkomend verschijnsel, *Fonds. Landbouw Export Bureau 1916–1918 (Wageningen)* **16**:11–13.

Wieringa-Brants, D. H., 1973, Infection of *Pythium* spec. *in vitro* with C¹⁴-labeled tobacco mosaic virus, *2nd Int. Congr. Plant Pathol. (Minneapolis) Abstr.* No. 0913.

Wood, H. A., 1973, Viruses with double-stranded RNA genomes, *J. Gen. Virol.* **20**:61–85.

Wood, H. A., and Bozarth, R. F., 1972, Properties of virus-like particles of *Penicillium chrysogenum:* One double-stranded RNA molecule per particle, *Virology* **47**:604–609.

Wood, H. A., and Bozarth, R. F., 1973, Heterokaryon transfer of virus-like particles associated with a cytoplasmically inherited determinant in *Ustilago maydis, Phytopathology* **63**:1019–1021.

Wood, H. A., Bozarth, R. F., and Mislivec, P. B., 1971, Virus-like particles associated with an isolate of *Penicillium brevi-compactum, Virology* **44**:592–598.

Wood, H. A., Bozarth, R. F., Adler, J., and Mackenzie, D. W., 1974, Proteinaceous virus-like particles from an isolate of *Aspergillus flavus, J. Virol.* **13**:532–534.

Woods, D. R., and Bevan, E. A., 1968, Studies on the nature of the killer factor produced by *Saccharomyces cerevisiae, J. Gen. Microbiol.* **51**:115–126.

Yamashita, S., 1974, Viruses of fungi, *Kagaku to Seibutsu* **12**:2–14.

Yamashita, S., Doi, Y., and Yora, K., 1971, A polyhedral virus found in rice blast fungus, *Pyricularia oryzae* Cavara, *Ann. Phytopathol. Soc. Japan* **37**:356–359.

Yamashita, S., Doi, Y., and Yora, K., 1973, Intracellular appearance of *Penicillium chrysogenum* virus, *Virology* **55**:445–452.

Yamashita, S., Doi, Y., and Yora, K., 1975, Electron microscopic study of several fungal viruses, *Proc. 1st Intersectional Congr. Int. Assoc. Microbiol. Soc.* **3**:340–350.

Yarwood, C. E., and Hecht-Poinar, E., 1973, Viruses from rusts and mildews, *Phytopathology* **63**:1111–1115.

CHAPTER 3

Cyanophages and Viruses of Eukaryotic Algae*

Louis A. Sherman

Division of Biological Sciences
University of Missouri—Columbia
Columbia, Missouri 65201

and

R. Malcolm Brown, Jr.

Department of Botany
University of North Carolina
Chapel Hill, North Carolina 27514

1. INTRODUCTION

The isolation of a virus that infects the blue-green algae (Safferman and Morris, 1963, 1964a,b) was looked upon with great interest by researchers in a number of fields. The driving force behind Safferman's search for what we shall call the cyanophages was an interest in underlying factors responsible for algal degeneration. Motivated by an increased concern over eutrophication and associated algal blooms, ecologists thought that the cyanophages may exercise some control over behavioral patterns of algae in nature. Plant virologists looked to the cyanophages as a model system to help resolve discrepancies which had arisen in the study of virus–host interactions. Those interested in photosynthesis and blue-green algal metabolism saw the cyanophages

---

* This review is dedicated to Dr. Kenneth M. Smith, Pioneer Plant and Insect Virologist, and posthumously to Dr. R. N. Singh, who died on March 9, 1977.

as a potentially useful tool in their studies. Finally, researchers in many diverse disciplines realized that the discovery of the cyanophages opened up the possibility of transduction, whereby the viruses could be utilized in a multitude of genetic experiments.

The work on the cyanophages which has been performed during the intervening 14 years accurately reflects the diversity of interests mentioned above. This diversity in turn reflects the rather unique position of the blue-green algae in the biological world. They are prokaryotic organisms and, as such, resemble the bacteria in many regards. However, the blue-greens carry out an aerobic photosynthesis which is virtually identical to that of eukaryotic plants and algae. And, until recently, all blue-green algae were considered to be obligate photoautotrophs. Nonetheless, evidence has accumulated over the past few years which indicates that the blue-green algae and bacteria are very much a coherent group. This relationship has been confirmed on the basis of their nucleic acid biochemistry, fine structure, and cell wall chemistry (Echlin and Morris, 1965; Stanier *et al.*, 1971; Whitton *et al.*, 1971; Drews, 1973). Although we will continue to call them blue-green algae in this chapter, they are now sometimes referred to as *cyanobacteria*, a term which is likely to be rigorously adhered to in the near future.

For this reason, we feel that the term *cyanophages* is appropriate for the viruses which attack the cyanobacteria. One of the goals of this chapter will be to attempt to convince researchers unfamiliar with the algal viruses that cyanophages have much in common with bacteriophages. In fact, the inclusion of both cyanophages and eukaryotic algal viruses in the same chapter represents, in part, the historical anachronism which caused the blue-greens to be classified as algae. However, it is our hope that such a combination will be useful to those in the fields of photosynthesis, plant virology, and ecology who some day will want to utilize such systems.

The general objective of a review such as this chapter is often to attempt a comprehensive survey of the entire field and to bring order to a rapidly expanding discipline. Although this is a noble goal, the diversity and range of research into cyanophages and eukaryotic algal viruses preclude that as a possibility. We shall therefore limit the scope of this chapter, and select timely and significant events relative to prokaryotic and eukaryotic virus structure, molecular biology, and infectious cycles. We shall examine the basic molecular biological and metabolic mechanisms in more detail than the applied aspects of the topic. We are able to do this because of the existence of three other reviews on algal viruses during the past few years (Brown, 1972; Padan and Shilo, 1973; Safferman, 1973). These reviews have discussed the older cyanophage literature in detail, and have dealt with the ecological considerations in some depth. One of the reviews (Brown, 1972) has given a preliminary account of eukaryotic algal viruses; a more recent review (Lemke, 1976) gives more data on eukaryotic algal, fungal, and protistic viruses.

Therefore, because of the extensive and growing literature on the topic, we shall be concerned primarily with the replication of cyanophages and its dependence on host metabolism. We shall first survey the field and discuss the structural and typological characteristics of the phages which have been isolated to date. We shall then discuss the replication of a few lytic phages which have been well characterized. It will be necessary to make an important subdivision during our analysis of lytic phage replication; we shall see at that time that many significant differences exist between phages which infect unicellular as opposed to filamentous cyanophycean hosts. We will then proceed to a consideration of lysogeny in the blue-green algae, an area of research which is only beginning to enter the rapidly expanding phase. Finally, we shall present a status report on viruses and viruslike particles (VLP) which infect eukaryotic algae. It will soon be evident, unfortunately, that this topic is only in its infancy. Nevertheless, the recent characterization and isolation of RNA and DNA viruses from these eukaryotic hosts demonstrate that this field is destined to receive a great amount of attention in the coming years.

2. GENERAL PROPERTIES OF CYANOPHAGES

2.1. Classification, Typology, and Host Range

In their original studies on cyanophages, Safferman and Morris (1963) tested 78 algal species and obtained lysis on 11 strains. These susceptible organisms were in the genera *Lyngbya, Plectonema,* and *Phormdium;* the virus was thus code-named LPP-1, using the initials of the generic names of the hosts. As more and more cyanophages have been isolated, this classification has been retained, with two important additions: (1) arabic numerals designate serological subgroups, and (2) different isolates of a similar type are indicated by a suffix consisting of the first letter of the provenance (Safferman *et al.,* 1969*b*). Therefore, cyanophage LPP-2SPI refers to an LPP strain of the second serological group which was isolated at St. Paul, Indiana.

Unfortunately, any attempt at systematic classification of the cyanophages is beset with all of the difficulties inherent in the designation of the blue-green algae. The taxonomy of the blue-greens is in a confusing and archaic state, since classification remains based mainly on microscopic examination of field collections. Therefore, the generic names used to identify the cyanophages are redundant and, in many cases, meaningless. For example, the cyanophage SM1 (Safferman *et al.*, 1969*a*) was thought to lyse strains *Synechococcus elongatus* and *Microcystis aeruginosa* NRC-1; further studies have indicated that NRC-1 could not be a *Microcystis* strain and is most likely a misclassified *Synechococcus* (Stanier *et al.*, 1971). As often happens in science, it is quite likely that the cyanophages will be of great value in a more comprehensive classification of the blue-green algae. Stanier *et al.* (1971) have proposed an integrated classification of the unicellular blue-green algae based on physiological and morphological considerations of axenic cultures and have discussed the use of cyanophage infection as a taxonomic criterion. Similar classification of the filamentous blue-green algae is in progress (Kenyon *et al.*, 1972).

Serological and morphological properties have proven to be the most definitive means of typing the cyanophages. Soon after the isolation of LPP-1, it was realized that viruses morphologically similar to the original strain were ubiquitous in waste-stabilization ponds. Therefore, a careful study of the serological specificities and morphology of a dozen LPP isolates was undertaken by Safferman *et al.* (1969*b*). They found that the strains divided equally into two distinct subgroups, LPP-1 and LPP-2, since members of one group showed no cross-reaction with antiserum prepared against a member of the other group. This finding has been verified more recently in an interesting fashion. As will be discussed in a later section, strain LPP-2SPI is lysogenic for *Plectonema boryanum*. The lysogenic alga is immune to the LPP-2 cyanophage but not to the LPP-1 subgroup. A third serological subgroup of the LPP group has been discovered in Russia (Mendzhul *et al.*, 1975*a*). To date, only one serological subgroup has been characterized for the other cyanophage groups. These factors resemble very closely the situation within the bacteriophages; this is the first of many reasons that will lead us to conclude that cyanophages should be considered as a class of bacteriophages.

In the past 10 years, well over a hundred independent cyanophage isolates have been obtained from all over the world. Nonetheless, at this time the cyanophage host range must be thought of as being remarkably narrow. The LPP group is by far the most numerous type, with serologically similar isolates having been obtained in India (Singh and Singh, 1967), Russia (Mendzhul *et al.,* 1975a,b), and Israel (Padan *et al.,* 1967), as well as in the United States. In fact, it is becoming increasingly clear that the LPP viruses are widespread in nature. On the other hand, cyanophage infection of other genera, especially the unicellular strains, is much harder to document. The difference in the distribution of cyanophages of the LPP and AS groups is indicated by a recent study aimed at the isolation of cyanophages that infect unicellular hosts. Of eight waste-stabilization ponds tested, seven gave rise to LPP isolates, while only two showed evidence of cyanophages of the AS group (Sherman and Connelly, 1976). This indicates quite clearly that the study of cyanophages is still in a primitive stage and that generalizations must be made cautiously. Furthermore, it allows us to speculate that we have only seen the tip of the iceberg and that much information remains below the surface. This is particularly true in dealing with lysogeny in the blue-greens. It has now been established that lysogeny exists in *P. boryanum* (Padan *et al.,* 1972), although definitive information in the unicellular hosts is still lacking. This fact and the difficulty in isolating cyanophages for unicellular hosts may turn out to be related.

Table 1 summarizes many of the major types of cyanophages that have so far been isolated. This table clearly indicates the prevalence of viruses which attack filamentous algae, particularly those of the LPP group. It is also obvious from this table that each individual isolate has a rather confined host range. If the physical and biochemical characteristics of the host strains are examined in greater detail, this narrow host range becomes even more evident.

The host strains lysed by the LPP cyanophages were originally classified as three separate genera of nonheterocystous, filamentous blue-green algae. Study of the DNA base ratio (46% G + C) and buoyant density (1.705 g/ml) by Edelman *et al.* (1967) revealed that these values were very similar in all susceptible organisms. Utilizing DNA-DNA hybridization, Cowie and Prager (1969) were able to show that the DNAs from these species had a high degree of homology to one another. Therefore, the host range of the LPP viruses can be considered to consist of three very closely related genera. The classification of the filamentous blue-greens by Stanier and co-workers (Kenyon *et al.,* 1972) reflects these and other facts.

Similar arguments apply to the host strains, *Anacystis nidulans* and *Synechococcus cedrorum,* infected by the AS phages. Although these have historically been considered separate genera, recent studies

TABLE 1
Summary of the Source, Host Range, and Special Attributes of Known Cyanophages

Virus	Reference	Source	Host range	Notes
LPP group (filamentous hosts)				
LPP-1	Safferman and Morris (1963)	Waste-stabilization pond, Indiana	Lyngbya, Plectonema, Phormidium	Presently limited to hosts of the Oscillatoriaceae; The archetype for this group (hosts presently limited to the Nostocineae, Oscillatoriaceae)
LPP-2	Safferman et al. (1969b)	Waste-stabilization ponds from Florida, Indiana, Missouri, and Arkansas	Same as LPP-1	Morphologically similar to LPP-1 but different serologically
LPP-3 GIII	Mendzhul et al. (1975a) Padan et al. (1967)	Russia Fish ponds, Israel	Same as LPP-1 Same as LPP-1	Synonymous with LPP-1, now called LPP-1G
Long-tailed	Singh (1974, 1975)	Polluted waters, India	Plectonema boryanum	Polygonal head, 55 nm diameter; long, contractile tail; has a latent period of 2 hr; possibly lysogenic
Unnamed	Ueda (1965) (see also Jensen and Bowen, 1970)	Streams, Japan	Oscillatoria princeps	Resembles tobacco mosaic virus; transmission not proven
D1	Daft et al. (1970)	Streams and waste-stabilization ponds, Scotland	Lyngbya, Plectonema, Phormidium	Identical to LPP-1 virus; used novel "baiting" method for isolation of virus in situ
N group (filamentous hosts)				Presently limited to hosts of the Rivulariaceae and Nostocaceae
C1	Singh and Singh (1967) (see also Jensen and Bowen, 1970)	Polluted waters, India	Cylindrospermum	Evidence based only on plaque morphology and development; no structural evidence for virus
AR1	Singh and Singh (1967)	Polluted waters, India	Anabaenopsis raciborskii, A. circularis, Raphidiopsis indica	Evidence based only on plaque morphology and development; Raphidiopsis does not produce heterocysts (= member of the Rivulariaceae); virus not observed

Name	Reference	Source	Host	Comments
Unnamed	Granhall and von Hofsten (1969)	Soil from Sweden?	*Anabaena*	Good evidence for a virus attack, but not clear for the heterocystous nature of host
A2	Mendzhul *et al.* (1975*b*)	Russia	*Anabaena variabilis*	Beaded fibers attached to neck at the same capsid vertex as the tail
N1	Adolph and Haselkorn (1971)	Freshwater lake, Wisconsin	*Nostoc muscorum* (Iowa strain)	Presently limited to hosts of the Chroococcaceae
SM group (unicellular hosts)				
Unnamed	Goryushin and Chaplinskaya (1968)	Reservoirs in Russia	*Microcystis aeruginosa, M. pulverea, M. muscicola*	Probably the same as Safferman's SM1, although evidence presented is circumstantial
SM1	Safferman *et al.* (1969*a*)	Waste-stabilization pond, Indiana	*Synechococcus elongatis, Microcystis aeruginosa*	The archetype for this group, also studied and characterized in the most detail
S1	Adolph and Haselkorn (1973*b*)	Freshwater lake, Wisconsin	*Synechococcus* strain NRC1	Structure similar to that of bacteriophage; host is the strain originally called *Microcystis aeruginosa*, which is also infected by SM1 and SM2
SM2	Fox *et al.* (1976)	Freshwater reservoir, Nebraska	*Synechococcus elongatus, Synechococcus* strain NRC1	Structure similar to that of S1
AS group				Presently limited to hosts of the Chroococcaceae
AS1	Safferman *et al.* (1972)	Waste-stabilization pond	*Anacystis nidulans, Synechococcus cedrorum*	Presently limited to hosts of the Chroococcaceae; similar hosts to the SM1 group, but the virus is different; no cross-reactivity between SM and AS groups
AS1-M	Sherman and Connelly (1976)	Waste-stabilization pond, Missouri	Same as AS1	Morphologically identical to AS1, but serologically different

have uncovered numerous morphological, biochemical, and physiological similarities. This has allowed them to be classified in one typological group by Stanier *et al.* (1971), which parallels their susceptibility to the AS phages.

2.2. Morphology

The structural work so far performed on the cyanophages indicates that they resemble bacteriophages and not plant viruses. As shown in Table 2, the cyanophages characterized to date fall into morphological groups A, B, and C proposed by Bradley (1967) for bacteriophages. Since all of these isolates are known to contain a double-stranded DNA genome, these represent the expected morphologies in Bradley's classification. In the designation of Tikchonenko (1970), which is based on increasing complexity of the nucleic acid injection apparatus, the cyanophages would be classifed in groups III–V. Since some bacterial genera contain only a few of the morphological groups, the blue-green algae seem quite typical in this regard (*see* Chapter 4).

The most extensively characterized cyanophages are those of the LPP group (Goldstein and Bendet, 1967; Goldstein *et al.*, 1967; Luftig, 1967, 1968; Luftig and Haselkorn, 1967, 1968*a,b;* Sherman, 1970; Sherman and Haselkorn, 1970*a;* Adolph and Haselkorn, 1972*b;* Safferman *et al.*, 1969*b*). Representative examples of the LPP-1 and LPP-2 subgroups are shown in Fig. 1, where it can be seen that they appear to be morphologically identical. Nonetheless, the two types do appear to be structurally different as will be discussed in the next section.

The capsid of LPP-1 is hexagonal in projection, with an edge-to-edge diameter of 58.6 nm. Attached to one of the vertices is a short tail that is 20 nm long and 15 nm wide (Luftig and Haselkorn, 1968*a*). Analysis of the capsid structural proteins by Adolph and Haselkorn (1972*b*) determined that there are two major head proteins of molecular weight 44,000 (310 copies/virion) and 14,000 (450 copies/virion). Utilizing these data with the principles of regular virus construction presented by Caspar and Klug (1962), they speculated that the LPP-1 capsid most likely was an icosahedron constructed with a triangulation number of $T = 4$ or 9. However, without any correlative data, it was not possible to carry this analysis any further.

Morphologically, the LPP-1 and LPP-2 viruses look indistinguishable. However, quantitative size measurements of the two groups first indicated that there may be structural as well as serological dif-

TABLE 2
Morphological and Physical Characteristics of the Cyanophages

Parameter	Filamentous host			Unicellular host		
	LPP-1[a,b,c,d]	LPP-2[e,f,g]	N1[h]	SM1[i,j]	S1[k]	AS1-M[l,m,n]
Morphology						
Capsid						
Shape	Icosahedral	Icosahedral	Hexagonal	Icosahedral	Hexagonal	Hexagonal
Edge-to-edge diameter (nm)	58.6 ± 2	57.3 ± 2	61.4 ± 3	67 ± 1.8	50	90 ± 2
Tail						
Shape	Short, noncontractile	Short, noncontractile	Long, contractile	Very short collar with thin appendage	Long, noncontractile	Long, contractile
Length × width (nm)	20 × 15	20 × 15	100 × 16	—	140 × 5(?)	240 × 20
Morphological class[o] and bacteriophage relationship	C, T7	C, T7	A, T-even	C, T7	B, λ	A, T-even
Physical characteristics						
Sedimentation coefficient, $s_{20,w}$ (S)	526–550	490	539	820	353	754
Dry mass (daltons)	53.4 × 10⁶	50.9 × 10⁶	—	—	—	—
Buoyant density in CsCl (g/ml)	1.48	1.48	1.498	1.48	1.501	1.49
Mg^{2+} requirement	Yes (1 mM)	Yes (1 mM)	Yes	No	Yes	No
Temperature of inactivation (°C)	55	55	—	55	—	55
Stability						
pH range	5–11	5–11	—	5–11	—	4–10
Temperature range (°C)	4–40	4–40	—	4–40	—	—
Major protein species (kilodaltons)	44,14	42,11	37,14	40,25	39,11,10	90,45,22,16,15
Growth						
Latent period (hr)	6	6	7	32	Slow	8
Lytic cycle (hr)	14	14	14	48	—	12
Burst size (pfu/cell)	200–350	200–350	100	100	—	50

[a] Luftig and Haselkorn (1967, 1968a,b).
[b] Goldstein et al. (1967).
[c] Schneider et al. (1964).
[d] Sherman and Haselkorn (1970a,b).
[e] Safferman et al. (1969b).
[f] Sherman (1970).
[g] Adolph and Haselkorn (1972b).
[h] Adolph and Haselkorn (1971, 1973a).
[i] Safferman et al. (1969a).
[j] MacKenzie and Haselkorn (1972a,b).
[k] Adolph and Haselkorn (1973b).
[l] Safferman et al. (1972).
[m] Sherman et al. (1976).
[n] Sherman and Connelly (1976).
[o] Bradley (1967).

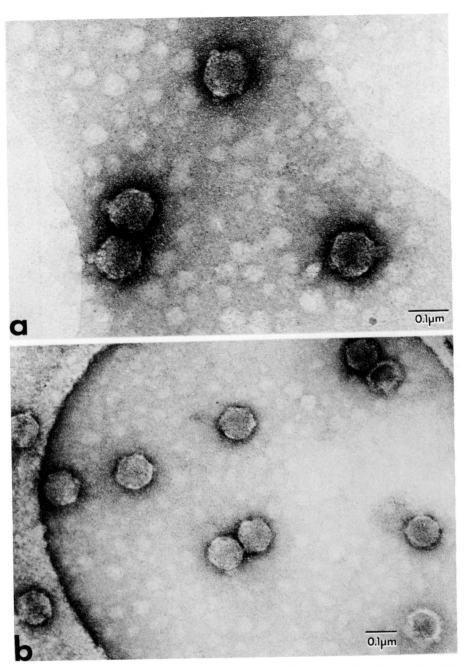

Fig. 1. Cyanophage LPP-1AT (a) and LPP-2GM (b) stained with 1% uranyl acetate. From Safferman *et al.* (1969a).

ferences between the two (Safferman *et al.*, 1969*b*; Sherman, 1970). Using the catalase calibration method of Luftig (1967), Safferman *et al.* (1969*b*) measured an edge-to-edge diameter for LPP-2 of 57.3 ± 2 nm and for LPP-1 of 58.6 ± 2 nm. This small difference was originally thought to be insignificant and due to systematic variation in the negative staining procedure. Nonetheless, the size distributions were statistically significant, and the structural data of Adolph and Haselkorn (1972*b*) also indicate that LPP-1 is more massive than LPP-2 (Table 2). Considering dry mass and sedimentation experiments, it would seem that LPP-1 is 6–8% larger than LPP-2. The structural proteins of the two phages are also slightly different; LPP-2 has two major proteins of 42,000 and 11,000 daltons present in the amount of 290 and 490 copies/virion, respectively. From these data, however, it is not possible to decide if the two viruses are constructed differently.

The cyanophage with the most intriguing structure has also been well characterized. This is phage N1, which infects the filamentous blue-green algae *Nostoc muscorum*. Besides a hexagonal capsid (61.4 nm diameter), electron micrographs show a long, contractile tail and flexible, beaded fibers attached to the neck (Adolph and Haselkorn, 1973*a*). Using optical diffraction, Adolph and Haselkorn (1973*a*) were able to document that structural changes accompany sheath contraction in a manner analogous to that in bacteriophages. The beaded fibers were found attached to the same capsid vertex as the tail. They presented the appearance of a chain of beads and were usually two in number, although some particles with three or four fibers were found. However, the functions of these fibers are still unknown.

Adolph and Haselkorn (1973*a*) also surveyed the effects of certain protein-denaturing agents such as urea, guanidine hydrochloride, detergents, pH extremes, and ultraviolet irradiation on the native viral morphology. They found that, for each agent tested, the first observable effect was triggering of a contraction of the sheath, while the most resistant viral substructure was the contracted sheath. From their investigations, they could arrange a hierarchy of increasing resistance to chemical degradation: capsid, tail core, tail sheath. Once more, this pattern is remarkably similar to that present in the T-even bacteriophages (To *et al.*, 1969).

Although they are few in number, the cyanophages which infect unicellular blue-green algae display a wide diversity of morphological types. The three best known within this class are SM1, S1, and AS1, all of which infect certain species of *Synechococcus*. SM1 has a class C morphology, although it is somewhat larger than the LPP viruses

(Table 2). Electron micrographs indicate that a very thin appendage protrudes from the capsid collar, and this may represent a tail (Safferman *et al.*, 1969a; MacKenzie and Haselkorn, 1972a). Cyanophage S1 (Adolph and Haselkorn, 1973b; also SM2, Fox *et al.*, 1976) is a class B virus with a long, contractile tail. The extent of physical characterization of SM2 and S1 is given in Table 2. However, these viruses grow slowly and their replicative cycles have not been studied in detail. Therefore, we will not discuss them greatly in the following sections.

The cyanophage which attacks unicellular algae and which will play a prominent role in the discussion of replication is AS1 (Safferman *et al.*, 1972). AS1 and the related strain AS1-M (Sherman and Connelly, 1976) are morphologically group A phage, with a large, hexagonal capsid (90 nm diameter) and a long, contractile tail (Fig. 2). This is the largest of the known cyanophages and possibly the most complex architecturally. Electron micrographs by the above authors indicate the presence of a base plate, short tail pins, and an internal core. AS1 superficially resembles the T-even phages, but there are a few major differences: (1) no tail fibers have yet seen on AS1; (2) the tail length of AS1 is much longer than that of T4; and (3) the capsid of AS1 more closely resembles an icosahedron than does that of T4, which is now thought to be a prolate icosahedron with a triangulation number of $T = 13$ (Branton and Klug, 1975; Bijlenga *et al.*, 1976). Neverless, AS1 must be considered the most complex cyanophage, and comparisons to the T-even phages would seem to be highly relevant.

2.3. Physical Characteristics

2.3.1. DNA

The cyanophages characterized to date all contain linear, double-stranded DNA. As shown in Table 3, the molecular weight of the DNA ranges from 23×10^6 for S1 to about 60×10^6 for SM1 and AS1-M. The %G+C of these nucleic acids also varies substantially, from 37% for N1 to 70% for S1. The good agreement between the G+C content calculated from buoyant density measurements and temperature melting experiments in all cases indicates that the viral DNAs probably do not contain large quantities of odd bases. Unfortunately for certain biochemical purposes, the G+C content of phage SM1 and AS1 DNA is almost identical to that of the host DNA. However, LPP-1 DNA can be readily resolved from that of *P. boryanum,* and great use has been

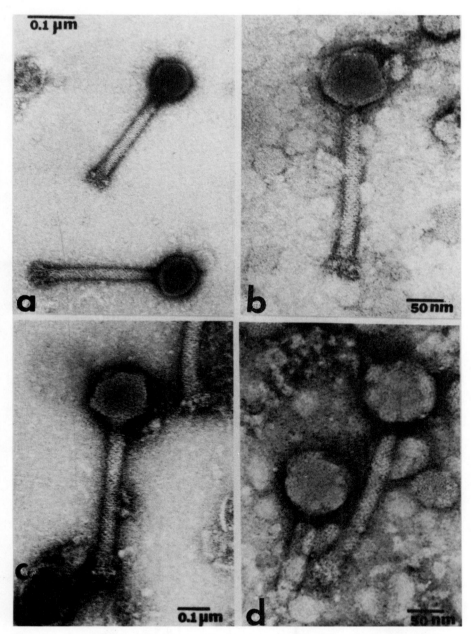

Fig. 2. Cyanophage AS1-M negatively stained with 1% uranyl acetate. From Sherman and Connelly (1976).

TABLE 3
Physical Chemical Characteristics of Cyanophage DNA

Parameter	Filamentous host			Unicellular host		
	LPP-1[a,b]	LPP-2[c]	N1[d]	SM1[e]	S1[f]	AS1-M[g,h]
Molecular weight	27×10^6	28×10^6	$41–45 \times 10^6$	$56–62 \times 10^6$	$23–26 \times 10^6$	57×10^6
Sedimentation coefficient, $s_{20,w}$ (S)	33.4	34.2	40.9	48	30.8	45.6
Contour length (μm)	13.2	—	~20	24.3	13.3	29.5
Buoyant density in CsCl (g/ml)	1.714	1.709	1.696	1.725	1.729	1.714
G + C content (%)	53	52	37–41	66	70–74	52–55

[a] Goldstein and Bendet (1967).
[b] Luftig and Haselkorn (1967).
[c] Sherman (1970).
[d] Adolph and Haselkorn (1971).
[e] MacKenzie and Haselkorn (1972a).
[f] Adolph and Haselkorn (1973b).
[g] Sherman and Connelly (1976).
[h] Sherman and Pauw (1976).

made of density gradient centrifugation to study DNA synthesis in this system.

With the exception of the parameters listed in Table 3, cyanophage DNA has not been studied in great detail. Luftig and Haselkorn (1967) did thermal denaturation analysis of LPP-1 DNA and found that it exhibited a single band in cesium chloride at a density 0.012 g/ml higher than that of the native DNA. They also attempted to separate the denatured DNA into individual strands by binding to poly(I,G). However, no separation was achieved by this technique. Although this finding contrasts with the behavior of T4 or λ DNA, this type of experimentation was not pursued in detail, and it has not been attempted with any other cyanophage DNA.

The amount of genetic information contained by the cyanophage genomes ranges from that of T7 to about one-half that of T4. The cyanophages, therefore, have the capacity to code for between 40 and 100 proteins. This can account for all of the structural proteins as well as enzymes involved in host degradation, phage DNA, and protein synthesis, and possibly host metabolism. There is an anomaly in regard to AS1-M, however. The size of the genome seems small with respect to the capsid diameter, especially relative to T4. Although AS1-M and T4 have almost the same-size head and what appears to be a similar degree of architectural complexity, AS1-M has only one-half the DNA. One possibility is that the AS1-M DNA can readily shear into two equal pieces so that the actual size of the genome is close to that of T4. On the other hand, it may be that T4 is a much more complex virus and that the cyanophage need not code for as many proteins.

2.3.2. Proteins

The structural proteins for the six viruses listed in Table 2 have been analyzed by SDS gel electrophoresis. The total complement of these structural proteins represents about 15% of the coding capacity of SM1 (MacKenzie and Haselkorn, 1972a) to about 50% for LPP-1, LPP-2 (Adolph and Haselkorn, 1972b), and AS1-M (Sherman and Pauw, 1976). The major protein species, presumably those of the capsid, are listed in Table 2. Again, these fall in the range of major head proteins found in bacteriophage capsids.

Because of the absence of conditional-lethal mutants in these systems, the structural parts of the virus have not generally been characterized. An exception is the LPP viruses, where a lack of Mg^{2+}

ions will cause the head and tail to separate. A number of workers have made use of this technique to study LPP-1 and LPP-2 in more detail (Luftig and Haselkorn, 1967, 1968a; Sherman, 1970; Sherman and Haselkorn, 1970c; Adolph and Haselkorn, 1972b). The major head proteins of LPP-1 have molecular weights of 44,000 and 14,000, while a major tail protein has a molecular weight of 70,000–80,000. No internal proteins have yet been discovered in any system, nor have structural enzymatic activities been detected (e.g., Adolph and Haselkorn, 1973a).

2.3.3. Other Physical Properties

Some basic physical parameters of the cyanophages are also listed in Table 2. These factors are probably closely related to the ecological significance of the cyanophages, although this has not been explicitly proven. However, the stability of cyanophages at alkaline pH is in accord with the pH range of optimal growth for the blue-green algae; this can be contrasted to the bacteriophages, which are generally stable only from pH 5 to 8 (Adams, 1959). The actions of AS1-M in this regard are interesting. It is stable only up to pH 10, although an exponential culture of *S. cedrorum* often reaches pH 11. Therefore, to ensure stability, it has been found beneficial to buffer the culture slightly upon infection (Sherman et al., 1976).

The cation requirements for cyanophages are variable. The LPP phages have an absolute requirement for at least 1 mM Mg^{2+}, while AS1 is stable in the absence of cations (Safferman et al., 1972). Some of the phages, especially AS1, require sodium chloride for optimal absorption. However, even this parameter is variable, since the closely related AS1-M does not have such a stringent requirement (Sherman et al., 1976).

3. REPLICATION OF CYANOPHAGES IN A UNICELLULAR HOST

The virus–host interations characteristic of cyanophage infection appear to be confined to two basic classes, which in turn seem closely linked to the host morphology. The cyanophages such as LPP-1, which infect filamentous blue-green algae, cause certain ultrastructural changes in the host cell (K. M. Smith et al., 1966a,b, 1967). This is typified by an invagination of the photosynthetic lamellae, leaving a virogenic stroma where the phages ultimately replicate. (However, there is an important exception which we will discuss in Section 6.) This replication cycle is usually complete by 14–16 hr after infection. Those cyanophages that infect unicellular blue-green algae, on the other hand, do not greatly disturb the host ultrastructure. Replication of SM1 in *Synechococcus* sp. takes place entirely within the nucleoplasm and no major morphological changes are seen (MacKenzie and Haselkorn, 1972*b*). The replication cycle of SM1 as well as S1 (Adolph and Haselkorn, 1973*b*) is, unfortunately, quite long, with complete lysis taking at least 48 hr.

These facts point out a dilemma confronting cyanophage researchers. Infection of the filamentous algae is relatively "fast," but the filamentous nature of the host makes certain quantitative experiments extremely difficult. Many of the statistical approaches used successfully in bacteriophage systems cannot be strictly applied when filamentous cells are involved. More importantly, the mere existence of long trichomes may influence viral development; understanding this interaction consumed a great deal of effort soon after the isolation of LPP-1 (K. M. Smith *et al.*, 1966*a,b*, 1967; Brown *et al.*, 1966; Padan and Shilo, 1973). Quantitative aspects of viral growth can be modified by the number and character of cells in the trichome; this assumption was sustained by the finding that plasmadesmata-like bridges link blue-green algal cells (Lang, 1968) and that substances are transferred from cell to cell in trichomes (Wolk, 1968). There have been indications that higher plant viruses can spread from cell to cell via plasmadesmata (Kitajima and Lauritis, 1969). The variability in host filament length has also been blamed for the nonlinear absorption kinetics of LPP-1 on *P. boryanum* (Goldstein *et al.*, 1967).

One solution to this dilemma was the isolation of a short trichome (one to six cells/filament) strain of *P. boryanum* by Padan and Shilo (1969). This permitted a number of biochemical experiments to be performed more appropriately. However, this system still had certain drawbacks, including the fact that the trichomes grew to normal length in time. A more suitable solution may have been provided by the isolation of AS1 by Safferman *et al.* (1972). AS1 (and the related AS1-M) infects the single-celled algae *Anacystis nidulans* and *Synechococcus cedrorum* and completes its replication cycle in 14 hr. Therefore, AS1 would seem to provide the best opportunity to date for the study of the biochemical, physiological, and genetic interactions of a cyanophage-algal system.

For this reason, we shall begin our discussion of cyanophage replication with AS1. Since it has been characterized more fully, we

shall most often refer to AS1-M. This in part reflects our value judg-
ment that this is the best model system available, and that, as with bacteriophages, it would be beneficial to focus on a small number of phage
systems. We will then discuss the replication of the cyanophage LPP-1,
which infects a filamentous alga and compare and contrast the two. We
will then move on to the interaction of phage replication and host
metabolism, where, once again, a dichotomy between unicellular and
filamentous algae will be evident.

3.1. Cyanophage Growth

One-step growth and intracellular growth curves for the infection
of *S. cedrorem* by AS1-M are shown in Fig. 3. The growth pattern is
typified by an eclipse period of 3 hr, a latent period of 6–8 hr, a total
lytic cycle of 12 hr, and a burst size of 40–55 plaque-forming units
(pfu)/ml (Sherman *et al.*, 1976). The growth of AS1 on the same host is
similar although somewhat slower (Safferman *et al.*, 1972). The results
described in Fig. 3 were obtained at high multiplicity of infection
(MOI = 15) and were taken from the EM experiment to be described
later. At this MOI, lysis-from-without is evident, and this causes a
slight decrease in the overall burst size (40 pfu/ml). However, at MOI
from 0.1 to 10.0, under a variety of physiological conditions, a burst
size of 50–55 pfu/ml was routinely obtained; the eclipse and latent
periods varied only slightly.

The only other cyanophage infecting a unicellular alga which has
been studied in detail is SM1. In this case, growth is substantially
slower. This phage shows a latent period of 32 hr, which is followed by
a rise period of another 16 hr. This ultimately culminates in the release
of 100 pfu/cell (MacKenzie and Haselkorn, 1972*b*). This slow growth,
coupled with the unusual physiological conditions needed to obtain
plaques, has made SM1 rather unsuitable for further biochemical
experimentation.

3.2. Ultrastructure of Infection

The prokaryotic nature of the blue-green algae is apparent in Fig.
4. The central region of the cell is the nucleoplasm, which contains
DNA, ribosomes, and structures called polyhedral bodies. This area is
surrounded by a series of concentric lamellae which contain all of the
chlorophyll, the photosynthetic electron transport chain, and the

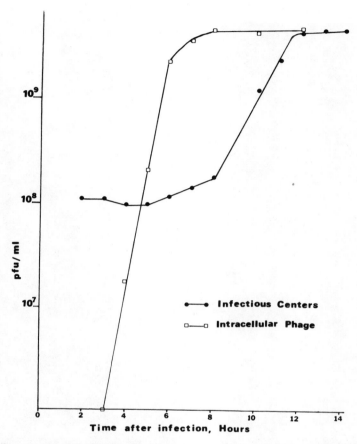

Fig. 3.   One-step growth (●) and intracellular growth (□) curves of AS1-M infection at MOI of 15. From Sherman *et al.* (1976).

photosystems which participate in the photosynthetic light reactions (Susor and Krogmann, 1964; Wolk, 1973). The cytoplasm is bounded by a plasma membrane, which in turn is surrounded by a complex cell wall. The tripartite construction of the cell wall has been described morphologically by Allen for *Anacystis nidulans* (1968a,b,c) and is now being analyzed chemically by Drews (1973, and personal communication). Located between the lamellae are a variety of other inclusions such as α granules (polyglucoside granules, Chao and Bowen, 1971), the lipid-containing β granules (Lang, 1968), and the peptide-containing cyanophycin granules (Simon, 1971). The ultrastructure of blue-green algae has been reviewed by Lang and Whitton (1973) and Drews (1973).

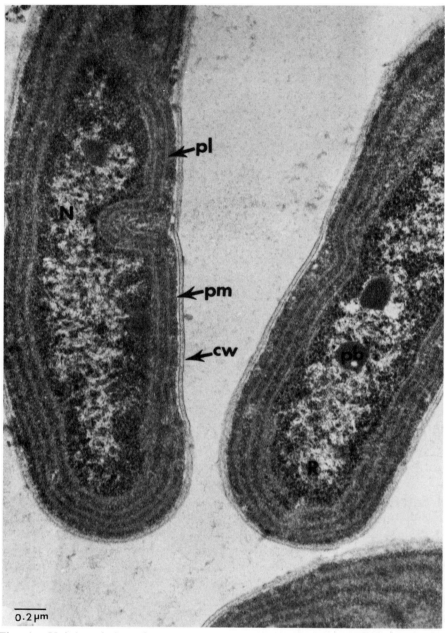

Fig. 4. Uninfected *Synechococcus cedrorum*. N, Nucleoplasm; R, ribosomes; pl, photosynthetic lamellae; pm, plasma membrane; cw, cell wall. From Sherman *et al.* (1976).

In order to study the morphological changes which occur after AS1-M infection, Sherman *et al.* (1976) prepared samples for electron microscopy hourly during the latent period. They found that the most important changes were noticeable at 2 hr (during the eclipse period), and at 4 hr and at 6 hr (which correspond to the early and middle periods, respectively, of intracellular growth). The first perceptible alteration in cellular morphology can be seen at 2 hr after infection (Fig. 5a). In the uninfected cells, the central nucleoplasm is of rather even density, with ribosomes and DNA distributed throughout this area. The nucleoplasm in 2 hr cells, on the other hand, appears almost electron transparent. Although DNA, as represented by fibrillar material, can still be seen in the nucleoplasm, the amount present seems to be less than that visible in uninfected cells. This correlates well with the biochemical finding that host DNA is rapidly degraded after AS1-M infection (Sherman and Pauw, 1976).

The next major modification within the cell is quite evident by 4 hr after infection (Fig. 5b), when a dramatic increase in fibrillar material can be seen. We can speculate that this is caused by the synthesis of phage DNA; once again, the EM observations correlate nicely with biochemical measurements which indicate that synthesis of phage DNA begins at 2.5–3 hr after infection (Sherman and Pauw, 1976). The size of this vegetative DNA pool continues to increase until 6 hr after infection, when a large number of phage particles begin to form (Fig. 6).

One other aspect of Fig. 5 and 6 is worthy of note. In these infected cells, the photosynthetic lamellae are no longer completely surrounding the nucleoplasm. This resembles the situation we will discuss later in regard to LPP-1 infection, in which the lamellae invaginate and leave a physically separate compartment (called the virogenic stroma) where the phages later develop. In AS1-M infection, the disruption of lamellae seems to be due to an actual breakage of the membranes which is apparent in about one-third of the cells by 6 hr. Other cells were seen in which only the inner lamellae were broken and were visible in the nucleoplasm as a free spiral. Despite this lamellar perturbation, phage development always seems to occur in what was originally the nucleoplasm.

The accumulation of vegetative phage DNA and the appearance of phage precursors increases greatly by 6 hr after infection, where the major events of phage morphogenesis can be seen with great clarity. Electron micrographs that depict a variety of morphogenetic stages at 6 hr are shown in Figs. 6–9. The salient structures seen at this time can be summarized as follows: (1) small vesicular structures which are pre-

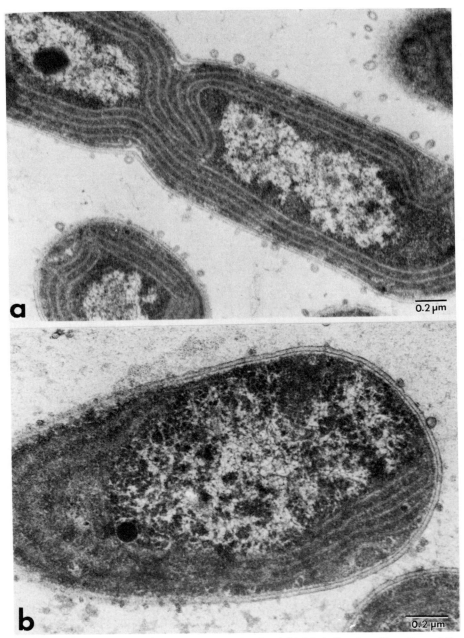

Fig. 5. (a) *S. cedrorum* at 2 hr after infection by AS1-M, indicating disruption of the nucleoplasm. (b) Four-hour-infected *S. cedrorum*. Acculumulation of fibrils that presumably represent vegetative phage DNA is apparent at this time. The hexagonal particle is one of the few filled heads visible at this time. From Sherman *et al.* (1976).

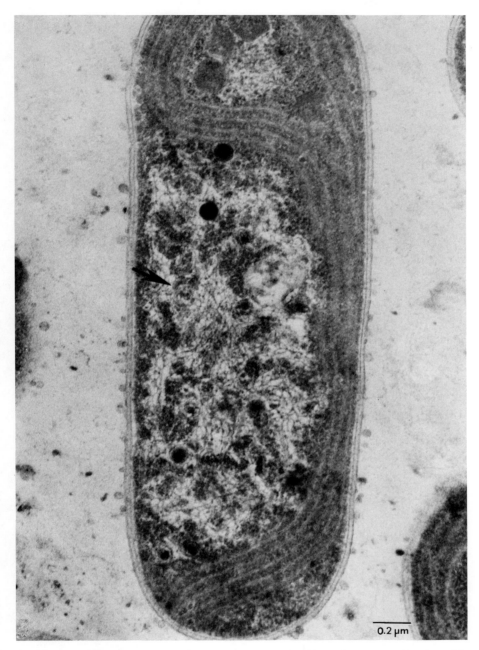

Fig. 6.   AS1-M-infected *S. cedrorum* at 6 hr. Phage DNA has accumulated at this time, and a number of partially filled heads are seen. The arrow is pointing to a structure that may represent a capsid during assembly. From Sherman *et al.* (1976).

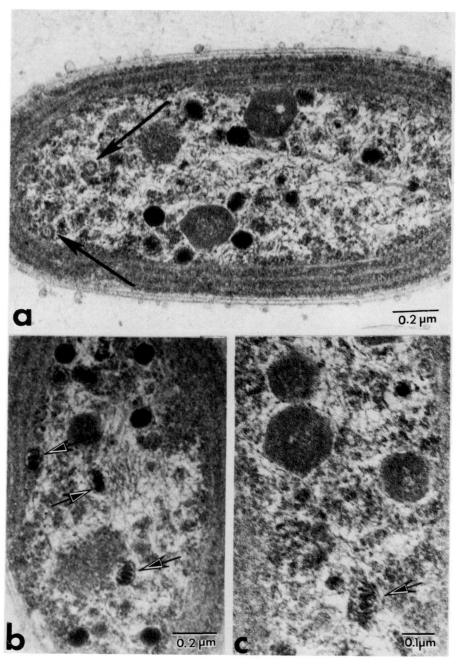

Fig. 7. *S. cedrorum* 6 hr after infection by AS1-M. (a) The arrows refer to structures that may be unfilled head intermediates. (b) A number of "helical" structures can be seen (arrows); these may be heads in the process of packaging DNA. (c) Another helical structure (arrow) which has fibrils extending outward from the particle. From Sherman *et al.* (1976).

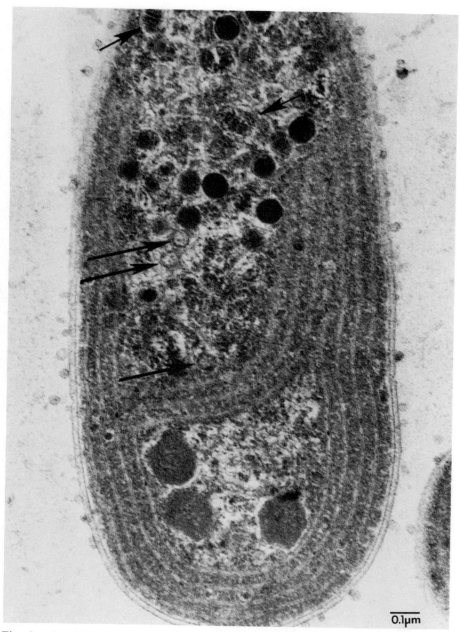

Fig. 8. *S. cedrorum* at 6 hr after infection. This micrograph shows a variety of intracellular phage particles, including the proteinaceous vesicles which may represent head precursors (large arrows), the helical structures (small arrows), and hexagonal, filled heads. From Sherman *et al.* (1976).

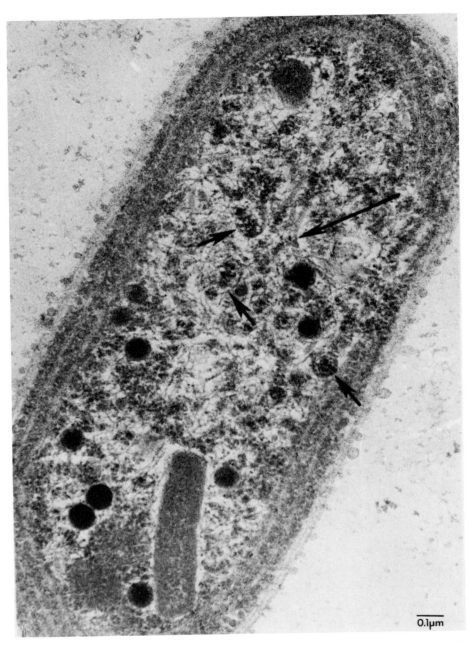

Fig. 9. Six hours after infection. The whole range of interimediates may be seen in this micrograph, including a variety of structures presumably in the process of DNA packaging (small arrows). The particle with the slightly curved tail (long arrow) may be in the final stages of assembly of the head to the tail. From Sherman *et al.* (1976).

sumably proteinaceous head precursors, (2) "helical" structures which may represent the phage DNA being packaged into the capsid, (3) darkly staining hexagonal particles which are most likely filled heads, and (4) complete phage containing both a filled head and tail.

This first structure has been termed proteinaceous because of the resemblance to τ particles seen in T4 infection (Bijlenga et al., 1973); however, there is no information on the chemical composition of these vesicles. These purported unfilled head precursors can be seen in Figs. 7a and 8 (long arrows). They are smaller than filled particles and they are round rather than polygonal. In a number of places, particularly Fig. 6 (arrow), structures that may represent a stage in the formation of these vesicles are visible.

The so-called helical structures are the most striking features in most 6-hr cells; they likely represent the packaging of DNA into the capsid (Figs. 7b,c, 8, and 9; small arrows). In certain cases, especially the high-magnification Fig. 7b,c, the darkly staining fibrils appear to be connected to the vegetative DNA pool. Viewing these many examples, there seems to be a progression from the rather irregularly shaped, loosely packed structures seen in Fig. 7c to the much more compact, elliptical particles seen in Figs. 7b, 8, and 9. None of these helical structures appears to be attached to tails.

From the sixth hour after infection on, the most obvious intracellular particles are the dense, hexagonal structures that resemble filled heads. The average edge-to-edge diameter of 150 measured particles was 90.0 ± 5 nm, which is similar to that of isolated AS1-M (Sherman and Connelly, 1976; Sherman et al., 1976). In most cases, these particles are seen in the nucleoplasm unattached to tails; however, at the same time, free tails are seen in many cells (Fig. 9). Finally, completed phages can be seen beginning at 6 hr, but especially by 8 hr after infection. These phages are seen throughout the cell, although there is a tendency to find the head oriented closer to the photosynthetic lamellae.

The precursor–product relationship between the empty vesicles and helical structures and between the helices and the hexagonal particles was studied more quantitatively. The electron micrographs indicate that the number of empty vesicles is maximal at 5 hr, the helices are maximal at 6 hr, and the dense hexagons are predominant at 8 hr. However, it is difficult to quantitate these data firmly because of the possibility of artifact and the moderate degree of synchrony during infection. This type of analysis is best done with the aid of conditional-lethal mutants in specific functions involved in the packaging of the DNA.

The process by which AS1 and AS1-M adsorb to the host is now becoming clear. Electron micrographs of sectioned cells that were infected by AS1 (Pearson *et al.,* 1975) and AS1-M (Sherman *et al.,* 1976) indicated that the phages attached to the cell wall by the distal part of the tail. This was verified by high-resolution scanning electron micrographs of AS1-M-infected *S. cedrorum,* which distinctly showed phage attachment to the outer cell wall (Sherman, unpublished observations). More recently, Samimi and Drews (personal communication) have been studying the biochemistry of adsorption. They found a very strong correlation between the host range of cells susceptible to AS1 infection and the composition of lipopolysaccharides (LPS) in the cell. The LPS of typical susceptible *Synechococcus* strains has a mannose polymer, and fucose, glucose, galactose, 3-*O*-methyl-D-mannose, and 4-*O*-methyl-D-mannose as the characteristic sugar components. LPS isolated from susceptible (but not from nonsusceptible) cells inactivates the phages, and a LPS–receptor complex has now been isolated and is being characterized. Importantly, highly purified receptor material can interact with AS1 and cause the tail to contract and the head to be emptied of DNA. Once again, the process is turning out to parallel bacteriophage adsorption of bacteria.

A very similar pattern of phage development was reported by Pearson *et al.* (1975) for AS1 infection. AS1 replication takes place in the nucleoplasm, and some of the intracellular precursors discussed above were seen as well. However, the replication cycle of AS1 (16 hr) was substantially slower than that of AS1-M and those authors did not see intracellular phage development until 9 hr after infection. In the absence of intracellular growth data on AS1, meaningful comparisons of the temporal appearance of precursors is difficult to make. Nonetheless, the precursors are a readily discernible feature of both AS1 and AS1-M infection. Nearly identical comments can be made in regard to SM1 infection, which has an intracellular growth cycle of 32 hr (MacKenzie and Haselkorn, 1972*b*). Once again, however, viral development is in the nucleoplasm.

3.3.   Assembly of AS1-M

Based on the ultrastructural observations, a speculative model for the assembly of AS1-M can be proposed. In the absence of conditional-lethal mutants that are blocked in various stages of phage assembly, this model is necessarily crude. Nonetheless, a working hypothesis based on the more detailed assembly pathways that have been obtained with bacteriophages would not seem out of order.

The first step in viral infection is the degradation of the host nucleoplasm at 2 hr, which is a manifestation of host DNA breakdown. Large amounts of what is presumably vegetative phage DNA accumulate by 4 hr. Phage infection does not affect the rest of the cell to any great extent; in fact, we will later indicate that photosynthesis continues normally throughout the latent period.

The appearance of numerous intermediates in head assembly at 6 hr can be explained utilizing recent models for T4 and λ development. The vesicles seen in Figs. 7 and 8 resemble the T4 τ particles seen in gene 24 mutant infection (Bijlenga et al., 1973) that have recently been shown to be intermediates in T4 head formation. There are also many striking similarities between the structures seen here and those seen by Simon (1972) after infection by T4 mutants. However, in both size and shape, these empty structures may have the same precursor relationship that petit forms of λ have to complete λ (Kellenberger and Edgar, 1971; Kaiser et al., 1975). If these structures are preformed head intermediates, then the most likely explanation for the helical particles is that they are in the process of packaging the DNA into the head. This model for the assembly of the AS1-M head would then be formally equivalent to that of T4 (Laemmli and Favre, 1973; Laemmli et al., 1974a,b) or to that of λ (Kaiser et al., 1975). After the head assembly has been finished, the hexagonal particles, representing filled heads, are seen unattached to tails. Phage assembly is then completed with the attachment of the tail to the head, again similar to the case of T4.

3.4. Protein Synthesis in AS1-M Infection

The biochemical mechanisms involved in AS1-M replication have not yet been worked out in great detail, but a start has been made. The temporal information is in accord with EM observations; together, the data will allow us to propose an overall life cycle for AS1-M.

The time course of protein synthesis after AS1-M infection was determined by analyzing pulse-labeled extracts of infected cells on SDS-acrylamide gels using both carrier-free ^{35}S and [^{14}C]amino acids (Sherman and Pauw, 1976). From these patterns, it is obvious that the cessation of host protein synthesis begins almost immediately and is complete by 4 hr after infection (Fig. 10). The pattern of those proteins which appear only after phage infection is quite complex, which is

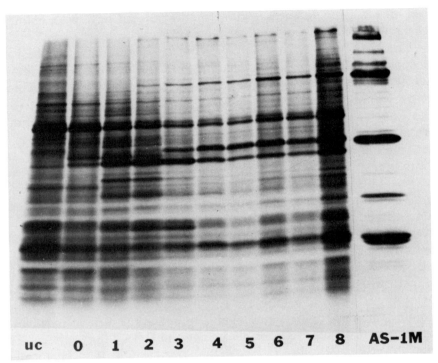

Fig. 10. SDS-polyacrylamide gel patterns of AS1-M structural proteins and proteins synthesized during infection. Samples were labeled with [^{14}C]amino acids for 1 hr and run on 10–15% acrylamide slab gels. Autoradiograms were made by exposure to Kodak Royal Plus X-ray for 3 days. The gels are as follows: uc, uninfected control; 0, 1, 2, etc., pulse begins at that time for 1 hr; AS1-M, phage structural proteins.

typical for a phage of this size. For convenience, the proteins have been divided into three distinct classes: early, middle, and late. Early proteins are those whose synthesis begins 0–3 hr after infection and which are shut off by 4 hr; synthesis of middle proteins begins at about 3 hr and continues throughout the infective cycle; late proteins begin synthesis at 3–4 hr, continue until lysis, and are the phage structural proteins.

During the first 3 hr, a number of early proteins are synthesized (Fig. 11), with molecular weights of between 10,000 and 70,000. In all cases, these bands are effectively shut off or are below the limit of resolution by 4 hr. Beginning at about 3 hr, the synthesis of a large number of middle proteins begins (Fig. 11, dashed lines). These proteins manifest much more variability than either of the other two classes; their start and stop times do not appear precisely synchronized. The seven middle proteins so far resolved have molecular weights ranging between 82,000 and 28,000. Finally, beginning at 3 hr, the synthesis of the structural proteins begins. Since the phenomenon of post-translational cleavage is so well documented in phage systems (e.g., Laemmli, 1970; Laemmli et al., 1974b), a number of pulse-chase experiments have been performed to see if such events take place during AS1-M infection. All such experiments have so far yielded negative results, despite the use of very sensitive techniques (Sherman and Pauw, 1976; Sherman, unpublished observations). To date, no example of post-translational cleavage has been confirmed in cyanophage infection.

An important problem in understanding the biochemical processes of the phage life cycle concerns the role of phage-directed protein synthesis in the turn-on of the late classes of proteins. In the case of T4 infection, it has been proposed that immediate early synthesis can take place using only the host transcriptional machinery, while delayed early and late RNA require phage-coded factors for proper synthesis (Travers, 1970; Horvitz, 1973; Mattson et al., 1974; Stevens, 1974).

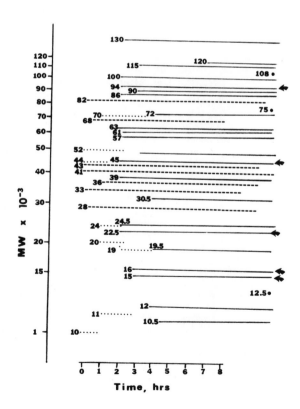

Fig. 11. Scheme for protein synthesis after AS1-M infection of *S. cedrorum.* Densitometer tracings of four experiments like that in Fig. 10 were analyzed for band position and molecular weight and are represented here at the times the bands appear during the pulses. Early proteins (····); middle proteins (----); late structural proteins (——). The arrows refer to the major phage structural proteins in Fig. 10. The dots refer to structural proteins detected in purified phage preparations but not in pulses. From Sherman and Pauw (1976).

Although the situation is poorly understood for AS1-M infection, a start has been made through the judicious use of the antibiotic rifampin. Rifampin (Rif) is known to stop RNA transcription in bacteria by binding to the β subunit of the DNA-dependent RNA polymerase (Travers, 1971). Although the exact mechanism of action is not known, Rif has a similar effect on RNA synthesis in blue-green algae (Sherman and Pauw, 1976). In the experiment summarized in Table 4, Rif was added at various times after infection, the cells were pulse-labeled for 1 hr with ^{35}S, and the labeled extracts were analyzed by SDS gel electrophoresis (Sherman and Pauw, 1976). For convenience, only the middle and late protein classes were tabulated.

The data indicate that Rif has a pronounced effect on later phage protein synthesis. Two important conclusions arise from the experiment: (1) synthesis of the middle proteins requires phage-directed RNA transcription that begins after 2 hr and is complete by 4 hr after infection; (2) synthesis of late proteins requires transcription that begins after 3 hr and is complete by 5 hr after infection. From this type of experiment, it is impossible to conclude whether Rif blocks the syn-

TABLE 4
Effect of Rifampin on Middle and Late Proteins Synthesized after AS1-M Infection of *Synechococcus cedrorum*

Rifampin added at (hr)[a]	Protein class	1-hr pulse[b] with $H_2{}^{35}SO_4$ beginning at		
		3 hr	4 hr	6 hr
2	Middle	−		
	Late	−		
3	Middle		+	+
	Late		−	−
4	Middle			++
	Late			(a few bands missing)
5	Middle			++
	Late			++

[a] Rifampin (100 μg/ml) was added to an infected culture at the time indicated. After the appropriate incubation time, the cells were washed, resuspended in low-sulfate medium, and pulsed with 70 μCi/ml $H_2{}^{35}SO_4$ for 1 hr. The pulsed cells were treated as described in Laemmli (1970), run on 10% SDS-polyacrylamide gels, and autoradiographed.

[b] ++, Normal or near normal gel pattern; +, protein synthesis somewhat decreased in quality and quantity; −, deficient in synthesis of the protein class. From Sherman and Pauw (1976).

thesis of a message for a factor needed to turn on later genes or whether it blocks the synthesis of the protein message directly. For simplicity, we will speak in terms of factors, although the alternative explanation cannot be ruled out. Nonetheless, the above experiments reveal a temporal control of protein synthesis in AS1-M infection that strongly resembles that found in bacteriophage.

3.5. DNA Synthesis in AS1-M Infection

What little is known about AS1-M DNA synthesis can be seen in Figs. 12 and 13. The fate of [^{3}H]uracil-labeled host DNA after infection is evident in Fig. 12. It is obvious that degradation of the host DNA begins almost immediately, for by 1 hr after infection about 15% of the radioactivity has been solubilized. The process continues until about 2.5–3 hr, when some 40% of the DNA has been degraded to acid-soluble form. This corresponds closely to the time at which the nucleoplasm is found disrupted in the EM. The label is slowly incorporated into newly synthesized phage DNA (Sherman and Pauw, 1976) until at 7 hr (near the end of the latent period) all of the label is again in acid-precipitable form. The data in Fig. 13 imply that AS1-M DNA synthesis begins at about 3 hr, and then continues linearly throughout the

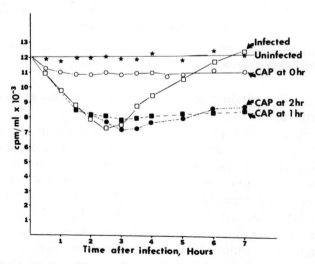

Fig. 12. Host DNA breakdown and the effect of chloramphenicol (CAP) on this process. The arrows refer to the times of addition of 100 μg/ml CAP. From Sherman and Pauw (1976).

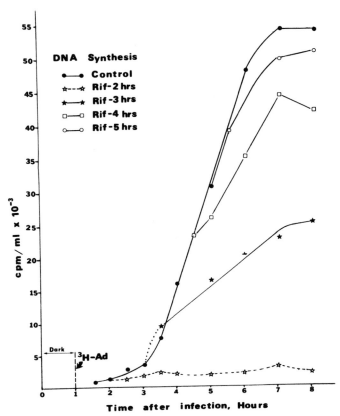

Fig. 13. Effect of rifampin (Rif) on AS1-M DNA synthesis. At the times indicated, 100 μg/ml Rif was added. From Sherman and Pauw (1976).

latent period. This finding agrees well with the appearance of intracellular fibrils seen in the EM.

These experiments also allow a determination of the capacity of phage-infected cells to synthesize phage DNA and to degrade the host genome. Furthermore, they show that postinfection protein synthesis is necessary for these functions to take place. As shown in Fig. 12, only the addition of chloramphenicol near the time of infection greatly affects host DNA breakdown; the addition of the antibiotic at 1 hr still permits optimal degradation. This strongly suggests that the phage-directed nucleases responsible for this function are early proteins. Similarly, the addition of Rif (Fig. 13) or chloramphenicol (Sherman and Pauw, 1976) between 2 and 5 hrs after infection prevents normal phage DNA synthesis. These results indicate that protein synthesis after infection is required for phage DNA synthesis and that transcription and translation of the requisite enzymes occur between 2 and 5 hr after infection. Therefore, those enzymes involved in DNA synthesis are middle proteins.

3.6. The AS1-M Replication Cycle

The information presented in the previous sections allows us to sketch a brief outline of the life cycle of AS1-M (Pearson *et al.*, 1975; Sherman *et al.*, 1976; Sherman and Pauw, 1976). The phage attaches to the algal cell wall and, via contraction of the tail, injects its DNA into the cell. Almost immediately after injection, a class of early proteins is synthesized; these proteins are turned on during the first few hours after infection and are turned off by 4 hr. Phage-directed DNase(s), which degrades the host DNA, appears to be an early protein. The cellular capacity for host DNA degradation is optimal by 2 hr, at which time a severe disruption of the host nucleoplasm is seen in the EM. This DNase activity is the only function that can definitely be considered early. Another early function may be a factor necessary for the initiation of the middle proteins. Altogether, a total of seven early proteins have been resolved on polyacrylamide gels. One other may be an enzyme involved in the breakage of the photosynthetic lamellae that are seen in the EM at this time.

Beginning at about 3 hr after infection, the middle proteins are synthesized, a process which continues until lysis. These proteins compose a large class which must be involved in many functions. Primary among these functions is phage DNA synthesis. The capacity for DNA synthesis rises to a maximum between 2 and 5 hr, while DNA replication itself begins at about 3 hr after infection. This synthesis of vegetative phage DNA is seen clearly in the EM at 4 hr after infection as an extensive matrix of fibrils. Another middle function might be an initiation factor necessary for the synthesis of late proteins, since, once again, this functional capacity increases between 3 and 5 hr.

The synthesis of viral structural proteins, the late class, begins somewhat before 4 hr after infection. The first signs of progeny phage production are seen at 3–4 hr in the intracellular growth experiment and at 4 hr in the EM, which is consistent with the polyacrylamide gel patterns. By 6 hr after infection and with DNA and protein synthesis in full swing, many examples of phage assembly can be visualized in the EM. We have speculated in Section 3.3 that assembly consists of formation of the phage head, followed by packaging of the DNA to yield a complete capsid; attachment of the tail then completes the assembly of the mature phage. Finally, the cell breaks open and releases the contents to the medium (Sherman *et al.*, 1976).

4. REPLICATION OF CYANOPHAGES IN A FILAMENTOUS HOST

Since most of the cyanophages isolated to date infect the filamentous blue-green algae, the replication of these viruses has been studied in some detail. Much of the information relating to the development of LPP-1 has been summarized in a comprehensive review by Padan and Shilo (1973). Therefore, we will not discuss all aspects of LPP-1 replication, but will confine our remarks to major points that can be compared and contrasted with AS1-M.

4.1. Ultrastructure of Infection

The most noticeable difference in the two systems is the place where progeny phage particles are formed (Fig. 14). LPP-1 infection causes an invagination of the photosynthetic lamellae, forming a space called the virogenic stroma between the lamellae and the plasma membrane where the phage later develop (Brown *et al.*, 1966; K. M. Smith *et al.*, 1966a,b; 1967; Sherman and Haselkorn, 1970a). This invagination occurs very early in the replicative cycle, and is apparent in most cells by 4 hr after infection. This process has so far been documented only in LPP-1 infection of *P. boryanum*. Interestingly, a somewhat more complicated pattern emerges during replication of the temperate phage, LPP-2SPI, in the same host. The reason for the lamellar invagination is not known; however, it is dependent on viral protein synthesis (Sherman and Haselkorn, 1970a) and closely parallels host DNA breakdown (Fig. 18).

The appearance of viral heads in the EM begins at about 4 hr, which corresponds to the end of the eclipse period in the intracellular growth curve (Padan *et al.*, 1970; Sherman and Haselkorn, 1970a). By the end of the latent period at 7 hr, most cells contain virogenic stroma filled with viral capsids (Fig. 14). Electron micrographs of sectioned material have not yielded any information on presumptive phage

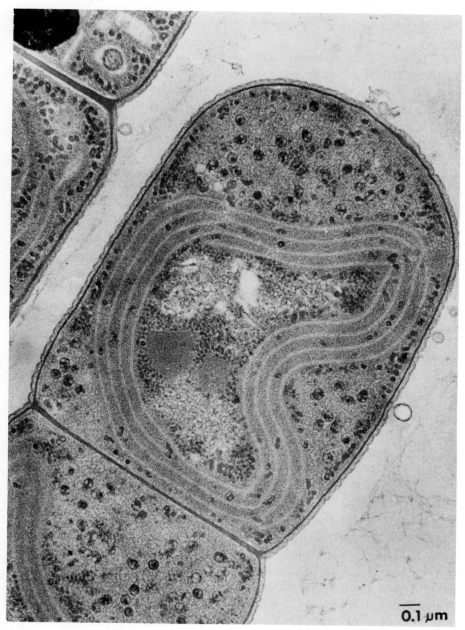

Fig. 14. *P. boryanum* at 10 hr after infection by LPP-1. From Sherman and Haselkorn (1970*a*).

precursors, so the assembly process is largely a mystery. Nonetheless, other techniques have been useful in helping to better understand this process.

The *in situ* lysis technique of Kellenberger *et al.* (1968) was used by Ginzberg (1973) to study the various morphological forms visible during LPP-1G infection. This technique permits the study of intracellular phage particles directly after staining of algal spheroplasts that have been burst on an EM grid. This approach yielded data for the appearance of phage precursors that were in accord with those provided by the sectioned material. At 4 hr, about 1% of the cells contained phagelike particles while by 7 hr this increased to almost 80% of the population. The morphological structures visible by this technique included (1) doughnut-shaped particles; (2) empty "capsids," both round and hexagonal; (3) tails; and (4) mature phage particles. Reaction of the *in situ* preparations with ferritin-labeled antibodies against mature virus showed that the empty, round structures contained viral antigens (Ginzburg, 1973; Padan and Shilo, 1973), indicating that they were phage-related structures. These data are again consistent with the idea that cyanophage assembly proceeds in a fashion similar to that of bacteriophage, although most of the details still must be determined. The close resemblance of LPP1 to T7 in morphology, genome size, and biochemical control (Sherman and Haselkorn, 1970*a,b,c;* Luftig and Haselkorn, 1968*b*) would indicate that the two phages may also have similar modes of assembly.

4.2. Protein Synthesis in LPP-1 Infection

The pattern of protein synthesis following LPP-1 infection is generally the same as that which we have discussed in regard to AS1-M (Sherman and Haselkorn, 1970*c*). As indicated in Fig. 15, three classes of proteins (termed early, middle, and late) are produced after infection. Utilizing a variety of antibiotics, Rimon (1971) has shown that mRNA synthesis and protein synthesis are required throughout the latent period for optimal phage yield. The temporal appearance of the protein classes is in accord with the growth and EM data, since large quantities of structural proteins are detectable at 4 hr after infection.

The functions of some of the nonstructural proteins will again be discussed from the perspective of DNA synthesis in the next section. However, the strong similarity in the time course of protein synthesis

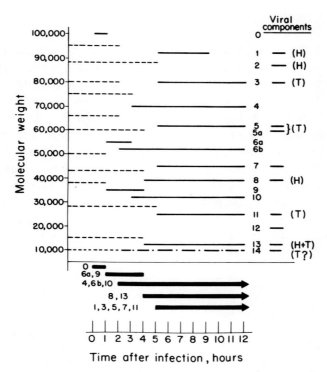

Fig. 15. Scheme for protein synthesis after infection of *P. boryanum* with LPP-1. Dashed line indicates host proteins; solid line represents viral proteins. Virus-coded proteins are numbered on the right; those proteins found in the mature virion are designated by horizontal bars, and their locations in heads or tails are indicated. The temporal classes are indicated on the bottom: Early proteins, 0, 6a, and 9; middle proteins, 4, 6b, and 10; late proteins, 1, 3, 5, 7, 8, 11, and 13. From Sherman and Haselkorn (1970c).

between LPP-1 and T7 (Studier and Maizel, 1969) calls for a speculative comparison. The gene 1 product of T7 codes for a virus-directed transcription factor (molecular weight 100,000) that is required for transcription of early and late mRNA (Summers, 1970). This gene product is produced immediately after infection. The first viral protein synthesized after LPP-1 infection has a molecular weight of 100,000 and is labeled in detectable amounts only during the first hour. It is tempting to suggest that this is a virus-directed transcription factor and that a transcription control mechanism similar to that of T7 exists for LPP-1. However, the finding of Rimon (1971) that RNA synthesis is Rif sensitive throughout infection could eliminate this as a viable hypothesis.

4.3. DNA Synthesis

The ability to readily separate LPP-1 and *P. boryanum* DNA on cesium chloride equilibrium gradients has yielded a more complete picture of LPP-1 DNA replication than that available to date for AS1-M. Although the buoyant densities in cesium chloride of the viral and host DNAs are similar (*P. boryanum*, 1.706 g/ml; LPP-1, 1.713 g/ml), the two species could be resolved using both analytical and preparative centrifugation (Sherman and Haselkorn, 1970*b*). These experiments show clearly that the bulk of the viral DNA synthesis begins at 6 hr, and continues throughout infection at a high level. It can be seen in Fig. 16 that about 20% of the label in a pulse between 1 and 2 hr goes into phage DNA, while at 7–8 hr over 70% of the label enters LPP-1 DNA. The nature of this early incorporation of label is still poorly under-

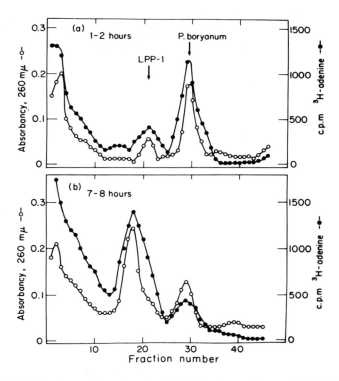

Fig. 16. Equilibrium centrifugation in CsCl density gradients of ³H-labeled DNA from LPP-1-infected *P. boryanum*. Cells prelabeled with [³H]adenine were infected at MOI of 10 and DNA was isolated at various times. Each DNA preparation was centrifuged at 25°C for 70 hr at 33,000 rev/min. From Sherman and Haselkorn (1970*b*).

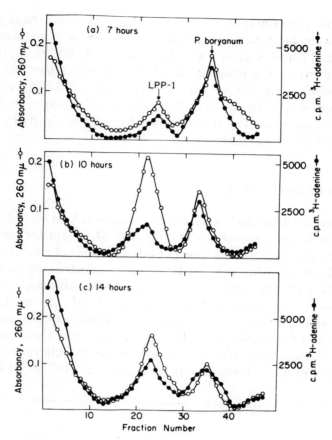

Fig. 17. Incorporation of host DNA breakdown products into LPP-1 DNA. Protocol was as described in Fig. 16. From Sherman and Haselkorn (1970*b*).

stood. It cannot be explained entirely as precocious DNA synthesis in an asynchronously infected population, since it is twice as extensive as would be expected from measurements of net DNA synthesis in the analytical ultracentrifuge (Sherman and Haselkorn, 1970*b*). Two possible sources of this active prereplication incorporation are recombination and repair. Recombination is likely because these experiments were performed at high multiplicity of infection. Repair is suggested because blue-green algae are very resistant to both ultraviolet and X-rays and contain very active repair systems (Wu *et al.,* 1967).

LPP-1 infection also causes extensive degradation of the host DNA commencing 1 hr after infection. This function requires virus-directed protein synthesis, and the cellular capacity for host DNA degradation is maximal at 3 hr. It was also shown that the degradation products were later reincorporated into the newly synthesized LPP-1 DNA (Fig. 17 and Sherman and Haselkorn, 1970*b*). Similarly, phage DNA replication is a phage-directed, protein-dependent process. The enzymes involved in DNA synthesis are produced 3–6 hr after infection, which would imply that they are middle (early in the older designation) proteins. The time course of the known early functions in LPP-1 synthesis is summarized in Fig. 18.

4.4. Life Cycle of LPP-1

The events described in the previous sections lead to a life cycle for LPP-1 that is quite similar to that of AS1-M except for the first few hours after infection. In this time period, a number of interesting occurrences take place. Almost immediately after infection, the early phage-directed proteins are synthesized. These proteins ultimately lead to the breakdown of the host DNA, invagination of the photosynthetic lamellae, and cessation of CO_2 photoassimilation (see Section 5). Based on the temporal coincidence of these phenomena and the limited coding capacity of the LPP-1 genome, we would like to believe that they are more than causally related (see also Padan *et al.*, 1971*b*). It is possible

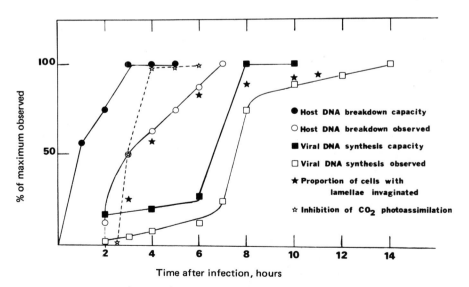

Fig. 18. Time course of early functions in LPP-1 infection of *P. boryanum*. Data from Sherman and Haselkorn (1970*a,b*) and Padan and Shilo (1973).

that degradation of the host DNA causes the invagination simply on a mechanical basis. The raised levels of nucleotides in the cell may explain the cessation of carbon dioxide photoassimilation, since these are known to inhibit this reaction in bacteria (Johnson, 1966). Unfortunately, host DNA degradation also occurs during AS1-M infection without the concomitant perturbations that take place in LPP-1 infection. These processes and anomalies are certainly deserving of further attention.

Once the lamellar invagination is complete, a new compartment is established within the cell that is thought to be physically distinct from the nucleoplasm. This virogenic stroma is the site of phage development. Beginning at about 4 hr after infection, viral structural proteins are produced, followed soon thereafter by the LPP-1 DNA.

5. CYANOPHAGE REPLICATION AND HOST METABOLISM

One of the main reasons for interest in the cyanophages is that the host blue-green algae are photosynthetic organisms. This is the common thread that underlies the four approaches to the field that were mentioned earlier. It has been thought that this virus–host system could be used as a model for infection in photosynthetic organisms as well as a tool for delving into the photosynthetic mechanism and the metabolism of the algae. In turn, this would have ramifications in the fields of ecology and bacteriophage biochemistry. The mechanism of photosynthesis and the nature of the metabolic machinery in blue-green algae are complex topics which we can not adequately review here. However, these subjects have been summarized recently (Holm-Hansen, 1968; Stanier *et al.,* 1971; Carr and Whitton, 1973; Wolk, 1973; Govindjee, 1975), so we will limit our discussion to those topics pertaining directly to cyanophage replication.

The presently accepted scheme for photosynthesis is a chlorophyll-mediated, light-driven mechanism which ultimately results in the production of ATP, the reduction of NADP, and the evolution of oxygen. Furthermore, there are two functionally and structurally distinct photosystems (PS) that act in series and that are connected by the photosynthetic electron transfer chain. Short-wavelength light is absorbed by PSII, which forms a strong oxidant leading to O_2 evolution, and a weak reductant. This process is strongly inhibited by the herbicide DCMU. Long-wavelength light is absorbed by PSI, which forms a strong reductant that can reduce NADP to $NADPH_2$, and a weak oxidant. These processes can be coupled to phosphorylation in one of two ways. In noncyclic photophosphorylation, the formation of ATP is coupled with electron transfers that accompany the conversion of primary oxidant and reductant to O_2 and $NADPH_2$. In cyclic phosphorylation, ATP formation is coupled with an electron shunt about PSI; no O_2 is evolved, nor is NADP reduced. This process is insensitive to DCMU and requires only far-red light. Both types of phosphorylation have been documented in blue-green algae (Jones and Meyers, 1964; Teichler-Zallen and Hoch, 1967; Wolk, 1973).

To a great extent, photosynthesis is the only important energy-producing mechanism in the blue-green algae, and these organisms have been considered obligate photoautotrophs. However, research in the past few years has shown that many strains of blue-greens have heterotrophic capabilities (Fay, 1965; Hoare et al., 1969, 1971; Kenyon et al., 1972; Khoja and Whitton, 1971; Pearce and Carr, 1969; A. J. Smith et al., 1967; Stanier, 1973). This has been demonstrated by growth in the dark on exogenous carbon sources and by growth in the light in the presence of DCMU (Kenyon et al., 1972). Quite interestingly, metabolic differences between unicellular and filamentous blue-green algae are becoming evident (Smith, 1973; Carr, 1973; Stanier, 1973).

5.1. Cellular Metabolism and Phage Development in the Filamentous Blue-Green Algae

5.1.1. Photosynthesis during Infection

The most important effect of LPP-1 infection on host metabolism is a rapid and complete cessation of CO_2 photoassimilation (Ginzberg et al., 1968). This phage-induced inhibition is dependent on phage-directed protein synthesis (Ginzberg et al., 1976), starts 2.5–3 hr after infection, and is complete by 5 hr. Therefore, cessation of CO_2 photoassimilation begins at about the same time as the invagination of the photosynthetic lamellae and about 1 hr after the commencement of host DNA degradation.

Infection with a single phage particle per cell is sufficient to inhibit CO_2 fixation fully, and neither the onset nor the magnitude of the cyanophage-induced inhibition depends on input multiplicity, as shown by Ginzburg et al. (1968). These authors also showed that this cessation of CO_2 fixation was different from that found in virus infection of plants (Zaitlin and Jagendorf, 1960) because nitrogen supplementation before or during infection had no influence on the inhibition. In the plant system, it had been found that most photosynthetic functions were normal if the nitrogen supply was sufficient.

Similar inhibition of CO_2 fixation was found to occur in *Nostoc* cells after infection by N1 (Adolph and Haselkorn, 1972a). The effect is not so rapid and commences at about 5 hr after infection. Nonetheless, the general similarity of this phenomenon and others to be discussed later indicates that infection of N1 and that of LPP-1 in filamentous hosts have many points in common.

If LPP-1 infection shuts off CO_2 fixation, what happens to the rest of the photosynthetic mechanism? This question has been answered in a detailed fashion by Ginzberg et al., (1976), who have shown that all other photosynthetic capacities are normal (Table 5). It is obvious that, despite the invagination of the photosynthetic membranes, the physical and functional attributes of the membranes are unchanged. The chlorophyll content remains the same, as does cyclic and noncyclic

TABLE 5

Photosynthetic Capacity in *Plectonema boryanum* with Cyanophage LLP-1G[a]

Parameter	Noninfected	Infected
CO_2 photoassimilation		
Intact cells	27.0	2.5
Cell extract (ribulose-1,5-diphosphate carboxylase)	8.4	8.4
Photophosphorylation in lysed spheroplasts		
Noncyclic (P/2e)	1.00	0.9
Cyclic (phenazine methosulfate)	1000	1000
O_2 evolution		
Intact cells	828	810
Lysed spheroplasts	82	66
Pigments in cell extract		
Chlorophyll	12	11
Phycocyanin	172	140
Total ATP in cell extract	0.6	0.6

[a] *Plectonema* cells were infected with cyanophage LPP-1G at input multiplicity of 10 plaque-forming units/cell; photosynthetic parameters were determined 6 hr after adsorption. Values are expressed by the following units: CO_2 photoassimilation in μmol CO_2/mg cell protein/hr (intact cells); μmol CO_2/mg protein/hr (cell extract); noncyclic photophosphorylation in μmol ATP/μatom O (P/2e); cyclic photophosphorylation in μmol ATP/mg chlorophyll/hr; O_2 evolution in μmole O_2/mg chlorophyll/hr; pigments in μg pigment/mg cell protein; total ATP in μg ATP/mg cell protein. From Ginzberg et al. (1976).

photophosphorylation, PSI and PSII activity, and O_2 evolution. These results indicate that electron flow, photochemistry, and phosphorylation are unaffected by infection.

Why, then, does CO_2 fixation cease? The answer must lie at some terminal step in the overall process. Ginzberg *et al.*, (1976) found that the activity of RUDP carboxylase, the main carboxylating enzyme, is the same before and after infection. However, they did not check the other Calvin cycle enzymes, nor did they study other parameters, such as pH or cofactors, on enzyme activity. Therefore, the functional state of the carboxylating enzymes must still be considered an open question. Nonetheless, a plausible mechanism for the inhibition of CO_2 fixation which does not require enzyme activity can be considered. This mechanism suggests that cessation of CO_2 fixation is due to competition of the viral biosynthesis of DNA and protein for ATP, NADPH, and other metabolic intermediates normally involved in CO_2 fixation. As a corollary, the phenomenon may be due to control of RUDP carboxylase at the level of substrates of the carboxylation reaction or by compounds within the infected cell (Ginzberg *et al.*, 1976). One trivial explanation is the inhibition due to increased nucleotide levels from host DNA degradation that was mentioned earlier. Whatever the nature of the mechanism of inhibition of CO_2 fixation, it is obvious that this is essential to the understanding of the metabolic control involved in the virus–host interaction.

One final question arises from this finding. If the Calvin cycle is not operating, how are ADP and NADP regenerated in the infected cell? Since viral synthesis can take place under anaerobic conditions and in the presence of DCMU (Section 5.2), it may be that biosynthetic reactions induced by the phage are responsible for this regeneration. There is presently no information at all with respect to virus-induced or virus-directed metabolic functions. However, evidence like that above would indicate that such events may occur. We will return to this topic in the next section.

5.1.2. Dependence of Phage Development on Cellular Metabolism

The picture that arises from the previous comments is that LPP-1 biosynthesis requires photosynthesis for energy production but not for the supply of carbon. Presumably, therefore, other sources are utilized to yield the building blocks for phage protein and DNA. In this section, we will discuss the possible ways in which the infected cell can obtain energy and carbon for phage replication. To do this, we will report on research that utilizes conditions which perturb the normal functioning of photosynthesis. Two compounds are particularly important in this regard, dichlorophenyl dimethylurea (DCMU) and carbonyl cyanide m-chlorophenyl hydrazone (CCCP). DCMU is a powerful inhibitor of oxygen evolution (Good and Izawa, 1964) which also blocks the production of the primary reductant of NADP (Susor and Krogmann, 1964). However, DCMU-treated cells are still capable of ATP production via the mechanism of cyclic phosphorylation (Avron and Neumann, 1968). CCCP is an uncoupler of photophosphorylation (Duane et al., 1965) and photosynthetic electron transport in blue-green algae (Susor and Krogmann, 1964). This compound is also capable of uncoupling oxidative phosphorylation (Heytler and Prichard, 1962).

From Table 6, it is obvious that phage replication can take place under conditions which impair normal photosynthetic functions (Padan et al., 1970, 1971a; Sherman and Haselkorn, 1971). Growth in the light in the absence of CO_2 yields a complete burst, in accordance with the finding that CO_2 fixation is shut off early after infection. In the presence of DCMU, a burst size of 75% of the control (Sherman and Haselkorn, 1971) for LPP-1 infection to 100% of the control (Padan et al., 1970) for LPP-1G infection was obtained. However, this treatment did slightly prolong the eclipse period. Therefore, neither noncyclic electron flow and phosphorylation nor NADP reduction is an absolute

TABLE 6
Yield of Cyanophages in Plectonema Cells Incubated under Different Experimental Conditions[a]

Experimental conditions		Yield[b] pfu/infected cell	Percent of maximal burst size under standard conditions
Dark	Aerobic	36	10
	Anaerobic	8	2
	Aerobic + CCCP (10^{-4} M)	0	0
Light	Aerobic	300–400	100
	Anaerobic	300–400	100
	Aerobic + DCMU (10^{-6} M)	300–400	100
	Anaerobic + DCMU (10^{-6} M)	300–400	100
	Aerobic + CCCP (10^{-4} M)	0	0

[a] From Padan and Shilo (1973).
[b] Yield of cyanophage was determined from intracellular growth curves for each experimental condition.

requirement for LPP-1 replication; they may only accelerate early functions. Infection in the dark under aerobic conditions permits burst sizes only about 10% of the control; however, this indicates that oxidative phosphorylation may play a metabolic role in the dark. The fact that anaerobiosis cuts the phage yield drastically in the dark supports this contention.

All of these results are consistent with the hypothesis that either photosynthetic or oxidative electron transport is required for phage growth. The fact that CCCP completely eliminates phage growth (Table 6) is in accord with this idea. Thus substrate-level phosphorylation or the existence of large, intracellular ATP pools does not seem to measurably contribute to phage growth under autotrophic conditions. Finally, anaerobiosis in the light does not seem to affect phage replication, which indicates that oxidative phosphorylation is relatively insignificant under these conditions.

The fact that phage development proceeds normally in the presence of DCMU implies that cyclic phosphorylation can supply all of the ATP needed for phage production. This, in turn, implies the possibility that only PSI is absolutely required for phage growth. This idea can be tested utilizing far-red light (~ 700 nm), which is absorbed only by PSI, in the presence of DCMU, which blocks PSII activity. In fact, these are the experimental conditions originally used to define cyclic phosphorylation. Ginzberg et al. (1976) have measured the growth of LPP-2SPI under these conditions and found that the burst size was identical to the control (white light and DCMU). These results strongly support the conclusion that the importance of photosynthesis for phage growth is as a supplier of ATP and that this requirement can be met via cyclic phosphorylation.

Another factor emerges from the ability of LPP-1 to grow in the presence of DCMU and in the absence of noncyclic electron flow. Under these conditions, reduced NADP is not produced, and the reducing power in infected cells must be low. However, since CO_2 fixation is inhibited and phage DNA utilizes breakdown products of the host DNA, the need for reducing power may be slight. This may also eliminate the need for the regeneration of NADP that was mentioned earlier.

Not only have we narrowed the light requirement of phage growth to PSI (far-red) light, but evidence has been obtained that such light is merely needed during part of the life cycle. Phage growth has been measured using complementary light–dark regimes. In the first series of experiments (Padan et al., 1970), infectious phage were monitored from infected cells exposed to light for 2–7 hr and then grown in the dark. It was determined that if light was provided throughout the eclipse period (about 4 hr) a maximal burst of phage could be produced, although at a slower rate. Illumination for 7 hr yielded results identical in rate and magnitude to those for continuously light-grown controls. Illumination for 2 hr followed by dark growth resulted in a much slower production of phage (30 hr latent period vs. 12 hr) and a yield about one-third that of the control. Therefore, the ATP requirement for viral synthesis can be satisfied by a 3–4 hr light period. However, light must also affect the release of the mature phage. Although intracellular phage can be produced in large quantities after only 1 hr of light (Padan *et al.*, 1971*a*), it is released very slowly after incubation in the dark. This may be due to inefficient synthesis of an early function such as a phage lysozyme.

The second type of experiment (Sherman and Haselkorn, 1971) measured infectious centers from infected cells first kept in the dark and then illuminated until lysis (Fig. 19). Under these conditions, extracellular virus appeared sooner and the burst size actually increased about twofold over that of the light-grown control. Although early functions such as degradation of the host DNA are slowed during dark

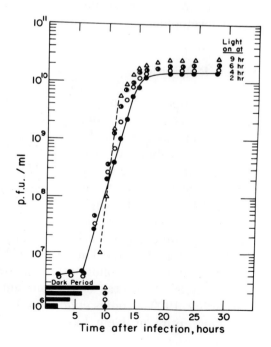

Fig. 19. Effect of duration of dark period on the growth of LPP-1 in *P. boryanum*. Duration of dark periods is indicated by bars in the lower left corner. Infectious centers were plated hourly. From Sherman and Haselkorn (1971).

growth (Sherman and Haselkorn, 1971), they do ultimately reach maximal levels. The results of these and the experiments of Padan *et al.,* (1970), when taken together, imply that no specific light requirement for either early or late functions exists. It is likely that growth in the dark allows for greater synchronization of the culture, and, by a faster inhibition of host functions, ultimately leads to a higher phage yield.

The other phage that infects a filamentous blue-green algae is N1; the dependence of N1 development on cellular metabolism resembles the pattern described for LPP-1 with slight modifications. In contrast to LPP-1, N1 growth in the presence of DCMU is reduced to 25% of the control (Adolph and Haselkorn, 1972*a*). Addition of CCCP severely restricts growth of N1, although the yield is still 2% of the control. This residual phage growth suggests that substrate-level phosphorylation or ATP pools may be able to supply sufficient energy for a limited amount of phage replication. Since the removal of light at any time throughout the lytic cycle reduces the phage yield, N1 is much more dependent on photosynthesis than LPP-1 (Adolph and Haselkorn, 1972*a*). It is difficult to determine how significant the differences between N1 and LPP-1 may be, since no attempts have been made to vary the physiological state of the host. As we shall see below, the exact growth history and physiology of the cell can play an important role in phage development.

5.1.3. Respiration

The importance of oxidative phosphorylation to the blue-green algae has been a most controversial subject during the past 10 years. The blue-greens are usually considered to be obligate photoautotrophs due to their inability to grow in the dark when supplied with exogenous carbon sources (see, e.g., Smith, 1973; Wolk, 1973; Stanier, 1973). This was thought to be due to their lack of important enzymes (A. J. Smith *et al.,* 1967), which would make oxidative metabolism impossible. Nonetheless, there have been reports over the years of heterotrophic growth of some algal strains (Cheung and Gibbs, 1966; Khoja and Whitton, 1971; Pan and Umbreit, 1972; Van Baalen *et al.,* 1971), and, in fact, oxidative phosphorylation has been demonstrated in intact cells and extracts of several blue-green algae (Biggins, 1969; Leach and Carr, 1969, 1970; Scholes *et al.,* 1969).

This entire problem has recently been reinvestigated by Stanier and co-workers (Stanier, 1973). They have shown that some strains of unicellular blue-greens are definitely faculative photoheterotrophs; that is, they are capable of growth in the presence of DCMU when supplemented with glucose. Of direct importance to our story, Padan *et al.* (1971*a*) have measured a large endogenous dark respiration in cells of *Plectonema boryanum*. They found that this dark respiration was highly dependent on the physiological state of the cell. The extent of this respiration was influenced by CO_2 fixation in the light and protein synthesis. Furthermore, White and Shilo (1975) have obtained heterotrophic growth of *Plectonema* using ribose, sucrose, glucose, mannitol, maltose, or fructose as substrate. The doubling time of these heterotrophically grown cells ranges from 5 days with ribose to 13 days with fructose. If these cells are illuminated with dim light (85 lux), the growth rate with ribose increases to nearly double the dark rate. In moderate light, growth takes place with ribose in the presence of DCMU.

Some of the earlier experiments with cyanophages had indicated the existence of a heterotrophic capability of infected cells (Padan *et al.*, 1970, 1971*a*; Sherman and Haselkorn, 1971). However, the data summarized in Table 7 are the most definitive to date (Ginzberg *et al.*, 1976). This experiment shows the ability of LPP-1G to develop in either

TABLE 7
Replication of Cyanophage LPP1-G in *Plectonema boryanum*[a]

Conditions during infection cycle		LPP-1G burst size (pfu/infected cell) and burst time (hours in brackets)	
		Photoautotrophic grown cells	Heterotrophic grown cells
Dark			
1. Aerobic	+glucose	300 [30]	300 [30]
2.	−glucose	30 [13]	300 [13]
Dark			
3. Anaerobic	+glucose		5 [13]
	−glucose		1 [13]
Light			
5. Aerobic	−glucose	300 [30]	—

[a] Logarithmic-phase *Plectonema* cells, grown either photoautotrophically or heterotrophically in the dark, were washed and suspended in the growth medium without glucose, and infected with LPP-1G cyanophage at an input multiplicity of 0.006 plaque-forming units/cell. After the adsorption and removal of unadsorbed cyanophages, the infected cells were incubated in conditions indicated in the table. Glucose was added to a concentration of 30 mM and anaerobiosis was performed by N_2 flushing. Phage burse size was obtained from intracellular growth curve. From Ginzberg *et al.* (1976).

phototrophically or heterotrophically grown *Plectonema* cells. Under aerobic dark conditions with glucose supplementation (line 1), both types of cells produced LPP-1G with the same efficiency as photo-autotrophically grown cells in the light (line 5). While the burst time in the dark in heterotrophically grown cells was identical to that of the light control, it took about 17 hr more to produce a burst in the dark in cells grown photoautotrophically. The relatively low cyanophage burst size in the dark from photoautotrophically grown cells (line 2) was enhanced tenfold in the presence of glucose during the infection cycle (line 1), while the viral burst size in heterotrophically grown cells did not depend on the presence of external glucose. Anaerobiosis inhibited phage development in both types of cells. Finally, the lag of 17 hr in the burst time of LPP-1G on transfer of light-grown cells to heterotrophic growth correlates well with the time required for induction of glucose uptake (Ginzberg *et al.,* 1976).

The results indicate that energy conversion by respiration is as sufficient as photosynthesis for sustaining viral growth. This is in accord with the observations of Stanier (1973) in regard to unicellular blue-greens. Therefore, we can now conclude that not all blue-green algae are obligate photoautotrophs. Some strains, however, seem to be obligate photoautotrophs in that oxidative processes cannot generate ATP at a rate sufficient to meet the needs of growth. The growth of cyanophages under different metabolic conditions may prove to be a useful analytical tool in finally unraveling these mysteries in blue-green algae.

5.1.4. Building Blocks for Phage DNA and Protein

Since viral replication in *P. boryanum* can occur without any net addition of carbon into the system, it would appear likely that the raw materials for phage protein and DNA are scavenged from the host macromolecules. This is indeed the case. As we have already mentioned, infection leads to an immediate breakdown of the host DNA; the breakdown products are then reincorporated into newly synthesized LPP-1 DNA (Sherman and Haselkorn, 1970b). Similar conclusions for the fate of host protein have been reached by Rimon (1971) and Padan and Shilo (1973). They found that algal proteins are the major precursors of phage proteins and that CO_2 assimilation does not contribute significantly, even before it is inhibited by infection. This was supported by similar results obtained by infection in the presence of DCMU.

5.1.5. Summary of the Relationship between Metabolism and the Replication of LPP-1

The pattern that arises from the relationship of cellular metabolism to viral development is one of extreme virulence. The number of infectious phage particles produced is the same when infection is performed in white light (PSI and PSII active), in far-red light plus DCMU (PSI only active), or in the dark in the presence of glucose. If the cells have previously been grown heterotrophically, the full phage burst is achieved even without glucose supplementation during infection. Therefore, growth of the host cell is not required for cyanophage replication. This places LPP-1 among the most virulent of bacteriophages, since development of most phage requires growing cells (Luria and Darnell, 1967). In this respect, this system differs from that of other phages which infect facultative autotrophic bacteria in which phage replication is sustained only under most stringent conditions. This is even true of another cyanophage–alga example to be discussed in the following section.

The unique dependence of LPP-1 replication on host cellular constituents represents an extreme degree of host–parasite interaction. The invading phage needs the cell to supply ATP and enzymes for various functions. It shuts off all host synthesis of macromolecules and then degrades the preexisting DNA and protein and uses the breakdown products as building blocks for its own protein and DNA. This extreme virulence would appear to offer a selective advantage to the phage, because the infective process would be less dependent on environmental factors. This may explain the worldwide distribution of the LPP viruses in quite diverse environments (Padan and Shilo, 1973; Safferman, 1973).

5.2. Cellular Metabolism and Phage Development in the Unicellular Blue-Green Algae

The dependence of the replication of cyanophage which infect unicellular blue-green algae differs markedly from that just described. Two systems have been described in some detail: SM1, which infects *Synechococcus elongatis* and *Synechococcus* NRC-1 (MacKenzie and Haselkorn, 1972c), and AS1, which infects *Synechococcus cedrorum* and *Anacystis nidulans* (Sherman, 1976). Since none of the *Synechococcus* strains studied by Stanier (1973) was a facultative photoheterotroph, these host strains must be considered to be metabolically distinct from the *Plectonema* strains that are infected by LPP-1. The reason for these differences is still obscure, because the cells contain significant levels of the key metabolic enzymes (Stanier, 1973) and in many other regards resemble those unicellular strains which are facultative heterotrophs. The results do indicate, however, that the absence of glucose permease in the *Synechococcus* strains may be the determining factor. Whatever the ultimate reason, these unicellular algae must be considered as obligate photoautotrophs. Since the results with AS1-M and SM1 are nearly identical, we will discuss them together.

The photoassimilation of CO_2 is not shut off during the period of intracellular growth. In AS1-M infection, all of the photosynthetic parameters measured remain almost unaffected until the end of the latent period at 8 hr (Sherman, 1976, and unpublished observations). CO_2 fixation, O_2 evolution, and photosystem I and II activity are still at or near 80% of the uninfected control level by 8 hr after infection. Removal of CO_2 causes a complete loss of cyanophage development (MacKenzie and Haselkorn, 1972c; Sherman, 1976).

The dependence on photosynthesis is complete. The entire apparatus must be functioning throughout the latent period for the full phage yield to be manifest (Sherman, 1976). The presence of DCMU, darkness, or many other photosynthetic inhibitors totally prevents phage growth. Removal of light at any time during the latent period lowers the burst size appreciably; placing infected cells in the dark even as late as 7 hr causes the burst size to be depressed (Fig. 20).

Nonetheless, the infected cells do contain a limited facultative heterotrophic capability. If infected cells growing in the dark or in the presence of DCMU are supplemented with 1% glucose, phage bursts of about 10% the control level are obtained (Sherman, 1976). Therefore, under these conditions, the infected cell is acting as a facultative heterotroph. This heterotrophic capacity is not brought about by infection. We have measured the incorporation of glucose into uninfected cells and infected cells in the presence or absence of chloramphenicol (to prevent phage protein synthesis), and, in all cases, the incorporation was the same (Sherman, unpublished observations). Therefore, this growth is not due to mechanical perturbations of phage infection or to the presence of phage-directed enzyme synthesis. This may indicate that the host has a small quantity of glucose permease but that the rate of ATP generation is normally not sufficient to meet the growth requirements of the cell (Stanier, 1973).

These results show that the dependence of phage growth on photosynthesis strongly parallels the host metabolic capabilities. Facul-

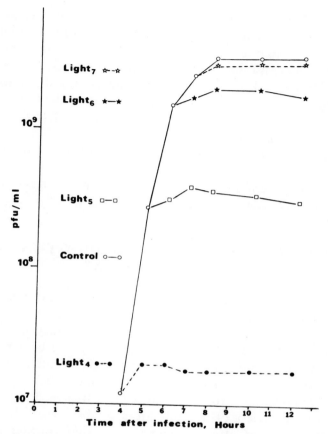

Fig. 20. Effect of darkness on the intracellular growth of AS1-M after infection of *S. cedrorum*. After infection, cells were grown in the light for the times indicated. Aliquots were then removed and grown in the dark for the remainder of the experiment. Intracellular phage were measured. From Sherman (1976).

tative heterotrophs allow phage development under conditions where photosynthesis is severely impaired, while the obligate photoautotrophs must have a functioning photosynthetic apparatus for phage to replicate. The two systems differ as to the ultimate sources of both energy and carbon. Unfortunately, cyanophages have been isolated only for a small number of different strains, and it is not known if virus infection is a typical characteristic of the blue-green algae. It is now more evident why cyanophages such as the LPP group are isolated so easily—the metabolism of the host gives them a distinct selective advantage. Nonetheless, cyanophages infecting the photoautotrophic unicellular algae have been found, and, with time, more examples should be isolated.

Finally, the study of the metabolic dependence of AS1-M infection (Sherman, 1976) has once again shown a similarity to that of bacteriophages. The sharp decrease in CO_2 fixation that commences after 8 hr resembles the cessation of O_2 uptake in T4-infected *Escherichia coli* (Mukai *et al.*, 1967). In the case of T4, O_2 uptake (used as a measure of oxidative metabolism) is seen to cease at or before the time of lysis. It is thought that this shutdown of metabolism actually triggers the irreversible destruction of the cell wall. The same series of events may take place during AS1-M lysis. It would be of interest to study the production of phage lysozyme after AS1-M infection as was done in T4 (Josslin, 1970) to see if a similar lytic mechanism operates during infection of a blue-green alga.

6. CYANOPHAGE LPP-2SPI AND LYSOGENY IN THE FILAMENTOUS BLUE-GREEN ALGAE

One of the driving forces behind cyanophage research has been the hope that the viruses could be used as vectors to transfer genetic information from cell to cell. This process of transduction would be of great significance in the genetic investigations of the blue-green algae, which are still in a quite primitive state. The poor status of genetics in the blue-greens has been due to some interrelated technical problems that have been overcome to some extent (Van Baalen, 1973). However, now that a wide variety of mutants are available, research is still hindered by the lack of effective means of sexual transfer of genetic information. Although transformation techniques are available (e.g., Shestakov and Khyen, 1970) and temporal genetic mapping has been used very effectively (Delaney *et al.*, 1976), a transduction system would be the best long-range solution for sophisticated genetic mapping. Therefore, a great deal of energy has been expended in the search for transduction and lysogeny in the blue-green algae (Padan and Shilo, 1973).

A number of reports on lysogeny of cyanophages have appeared in the past few years. Cannon *et al.* (1971) and Cannon and Shane (1972) claimed to have produced a lysogenic strain of *Plectonema* by treatment with chloramphenicol for 2 hr before infection. Addition of mitomycin C was shown to induce a hundredfold increase in virus production about 4 hr after treatment. However, immunity studies were not performed, and the purported lysogenic strain has not been used for other studies. Furthermore, the phenomenon is not reproducible; both Sherman (unpublished observations) and Haselkorn (personal communication) have been unable to repeat Cannon's findings. Until further corroborating evidence is produced, it must be concluded that these reports do not unequivocally demonstrate lysogeny in the blue-green algae.

The same must be said for all but one of the other reports of lysogenic cyanophages. Cannon has also claimed the isolation of lysogenic strains of *Nostoc* and *Anacystis nidulans* (personal communication), and a lysogenic strain of *Anabaenopsis* has been indicated (Singh *et al.*, 1969; Singh and Singh, 1972). Singh (1975) has also alleged to have demonstrated lysogeny in *Plectonema*. However, the data supporting these assertions are sketchy, and the lysogenic strains have not been used for subsequent studies. Therefore, it was difficult to decide if lysogeny did occur in blue-green algae, and, if so, how useful the phenomenon would be for genetic manipulations. This uncertainty has now been alleviated by the discovery that strain LPP-2SPI is lysogenic (Padan *et al.*, 1972). Since this phage is now being carefully characterized, we will discuss it as a model system for cyanophage lysogeny.

6.1. Lysogeny in *Plectonema boryanum* by LPP-2SPI

Cyanophage strain LPP-2SPI was originally isolated by Safferman, and was found to belong serologically and morphologically to the LPP-2 class (Safferman *et al.*, 1969*b*). However, careful study by Padan *et al.* (1972) showed that it gave rise to turbid plaques on *Plectonema*; therefore, they isolated clones from the center of such plaques and purified them. One of these clones, which they refer to as "SPI," was chosen for further study. SPI acted like a typical strain of *Plectonema,* but the culture continually liberated phage, and over 90% of a population was found to produce phage. Furthermore, the lysogenic strain was immune to superinfection of LPP-2 phage, but not to those of the LPP-1 group. Although the authors could not demonstrate induction with ultraviolet light, X rays, or mitomycin C, SPI appeared to be lysogenized by the cyanophage.

The lack of induction was overcome by a technique borrowed directly from bacteriophage research. Rimon and Oppenheim (1975) isolated a temperature-sensitive, clear plaque mutant of LPP-2SPI which lysogenized *Plectonema* in a stable fashion at 26°C. When the lysogenized cells were grown at 40°C, induction took place with the formation of progeny phage. With this mutant (SPI*cts*1) as the

essential tool, they were then able to begin the study of the biochemistry and physiology of lysogeny.

The kinetics of induction of SPIcts1 at 40°C is shown in Fig. 21.
After a lag of about an hour, intracellular phages are produced, and
lysis is completed by 3 hr after induction. However, it is obvious that
the burst of SPIcts1 after infection is about 10–100 times greater than
the burst after induction. Although the reason for this is obscure, it
may be related to the natural temperature sensitivity of SPIcts1, which
loses infectivity when incubated at 40°C (Rimon and Oppenheim,
1975). However, other possibilities such as incomplete assembly of
phage particles and metabolic interactions are being investigated.

The assembly site of LPP-2SPI after infection and after induction
has recently been investigated in the EM (Silverberg et al., 1977). In
contrast to the other LPP phages that have been studied, this strain
does not absolutely cause invagination of the photosynthetic lamellae
after infection. Furthermore, viral replication can occur either in a

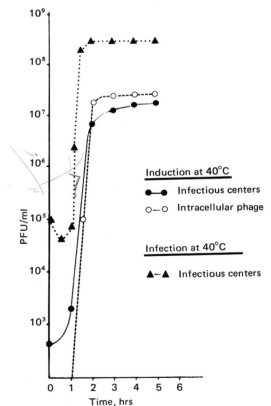

Fig. 21.   Phage production of LPP-
2SPI after infection and P1 (SPIcts)
after induction of P. boryanum at
40°C. Data taken from Rimon and
Oppenheim (1975).

virogenic stroma or in the nucleoplasm; quantitative measurements indicate that SPI has no preference for either site. Upon induction, viral replication is always found in the nucleoplasm and there is no evidence for invagination of the photosynthetic lamellae.

Rimon and Oppenheim (1975) have also studied the physiological requirements for induction utilizing many of the techniques and concepts that we discussed in Section 5. Their findings can be summarized as follows: (1) The optimal temperature for induction is 40°C. (2) A 1-hr heat treatment at 40°C is sufficient for complete induction, although temperature jump experiments modulate the kinetics of phage appearance. (3) Induction also requires at least 1 hr of light. (4) The heat and light treatments must be given concomitantly and they are required for an early step in the induction process. (5) Induction occurs in the absence of exogenous CO_2. (6) DCMU limits induction to less than 1% of the control level when added at or before the heat pulse, while CCCP completely prevents induction. (7) CO_2 photoassimilation continues unaffected after induction until the onset of lysis. The cessation of CO_2 fixation during infection and induction correlates very closely with extensive leakage of radioactively labeled material.

These results indicate that induction of a temperature-sensitive lysogen in the blue-green algae proceeds in a manner similar to that for bacteriophages, but with a major difference. The effect of heat on induction suggests that the mutated gene codes for a temperature-sensitive protein whose function is to repress prophage development. This protein is likely to resemble the product of the cI gene in λ which codes for the repressor protein (Szybalski et al., 1970). On the other hand, the necessity for a functional photosynthetic apparatus for induction is surprising. Infection by LPP-2SPI, as discussed in Section 5, can develop in complete darkness. The light-dependent step is early in the induction process, indicating that light (or the photosynthetic apparatus) may be involved in repressor inactivation, early gene expression, or even the excision of the prophage from the host chromosome. However, at the present time, no experiments have been performed that can help to differentiate among these posibilities.

The work with SPI induction has also raised questions which pertain to our discussion of phage development and host metabolism (Section 5). Like infection of the unicellular algae, replication of SPI after induction takes place in the nucleoplasm, is dependent on photosynthesis for part of the time, and does not cause the cessation of CO_2 fixation until lysis. This pattern of phage development separates SPI induction from the infection of filamentous hosts that have been studied to date. The finding of a close correlation between cessation of CO_2 fixation and cellular leakage may also be germane to our previous discussion. However, the precise nature of the correlation is intriguing. It may be that invagination of the photosynthetic membranes allows leakage to proceed more readily and that CO_2 fixation is a totally unrelated phenomenon, or it could be that leakage directly causes the cessation of CO_2 fixation. In this regard, it would be of interest to determine the effect of induction on the degradation of the host DNA to see if this parameter is involved as we had speculated earlier.

The low productivity of LPP-2SPI replication after induction and the failure to shutoff CO_2 fixation may also be related. In our earlier treatment of this subject, we indicated that it was to the benefit of the parasite to channel all of the cellular ATP and NADPH for its own use. In SPI induction, both processes occur simultaneously, possibly to the detriment of the phage. If this is indeed the case, it may be possible to increae the phage yield after induction by using "heterotrophic" cells (Ginzberg *et al.,* 1976), heating at 40°C for 1 hr in the light, and then allowing replication to proceed under heterotrophic conditions. This is obviously just one of many experiments that remain to be performed.

6.2. Temperature-Sensitive Mutants of LPP-2SPI

Probably the most successful contemporary approach to the solution of many of the problems that we have encountered is the use of conditional-lethal phage mutants (Epstein *et al.,* 1963). However, very few cyanophage mutations of this type have been obtained over the years, although host-range mutants are easily isolated. Until recently, no systematic search for temperature-sensitive (*ts*) mutations in many of the genes of a cyanophage had been carried out. This sad state of affairs has fortunately been changed with the isolation of a series of *ts* mutations of LPP-2SPI (Rimon and Oppenheim, 1974). Since that time, *ts* mutations in AS1-M have also been isolated (Sherman, unpublished observations), but the LPP-2SPI system is the only one to have been studied in detail.

Rimon and Oppenheim (1974) isolated 23 *ts* mutants which showed an efficiency of plating of less than 10^{-5} at 40°C and a drastically reduced yield of phage when grown at the nonpermissive temperature. Using standard streak tests, they found that the mutants fell into 14 distinct complementation groups. They then performed pairwise two-factor crosses with representatives from each complementation group and obtained the genetic map of LPP-2SPI shown in Fig. 22. Since no anomalies were found with the end markers, the map appears to be linear, and there is no evidence that the genome might be terminally redundant or circularly permuted.

In order to try to understand the functions of the 14 cistrons, Rimon and Oppenheim (1976) ran labeled extracts of cells infected at 26°C and 40°C on SDS-polyacrylamide gels. Using the more sensitive gel techniques now available, they obtained gel patterns which resolved 24 phage-induced protein species. Of these, 12 were phage structural proteins, while the other 12 were early and middle proteins (Table 8). The kinetics of synthesis of the phage-directed proteins after wild-type infection was studied by pulsing infected cells every 15 min with ^{35}S (Fig. 23). There is obviously a transient depression of host protein synthesis, which then resumes until about 2 hr. At about 45 min, the early proteins appear (P7, P13, P14, P19, and P20). These proteins are present at maximum quantities around 2 hr, after which they quickly decline. Synthesis of the late proteins begins at about 2 hr and then quickly reaches a maximum by 3 hr after infection.

When protein synthesis following infection with the *ts* mutants was observed, the gel patterns permitted the mutants to be divided into a number of different classes (Fig. 24). Many of the mutants showed pleiotropic effects on phage protein synthesis. The different classes of mutants, based on their influence on protein synthesis at 40°C, can be summarized as follows (Table 9): (1) Mutants L and M are unable to induce synthesis of any of the recognizable phage-induced proteins.

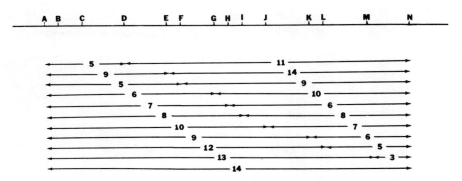

Fig. 22. Genetic map of cyanophage LPP-2SPI. The top line represents a tentative genetic map based on a set of two-factor crosses. The arrows indicate representative crosses of the two outside mutants (A and N) with various *ts* mutants and provides an indication of the scatter and additivity of the results. From Rimon and Oppenheim (1974).

TABLE 8
Molecular Weights of LPP-2SPI Coded Proteins[a]

Structural proteins		Nonstructural proteins	
Protein band No.	Molecular weight	Protein band No.	Molecular weight
P1	120,000	P7	72,000
P2	110,000	P9	54,000
P3	105,000	P11	51,000
P4	97,000	P13	47,000
P5	90,000	P14	38,000
P6	86,000	P15	36,000
P8	70,000	P17	28,000
P10	52,500(55%)	P18	26,000
P12	49,000	P19	23,000
P16	31,000	P20	21,500
P22	18,000	P21	18,000
P24	15,000(21%)	P23	17,000

[a] From Rimon and Oppenheim (1976). Phage proteins were separated by SDS gel electrophoresis and their molecular weights were determined by comparison to standard proteins. The figures in parentheses refer to the relative proportion of that protein based on cutting and weighing of peaks of Coomassie blue stained gels. Bands P10 and P24 are the two major structural proteins and represent 60–75% of the structural proteins.

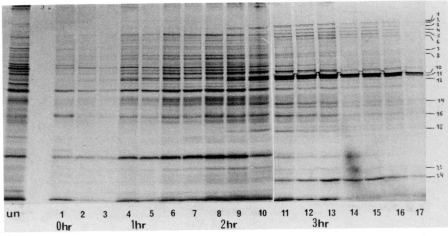

Fig. 23.   Kinetics of synthesis of LPP-2SPI-induced proteins after infection. Infected cells were pulsed for 15 min with ^{35}S and run on SDS-polyacrylamide gels. Slots 1–17 contain extracts of infected cells labeled consecutively for 15-min periods. (1) 0–15 min; (2) 15–30 min, etc. Phage-induced proteins are identified by numbers. From Rimon and Oppenheim (1976).

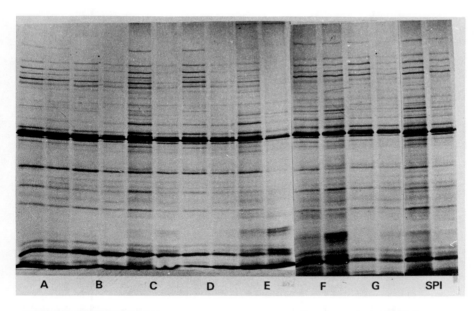

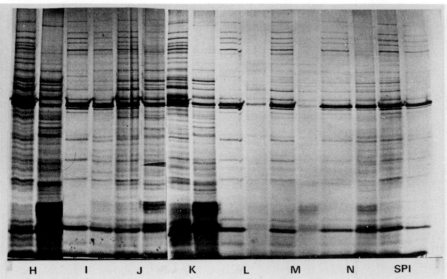

Fig. 24.   Protein synthesis by LPP-2SPI *ts* mutants. *Plectonema* cells were infected with mutants in the genes mapped in Fig. 22. After infection, the cells were labeled with ^{35}S and then divided into two parts: one was incubated for 6 hr at 26°C (top) and the other for 4 hr at 40°C (bottom). The letters stand for the corresponding cistrons. From Rimon and Oppenheim (1976).

TABLE 9
Protein Synthesis of LPP-2SPI *ts* Mutants[a]

Protein band	A	B	C	D	E	F	G	I	J	K	L	M	N	Proteins of the mature phage	Early proteins	Late proteins
P1	+	−	−	−	−	+	−	+	+	+	−	−	−	+		+
P2	+	+	+	+	−	+	+	+	+	−	−	−	−	+		+
P3	+	+	+	+	−	+	+	+	+	−	−	−	−	+		+
P4	+	−	−	−	−	+	+	+	−	+	−	−	−	+		
P5	+	+	+	+	−	+	+	+	−	+	−	−	−	+		+
P6	+	+	+	+	−	+	+	+	−	+	−	−	−	+		+
P7	+	+	+	+	+	+	+	−	+	−	−	−	+		+	
P8	+	−	−	−	−	+	+	−	−	+	−	−	−			
P10	+	+	+	+	+	+	−	+	−	+	−	−	+	+		+
P11	+	+	+	+	+	+	+	+	+	+	−	−	+	+		+
P12	+	−	−	−	−	+	−	+	−	+	−	−	−	+		+
P14	+	+	+	+	+	+	+	+	+	−	−	−	+		+	
P16	+	+	+	+	−	+	+	+	+	+	−	−	+	+		+
P17	+	+	+	+	−	+	+	+	+	+	−	−	+			+
P18	+	+	+	+	−	+	+	+	+	−	−	−	−	+		+
P22	+	+	+	+	+	+	+	+	+	−	−	+	+	+		+
P24	+	+	+	+	+	+	+	+	+	+	−	+	+	+		+

[a] A summary of the results obtained from infections with *ts* mutants at the restrictive temperature (Fig. 23 and Fig. 24) showing the presence (+) or absence (−) of virus induced proteins. From Rimon and Oppenheim (1976).

They do, however, suppress host proteins. These genes likely code for early functions. (2) Mutant K does not synthesize P14, which is one of the early proteins, and shows depressed synthesis of many of the late proteins, and would also appear to be an early gene. (3) Mutant H is defective at both temperatures and is unable to shut off host protein synthesis. Both mutants H and K fail to depress the major band around 20,000 daltons, which may represent a subunit of phycocyanin, the important accessory pigment in photosynthesis. These genes, therefore, are involved with early functions that lead to the suppression of host protein synthesis. They may be regulatory genes that are required to switch the transcriptional machinery from host synthesis to phage synthesis; (4) Mutants E, I, J, and N can synthesize the two major structural proteins (P10, P24), but are unable to synthesize a variety of other late proteins. These gene products may be necessary to activate specific late functions. (5) Mutants B, C, D, and G fail to express several structural proteins (P1, P4, P8, and P12) at the nonpermissive temperature. (6) Mutants A and F synthesize all of the recognizable proteins and may therefore be maturation-defective mutations.

This work is still in its infancy and much remains to be done. In particular, the effects of mutations on DNA synthesis, insertion, excision, and, of course, repression have not yet been studied. Nonetheless, the study that has been done indicates a number of the basic tenets of this system. The regulation of gene expression resembles that found for other temperate bacteriophages such as λ and P22 (Echols, 1972; Levine, 1972). The results presented here and in Section 4 are consistent with the following pathway for SPI gene expression: early genes $\rightarrow$ middle genes $\rightarrow$ late genes $\rightarrow$ phage morphogenesis $\rightarrow$ lysis (Rimon and Oppenheim, 1976). Furthermore, the tentative functional assignments of the 14 genes indicate a definite map clustering. It would appear that the genes at the right end of the map in Fig. 22 are involved in early functions, while those at the left end are responsible for structural proteins and the maturation process. Obviously, the positions of genes involved in DNA synthesis and repression are still unknown; however, the basic framework for further research in this area has now been built.

7. EUKARYOTIC ALGAL VIRUSES

A summation of viruses and viruslike particles (VLPs) of the eukaryotic algae is presented in Table 10. It is at once evident that a great majority of the descriptions are, in fact, only preliminary

TABLE 10
Viruses and VLPs of Eukaryotic Algae[a]

Host species[b]	Location of virogenic stroma	DNA, RNA, or VLP	Morphological characteristics	Infection and replication facts	Key references
CHAROPHYTA					
Chara corallina[1]	Cytoplasmic?	RNA	Tubular, 532 nm long, 18 nm wide, helical, with basic pitch of 2.75 nm	By artificial injection of virus into uninfected Chara cells; chlorosis and death in 10–12 days	Gibbs et al. (1975), Skotnicki et al. (1976)
CHLOROPHYCEAE					
Coleochaete scutata[2]	Intranuclear	unk (VLP)	Polyhedral, 41 nm diameter	unk	Mattox et al. (1972)
Stigeoclonium farctum[3]	Chloroplast?	unk (VLP)	Tubular? 16.3 nm in cross-section	unk	Mattox et al. (1972)
Radiofilum transversale[4]	Intranuclear	unk (VLP)	unk	unk	Mattox et al. (1972)
Uronema gigas[5]	Intranuclear	DNA	Polyhedral, 500 nm in diameter, with long tails (1000 nm)	unk	Mattox et al. (1972), Dodds et al. (1975), Dodds and Cole (1978)
Dichotomosiphon tuberosus[6]	Chloroplast	unk (VLP)	Tubular, striated, square in cross-section 30–33 nm; helical striations with 16-nm repeat pattern	unk	Moestrup and Hoffman (1973)
Oedogonium sp.[7]	Cytoplasm	unk (VLP)	Polyhedral, 240 nm in diameter	unk	Pickett-Heaps (1972)
Chlorella spp.[8]	unk	unk (VLP)	unk	unk	Tikhonenko and Zavarzina (1966)
Aulacomonas sp.[9]	Intranuclear	unk (VLP)	Hexagonal in cross-section, 200–230 nm in diameter, with tail 150–200 nm in length	unk	Swale and Belcher (1973)

Organism					Reference
Cylindrocapsa geminella[10]	Intranuclear, then cytoplasmic?	unk (VLP)	Polyhedral, 200–230 nm in diameter with a membranous coat	unk	Hoffman and Stanker (1976)
PRASINOPHYCEAE Platymonas sp.[11]	Intranuclear	unk (VLP)	Polyhedral, 51–58 nm in diameter	unk	Pearson and Norris (1974)
Micromonas pusilla[12]	Cytoplasmic	unk (VLP)	Polyhedral, 129 nm in diameter	unk	Pienaar (1976)
Pyramimonas orientalis[13]	Intranuclear	unk (VLP)	Polyhedral, two size classes: 60 and 200 nm	unk	Moestrup and Thomsen (1974)
Heteromastix sp.[14]	Intranuclear?	unk (VLP)	Polyhedral, 320 nm, with complex membranes	unk	Ott and Hommersand (unpublished material)
HAPTOPHYCEAE Chrysochromulina	unk	unk (VLP)		unk	Manton and Leadbeater (1974)
Hymenomonas carterae Strain 1[15]	Intranuclear	unk (VLP)	Polyhedral, 63 nm in diameter	unk	Pienaar (1976)
Strain 2[16]	Cytoplasmic	unk (VLP)	Polyhedral, 400 nm in diameter	unk	Cooper and Brown (unpublished material)
CHRYSOPHYCEAE Chrysophycean-like[17]	Intranuclear and cytoplasmic	unk (VLP)	Polygonal, 35 nm in diameter	unk	Tomas et al. (1973)
Dinobryon sp.[18]	Cytoplasmic	unk (VLP)	Small polyhedra, ca. 100 nm	unk	Carson and Brown (unpublished material)
Hydrurus sp.[19]	Cytoplasmic	unk (VLP)	Polyhedral, 52–59 nm	unk	Hoffman (unpublished materials)
EUSTIGMATOPHYCEAE Monodus sp.[20]	Intranuclear?	unk (VLP)	Polyhedral, 400 nm in diameter	unk	Ott, Huang, and Hommersand (unpublished materials)

(Continued)

TABLE 10. (Continued)

Host species[b]	Location of virogenic stroma	DNA, RNA, or VLP	Morphological characteristics	Infection and replication facts	Key references
CRYPTOPHYCEAE Cryptomonas sp.[21]	Intranuclear and cytoplasmic	unk (VLP)	Two particle size classes: small, polyhedral, 99 nm in diameter; large, stalked particle, 120 nm in diameter, surrounded by membrane	unk	Pienaar (1976)
DINOPHYCEAE Gyrodinium resplendens[22]	Cytoplasmic	unk (VLP)	Tubular arrays, with particles 35 nm in diameter	unk	Franca (1976)
PHAEOPHYCEAE Chorda tomentosa[23]	Cytoplasmic?	unk (VLP)	Polyhedral, 170 nm in diameter, surrounded by membrane	unk	Toth and Wilce (1972)
Pylaiella littoralis[24]	Intranuclear then cytoplasmic?	unk (VLP)	Hexagonal in section, 130–170 nm in diameter	unk	Markey (1974)
Ectocarpus fasciculatis[25]	Intranuclear	unk (VLP)	Polyhedral, 170 nm in diameter with distinct shell and tubular elements	unk	Clitheroe and Evans (1974), Baker and Evans (1973)
RHODOPHYCEAE Sirodotia tenuissima[26]	Cytoplasmic	unk (VLP)	Polygonal, 50–60 nm in diameter	unk	Lee, 1971
Porphyridium purpureum[27]	Intranuclear and cytoplasmic	unk (VLP)	Subspherical to polygonal, 40 nm in diameter	unk	Chapman and Lang (1973)

[a] Key: unk, unknown; VLP, viruslike particle.

[b] Notes:

[1] (a) Chara may not be considered as closely related to algae as to other plants such as Equisetum and Ephedra (see Bold, 1973). Certainly the Charophyta differ considerably in structure and reproduction from the major algal groups. (b) The virus is considered to be a tobamovirus distantly related to all but the curcurbit tobamovirus. (c) It represents the first instance of experimental transmission of a virus and infection of a eukaryotic algal cell. (d) Physicochemical data: $s_{20,w} = 230$ S; molecular weight 3.6×10^6; base ratio of G 24.5, A 28.0, C 20.0, U 27.5; 5% RNA; protein coat molecular 17.5×10^3.

[2] (a) Regular aggregations of spherical-hexagonal inclusions in heterochromatic regions of the nucleus. (b) From infected cultures (Indiana University Culture No. 610).

[3] Indiana University Culture No. 439.

[4] Indiana University Culture No. 1252; not illustrated but resembles the virus of *Coleochaete*.

[5] (a) This virus was first demonstrated by Mattox et al. (1972). (b) Recent unpublished observations of Dodds and Cole indicate the virus to be a double-stranded DNA virus. This is the first confirmed account of a DNA virus in a eukaryotic algal cell. The virus resembles a gigantic bacteriophage. The stages of infection, including virus attachment, have not been observed. See Fig. 25.

[6] Also interpreted as a unique form of striated microtubule in the chloroplast (see Hoffman, 1967).

[7] In developing germlings, the infection may be initiated in the wall-less state (= zoospore). From natural collections.

[8] The presence of viruses in *Chlorella* is questionable since the cultures were contaminated with bacteria. Published documentation is scarce and difficult to obtain.

[9] (a) *Aulacomonas* is a colorless flagellate, lacking a chloroplast. It resembles volvocalean algae in other aspects. (b) There is no tail plate or tail filament.

[10] (a) Viruses confined to single-cell germlings. See Fig. 26. (b) From cultured material. (c) A heat shock of 40°C, 6–24 hr, is necessary to induce VLP formation. The heat shock, applied to the zoospores, produces germlings in which 3–10% are infected with VLPs. Thus an apparently healthy culture may carry a latent virus infection.

[11] (a) This virus is similar to the herpes-type virus infecting oysters (see Farley et al., 1972). (b) Marine algae were postulated as vectors for transmission of viral infections to marine animals.

[12] From phytoplankton of San Juan Island, Washington, used enrichment cultures.

[13] (a) Two size classes of viral particles were found within the same cell. (b) From cultured material, the viral infection was minimal on the cultured population.

[14] From a marine sewage lagoon near Beaufort, North Carolina, observed only in a natural collection. See Fig. 27.

[15] Pienaar's strain was isolated from San Juan Island, Washington.

[16] The Cooper and Brown strain was the Plymouth strain No. 156, isolated by Adams in 1956. See Fig. 28.

[17] The unidentified chrysophycean alga was considered an endosymbiont in the dinophycean alga *Peridinium*.

[18] This viruslike particle may be responsible for rapid cell lysis when natural collections are brought into elevated temperatures of the laboratory, or in late spring when the water temperature increases. This may explain the mass-scale lysis of *Dinobryon* in major water supplies, giving rise to subsequent blooms of other algae. See Fig. 29.

[19] Found in cold, running streams. When brought into the laboratory, it undergoes rapid lysis. See Fig. 30.

[20] (a) Collected from eutrophic brackish waters near Beaufort, North Carolina. See Fig. 31. (b) The host could be xanthophycean or intermediate between the Xanthophyceae and Eustigmatophyceae (based on graduate research of Yuan-Shen Huang).

[21] (a) The stalked particles may have a tail. (b) The host cells with viral particles may be scenescent.

[22] (a) From clones isolated south of Lisbon, Portugal. (b) No observed modification of organelles by the VLPs.

[23] (a) VLPs found in 3-day-old spore of *Chorda*. (b) Spores with VLPs do not synthesize a cell wall.

[24] (a) VLPs found in the unilocular sporangia. (b) The VLPs resemble in size and shape those of the iridescent group. (c) Interesting dark-staining areas within the nucleus of the infected sporangium.

[25] (a) VLPs found in the plurilocular sporangia of wild material. (b) VLPs also found in developing zoospores; thus the VLPs may have been transmitted through the zoosporangium to the zoospore by subsequent cleavage.

[26] (a) The host is a culture from the Indiana University Culture Collection (No. LB 1499). (b) The virogenic stroma is prominent in the apical cell and appears to be transmitted to the daughter cells during division of the apical cell. (c) Isolated VLPs examined by negative staining, also.

[27] (a) The intranuclear particles (termed concentrosomes) may not be VLPs. (b) The host came from Indiana University Culture Collection (No. 161, *P. cruentum*).

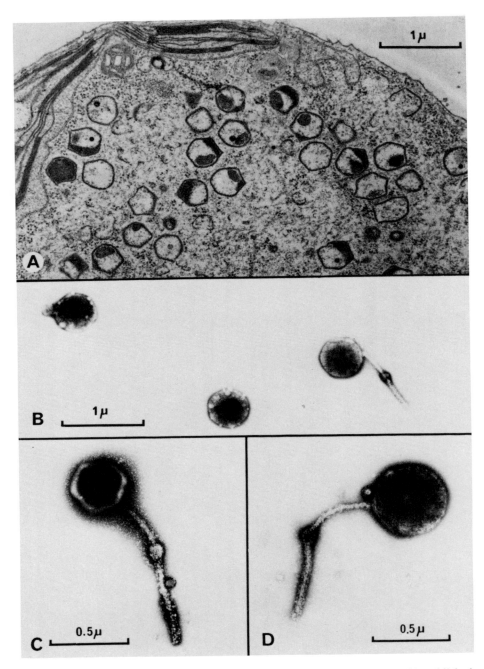

Fig. 25. The double-stranded DNA virus and its host, *Uronema gigas*. Unpublished micrographs, courtesy of J. Allan Dodds and Anabel Cole. A: Viruses *in situ*. Note (1) polyhedral particles 400 nm in diameter with five- and six-sided views, (2) densely stain-

accounts. Most of the published reports of viruslike particles in eukaryotic algal cells are single accounts and descriptive in nature. For this reason, the data are merely summarized in Table 10 and the reader is encouraged to consult the appropriate key references for more details (Figs. 25–32). In spite of the paucity of information on eukaryotic algal viruses (for the two reviews on this topic, see Brown, 1972, and Lemke, 1976), it is almost certain that many such viruses do exist. Only three reports will be examined in detail in this chapter since they have, in our opinion, added substantially to our knowledge of algal viruses.

A major breakthrough in the understanding of algal viruses has come from study of the virus which infects the green alga *Uronema gigas*. In 1972, Mattox and co-workers first described polyhedral particles in association with the degrading nucleus of *Uronema*. Subsequently, Dodds *et al.* (1975) reported that VLPs could be liberated from the cells of *U. gigas* into the culture medium. Although the VLPs were released in extremely low concentrations, it was possible to concentrate and partially purify the particles by differential centrifugation.

Dodds and Cole (1977, and personal communication) have succeeded in purifying the VLPs so that a compositional analysis could be made. Dodds and Cole report that the particles infecting *U. gigas* contain double-stranded DNA, which is the first conclusive evidence of DNA viruses among the eukaryotic algae. Because of the importance of this discovery, we shall briefly describe some of the properties of the *Uronema* virus. The virus is a large polyhedron, with the capsid appearing to be an icosahedron approximately 400 nm in diameter at pH 5.0 and 500 nm at pH 7.0. The particle is bounded by a membranelike outer coat and the interior of the virus is characterized by a densely packed, highly electron-dense material which is trypsin digestible (Fig. 25). Negative stain preparations of the purified virus indicate that approximately 10% of the particles exhibit a prominent tail about 1 μm in length. Tails also have been observed associated with the particles *in situ* (on the basis of ultrathin sections of infected cells of *Uronema*).

The following physicochemical properties of the *Uronema* virus have been determined by Dodds and Cole (1978) and Cole (personal

---

ing material adjacent to an edge but not all edges, (3) less densely staining material occupying a more central position in the particle, and (4) the particles within a matrix devoid of ribosomes and rich in fibrillar material. B: Partially purified particles in 2% PTA, pH 7.0. Some particles have a 1-μm-long tail which has a central swelling. Head diameter is about 500 nm and its outline is not angular. C: Tailed particle in 2% uranyl acetate, pH 4.4. Head is about 400 nm in diameter and is angular. D: Tailed particle in 2% PTA, pH 7.0. Head is swollen. The small body beside the head in D, and beside the tail in C, is a product of Beijerinck's medium.

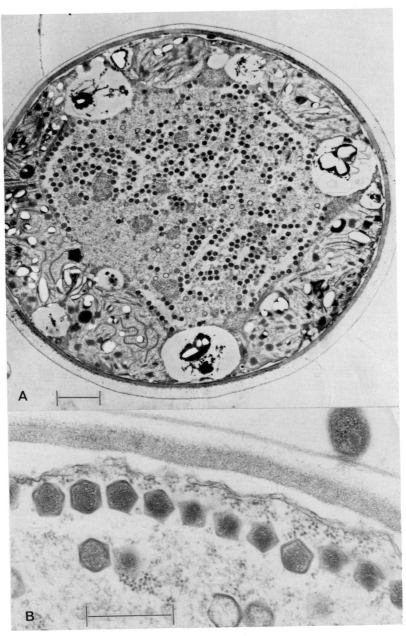

Fig. 26. VLPs in the host *Cylindrocapsa geminella*. A: Near-median, longitudinal section through a single-celled germling. Note the virogenic stroma, probably in the region formerly occupied by the nucleus. Bar equals 1 μm. Unpublished micrograph courtesy of Larry Hoffman and Larry Stanker. B: Row of VLPs in a cell early in the infection stage and before the development of distinct ribosomal aggregates. Note the distinct membrane around the polyhedra. Bar equals 0.5 μm. From Hoffman and Stanker (1976).

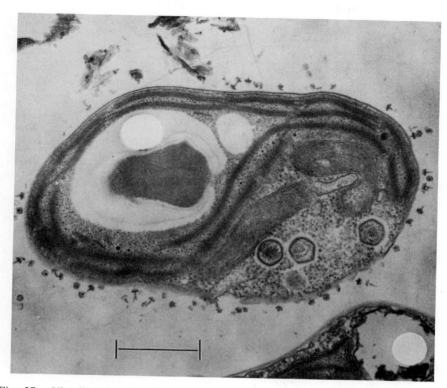

Fig. 27. Viruslike particles in the prasinophycean alga *Heteromastix* sp. Note the prominent scales which decorate the surface of the alga. Three viruslike particles are shown in the cytoplasm. Bar equals 1 μm. Unpublished micrographs courtesy of Donald Ott and Max Hommersand.

communication). The virus particle density is 1.30 g/ml as determined by sucrose density gradient analysis. The virus population is quite heterogeneous in density and cesium chloride degrades the particles. The sedimentation coefficient on the basis of an average of three runs is 6200 $\pm$ 200 S. The virus particles scatter light intensely and only about 20% of the absorbance at 260 nm is due to true absorbance of protein and nucleic acid. The $A_{260}/A_{280} = 1.20$ when corrected for light scattering. The *Uronema* virus particle composition is of great interest. The nucleic acid content, as mentioned above, is double-stranded DNA. This was determined by the following tests: a positive diphenylamine test, enzyme digestions of fixed and sectioned virus, acridine orange staining, buoyant density, and electron microscopy using the Kleinschmidt technique. The buoyant density in cesium chloride of purified DNA was 1.732 g/ml (based on an average of two runs), which

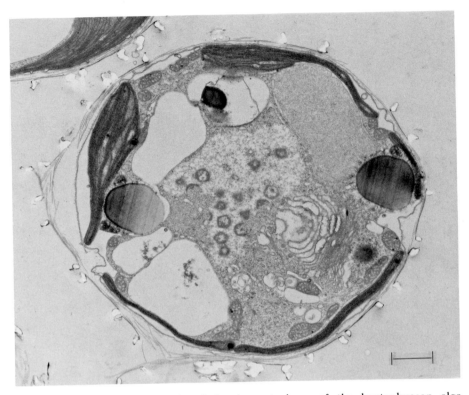

Fig. 28. Viruslike particles found in the cytoplasm of the haptophycean alga *Hymenomonas carterae*. Note the polyhedral particles below the nucleus and adjacent to the Golgi apparatus. Specialized scales known as coccoliths cover the cell surface. Bar equals 1 μm. Unpublished micrograph courtesy of Kay Cooper and R. Malcolm Brown, Jr.

is equivalent to a G+C content of 73.5%. Six viral proteins have so far been detected by SDS-polyacrylamide gel electrophoresis. They have the following molecular weights: 64,900, 54,600, 46,000, 41,500, 35,500, and 26,600 daltons.

A few comments should be made regarding the biology of this virus (Dodds and Cole, 1978, and personal communication). The type cultures of *U. gigas* from both Indiana University and Cambridge University Culture Collections are infected with the virus. Subcultures established from single zoospores always release some viruses, although at varying rates. To date, Dodds and Cole have failed to produce a virus-free culture by heat shock or ultraviolet irradiation; nevertheless, a vast majority of cells examined by electron microscopy are free of virus particles or any pathological appearance. The most susceptible stage in the algal life cycle for virus release seems to be the young germling at the two- to four-celled stage, shortly after zoospore attachment to a surface. From this first detailed account of a eukaryotic algal virus containing DNA, it is obvious that we are on the verge of an exciting and productive era of discovery with eukaryotic algal viruses.

A brief discussion on the status of our knowledge of the mode of the infection process deserves attention. The cell walls of most eukaryotic algae are many times thicker than those of the blue-green algae. Thus the modes of infection are not understood at present. However, it may be possible for the 1-μm tails of the *Uronema* virus to penetrate the host cell wall. The proposal for the necessity of a wall-less phase in the life history of the host (e.g., zoospore, gamete) certainly is germane to the infection question (Pickett-Heaps, 1972). Far more intriguing is the possibility that many if not most eukaryotic algal cells may be able to carry and transmit the virion from cell generation to cell generation without intervening lysis and cell degradation. Could these large VLPs behave as plasmids? What nutritional and environmental parameters could trigger the induction of the virulent phase? Interesting responses

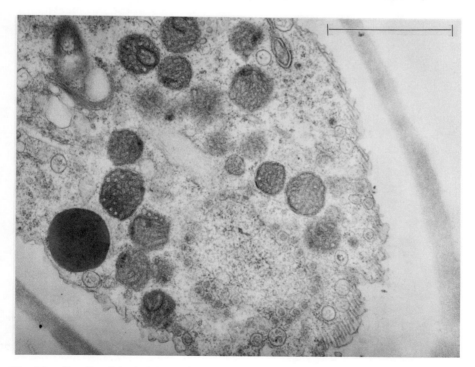

Fig. 29.   Small polyhedral VLPs in the chrysophycean alga *Dinobryon* sp. Bar equals 1 μm. Unpublished micrograph courtesy of Johnny L. Carson.

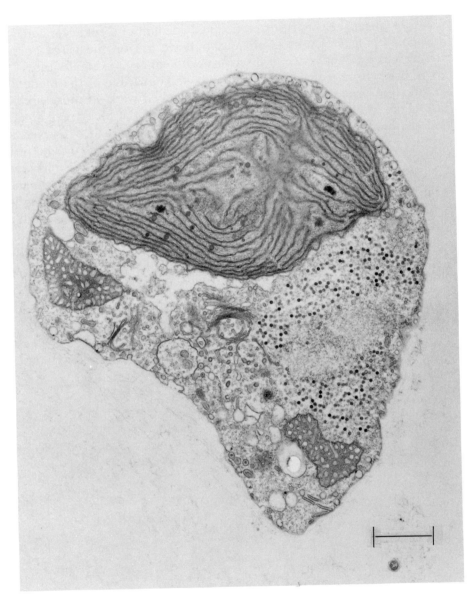

Fig. 30. Viruslike particles in a young zoospore of the chrysophycean alga *Hydruras*. Bar equals 1 μm. Unpublished micrograph courtesy of Larry Hoffman.

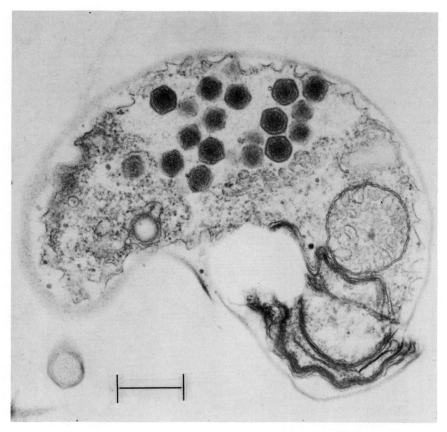

Fig. 31.   Large polyhedral VLPs in the eustigmatophycean alga *Monodus* sp. Bar equals 1 μm. Unpublished micrograph courtesy of Yaun Shen Huang and Max Hommersand.

to the latter question have come from studies of Hoffman and Stanker (1976), who isolated VLPs from heat-shocked zoospores of the green alga *Cylindrocapsa* (Fig. 26). The apparent necessity of a heat shock of 40°C for 6–24 hr is intriguing and in need of further analysis and study. This study also demonstrates the importance of the necessity to obtain purified virus from *Cylindrocapsa* and to establish *bona fide* virus-free strains for testing. (For EMs of several of these viruses see Fig. 27–31).

Turning to yet another system, details are beginning to emerge from the study of an interesting RNA virus which attacks *Chara corallina* (Gibbs *et al.*, 1975; Skotnicki *et al.*, 1976) (Fig. 32). This virus has been isolated from infected *Chara* populations, purified, then experimentally injected into virus-free giant internodal cells. Within 10–12

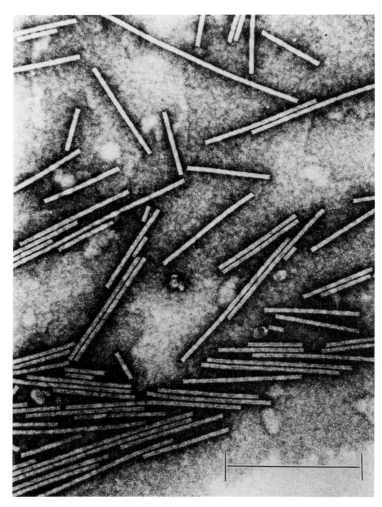

Fig. 32.   Negative stain preparation of an equal mixture of purified tobacco mosaic virus (short rods) and *Chara corallina* virus (long rods). Bar equals 500 nm. Courtesy of Adrian Gibbs.

days after infection, these cells undergo chlorosis and death. Cells injected with UV-irradiated viral sap remain green and appear healthy. The physicochemical data for the *Chara* virus are summarized in Table 10. Gibbs *et al.* (1975) and Skotnicki *et al.* (1976) consider the *Chara* virus to have so many properties in common with the tobamoviruses that they believe that these viruses should be grouped together. If the *Chara* virus is, in fact, a tobamovirus, it is the first case of such a virus found outside of the angiosperm plants. This point raises an important question regarding the taxonomic status of *Chara*. Bold (1973) and others believe that *Chara* and *Nitella* are so different on the basis of structure and reproduction that these representatives should be grouped into a separate division, the Charophyta. While it is not the purpose of this chapter to summarize and compare the taxonomic status, it should be reconsidered if *Chara* is sufficiently algalike to warrant continued classification of it as an alga. The fact that no RNA virus remotely resembling the *Chara* virus has yet been discovered among the designated algal classes (Bold, 1973) seems to substantiate this point. It seems to us that continued studies of viruses among the Charophyta might help to better establish the relationship of the Charophyta to primitive land plants (mosses, liverworts, ferns, etc.).

We should not forget brief mention of those algal classes in which VLPs have yet to be discovered at the time of this writing. These include only the Euglenophyceae (*Euglena*, euglenoid flagellates), the Bacillariophyceae (diatoms), the Chloromonadaophyceae (rare flagellates with some affinities with the Chrysophyceae), and possibly the Xanthophyceae (the yellow-green algae). In regard to the last class, we cannot yet rule out the possibility of a representative having a VLP infection. *Monodus* sp. (Fig. 31) (Hommersand and Huang, personal communication) may occupy a position intermediate between the Xanthophyceae and the Eustigmatophyceae. This merely points out the present confusion in many areas of eukaryotic algal taxonomy. The wide distribution of viruses among the classes of algae lends hope that future knowledge of these viruses and virus–host relationships will provide an indispensable tool for solving phylogenetic and taxonomic problems in the algae.

8. CONCLUSIONS

This status report on algal viruses has been mostly concerned with cyanophage replication in blue-green algae. Although our knowledge of viral infection in the eukaryotic algae is limited, the existence of such viruses is now well documented and a breakthrough on their propagation can be looked for in the near future. On the other hand, the cyanophage–alga system has been studied in some detail. In many respects, cyanophage replication parallels events that occur in bacterial infection, although even in this system much work remains to be done.

The most intriguing aspect of cyanophage replication is the different pattern that arises after infection of unicellular and filamentous hosts. Most importantly, this dichotomy is dependent on the nature of the host's metabolism. Cyanophages which infect cells capable of heterotrophic growth require only a ready supply of ATP, which can be supplied by either photosynthetic or oxidative phosphorylation. The building blocks for viral protein and DNA are scavenged from degradation products of the host's own macromolecules. Cyanophage infection of obligate photoautotrophs, on the other hand, requires a functioning photosynthetic machinery to supply both ATP and carbon for viral macromolecular synthesis.

A number of important questions still remain unanswered. First, what is the reason for the invagination of photosynthetic lamellae during LPP-1 infection? This invagination occurs at about the time of host DNA degradation, cessation of CO_2 assimilation, and leakage of cellular materials. However, the membranes are still highly functional despite the physical disruption. Some but not all of these events occur during infection of *S. cedrorum* by AS1-M and during induction of *Plectonema* lysogenic for LPP-2SPI. Therefore, a correlation of the four phenomena would be an important synthesis of early events in cyanophage replication. Since the photosynthetic lamellae remain functional during infection, it may be of interest to see if viral infection can be used in the isolation of photosynthetic membranes for studies in photosynthesis.

Second, what is the nature of the control mechanisms involved in the shutoff of host synthesis and the turn-on of phage-directed synthesis? We have alluded numerous times to the similarity of cyanophage replication to that of bacteriophages. Nonetheless, this has been studied only in a rather superficial way for LPP-1, LPP-2SPI, and AS1-M infection. In particular, research directly aimed at understanding the transcriptional and translation controls involved in phage replication has not yet been done. The isolation of *ts* mutants in LPP-2SPI should greatly aid in this regard.

The general nature of lysogeny in the blue-green algae is also an open question. Although claims of lysogeny have been made in a number of diverse species, the only *bona fide* example is the LPP-2SPI case. Here, induction of a temperature-sensitive mutant is triggered by a heat pulse which allows the whole lysogenic process to be carefully studied. However, the universality of the lysogenic phenomenon remains to be discovered. Furthermore, the use of such lysogenic strains in transduction studies has hardly begun. The future importance of the cyanophages in particular, and of blue-green algal genetics in general, may be dependent on an understanding of these questions.

There is one last similarity between cyanophages and bacteriophages. One of the driving factors behind the early research in bacteriophages was the hope that they could be used to control bacterial infections (Stent, 1963). In a like vein, the isolation of cyanophages was initially accomplished because of a desire to control algal blooms (Safferman, 1973; Padan and Shilo, 1973). The use of cyanophages to manipulate algal blooms remains an open and still promising field of endeavor, but much of the more recent work has revolved around the molecular biology and biochemistry of cyanophage replication. It may be that the ultimate importance of the cyanophages will, as with the bacteriophages, lie in the basic fields of genetics and molecular biology, and not in applied fields such as ecology.

Turning to the eukaryotic algae, it is obvious that our knowledge of viral infection of these organisms is miniscule at best. We have seen that most of the major algal classes have organisms which are infected by VLPs. Confirmation and characterization of a true algal virus infection have come only from the recent breakthroughs with the *Chara* virus (Skotnicki *et al.*, 1976) and the *Uronema* virus (Dodds and Cole, 1978, and personal communication). The *Chara* virus may not even be considered a true algal virus, based on the differences of the host (Bold, 1973). On the other hand, the discovery of a large double-stranded DNA virus in *Uronema* is truly indicative of the existence of viruses among the eukaryotic algae. This discovery should spur the successful search for the characterization of the many viruslike particles among the eukaryotic algae.

The necessity of a temperature shock to induce VLP formation in the green alga *Cylindrocapsa* (Hoffman and Stanker, 1976) is interesting and reminiscent of the lysogenic pathway in some of the blue-green algae (Rimon and Oppenheim, 1974, 1976). This points out one of the major problems facing investigators studying eukaryotic algal viruses: that the host may have the capacity to carry latent viral genomes which can spontaneously express virulence on stimulus from a number of nutritional and environmental parameters. Future research should be directed toward understanding the infection cycles of eukaryotic algal viruses. In this connection, we cannot conclude this chapter without a statement about unexplained and rapid populational changes of algae in water supplies and aquatic habitats. If viruses indeed do play an important role in mediating these populational fluxes of eukaryotic as well as prokaryotic algae, this field will then become increasingly important to researchers of aquatic ecology and public health.

To end on a philosophical note, it is not inconceivable to us that, as the discovery of eukaryotic and prokaryotic algal viruses proceeds, the accumulation and characterization of host–virus specificity data will become indispensable as a tool for understanding the phylogeny and systematics within this vast assemblage of diverse photosynthetic organisms (Klein and Cronquist, 1967). Ultimately, a better understanding of algal viruses may help us to more fully comprehend the evolutionary history of the land plants as well as the evolution of their numerous and often destructive viral diseases.

ACKNOWLEDGMENTS

The authors are indebted to those colleagues who have supplied unpublished information for this review or who provided figures and micrographs from previously published papers. We also would like to express our appreciation to Academic Press and The American Society for Microbiology for permission to use the published material. Although thanks are due many students for their efforts throughout the years, we would especially like to single out Ms. Rosemary Crane for her excellent work in the preparation of the manuscript.

The investigations of one of the authors (L. A. S.) were supported by grants from the University of Missouri Research Council and a grant from the USPHS (GM 21827).

9. REFERENCES

Adams, M. H., 1959, *Bacteriophages,* Interscience, New York.

Adolph, K. W., and Haselkorn, R., 1971, Isolation and characterization of a virus infecting the blue-green alga *Nostoc muscorum, Virology,* **46**:200.

Adolph, K. W., and Haselkorn, R., 1972a, Photosynthesis and the development of blue-green algal virus N-1, *Virology* **47**:370.

Adolph, K. W., and Haselkorn, R., 1972b, Comparison of the structures of blue-green algal viruses LPP-1M and LPP-2 and bacteriophage T7, *Virology* **47**:701.

Adolph, K. W., and Haselkorn, R., 1973a, Blue-green algal virus N-1: Physical properties and disassembly into structural parts, *Virology* **53**:427.

Adolph, K. W., and Haselkorn, R., 1973b, Isolation and characterization of a virus infecting a blue-green alga of the genus *Synechococcus, Virology* **54**:230.

Allen, M. M., 1968a, Simple conditions for growth of unicellular blue-green algae on plates, *J. Phycol.* **4**:1.

Allen, M. M., 1968b, Photosynthetic membrane system in *Anacystis nidulans, J. Bacteriol.* **96**:836.

Allen, M. M., 1968c, Ultrastructure of the cell wall and division of unicellular blue-green algae, *J. Bacteriol.* **96**:842.

Avron, M., and Neumann, J., 1968, Photophosphorylation in chloroplasts, *Annu. Rev. Plant Physiol.* **19**:137.

Baker, J. R. J., and Evans, L. V., 1973, The ship fouling alga *Ectocarpus.* I. Ultrastructure and cytochemistry of plurilocular reproductive stages, *Protoplasma* **77**:1.

Biggins, J., 1969, Respiration in blue-green algae, *J. Bacteriol.* **99**:570.

Bijlenga, R. K. L., Scraba, D., and Kellenberger, E., 1973, Studies on the morphopoiesis of the head of T-even phage. IX. τ-particles: Their morphology, kinetics of appearance and possible precursor function, *Virology* **56**:250.

Bijlenga, R. K. L., Aebi, U., and Kellenberger, E., 1976, Properties and structure of a gene 24-controlled T4 giant phage, *J. Mol. Biol.* **103**:469.

Bold, H. C., 1973, *Morphology of Plants*, 3rd ed., Harper and Row, New York.

Bradley, D. E., 1967, Ultrastructure of bacteriophages and bacteriocins, *Bacteriol. Rev.* **31**:230.

Branton, D., and Klug, A., 1975, Capsid geometry of bacteriophage T2: A freeze-etching study, *J. Mol. Biol.* **92**:559.

Brown, R. M., Jr., 1972, Algal viruses, *Adv. Virus. Res.* **17**:243.

Brown, R. M., Jr., Smith, K. M., and Walne, P. L., 1966, Replication cycle of the blue-green algal virus LPP-1, *Nature (London)* **212**:729.

Cannon, R. E., and Shane, M. S., 1972, The effect of antibiotic stress on protein synthesis in the establishment of lysogeny of *Plectonema boryanum, Virology* **49**:130.

Cannon, R. E., Shane, M. S., and Bush, V. N., 1971, Lysogeny of a blue-green alga, *Plectonema boryanum, Virology* **45**:149.

Carr, N. G., 1973, Metabolic control and autotrophic physiology, in: *The Biology of Blue-Green Algae* (N. G. Carr and B. A. Whitton, eds.), pp. 39–65, University of California Press, Berkeley.

Carr, N. G., and Whitton, B. A. (eds.), 1973, *The Biology of Blue-Green Algae,* University of California Press, Berkeley.

Caspar, D. L. D., and Klug A., 1962, Physical principles in the construction of regular viruses, *Cold Spring Harbor Symp. Quant. Biol.* **27**:1.

Chao, L., and Bowen, C. C., 1971, Purification and properties of glycogen isolated from a blue-green alga, *Nostoc muscorum, J. Bacteriol.* **105**:331.

Chapman, R. L., and Lang, N., 1973, Virus-like particles and nuclear inclusions in the red alga *Porphyridium purpureum* (Bory) Drew et Ross, *J. Phycol.* **9**:117.

Cheung, W. Y., and Gibbs, M., 1966, Dark and photometabolism of sugars by a blue-green alga: *Tolypothrix tenuis, Plant Physiol.* **41**:731.

Clithero, S. B., and Evans, L. V., 1974, Viruslike particles in the brown alga *Ectocarpus, J. Ultrastruc. Res.* **49**:211.

Cowie, D. B., and Prager, L., 1969, Cyanophyta and their viruses, in: *Carnegie Institution of Washington Year Book, 1968–69,* pp. 391–397, Carnegie Institution, Washington, D. C.

Daft, M. J., Begg, J., and Stewart, W. D. P., 1970, A virus of blue-green algae from fresh water habitats in Scotland, *New Phytol.* **69**:1029.

Delaney, S. F., Herdman, M., and Carr, N. G., 1976, Genetics of blue-green algae, in: *The Genetics of Algae* (R. A. Lewin, ed.), pp. 7–28, University of California Press, Berkeley.

Dodds, J. A., and Cole, A., 1978, Initial accounts of a DNA virus in *Uronema gigas, Virology,* (submitted).

Dodds, J. A., Stein, J. R., and Haber, S., 1975, Characterization of virus-like particles in a filamentous green alga, *Proc. Can. Phytopathol. Soc.* **42**:26.

Drews, G., 1973, Fine structure and chemical composition of the cell envelopes, in: *The Biology of Blue-Green Algae* (N. G. Carr and B. A. Whitton, eds.), pp. 99–116, University of California Press, Berkeley.

Duane, W. E., Sr., Hohl, M. C., and Krogmann, D. W., 1965, Photophosphorylation activity in cell-free preparations of a blue-green alga, *Biochim. Biophys. Acta* **109**:108.

Echlin, P., and Morris, I., 1965, The relationship between blue-green algae and bacteria, *Biol. Rev. (Cambridge)* **40**:143.

Echols, H., 1972, Developmental pathways for the temperate phage: Lysis vs. lysogeny, *Annu. Rev. Genet.* **6**:157.

Edelman, M., Swinton, D., Schiff, J. D., Epstein, H. T., and Zeldin, B., 1967, Deoxyribonucleic acid of the blue-green algae (Cyanophyta), *Bacteriol. Rev.* **31**:315.

Epstein, R. H., Bolle, A., Steinberg, C. M., Kellenberger, E., Boy de la Tour, E., Chevalley, R., Edgar, R. S., Susman, M., Denhardt, G. H., and Lielausis, A., 1963, Physiological studies of conditional lethal mutants of bacteriophage T4D, *Cold Spring Harbor Symp. Quant. Biol.* **28**:375.

Farley, A. C., Banfield, W. G., Kasmic, G., Jr., and Foster, W. S., 1972, Oyster herpes-type virus, *Science* **78**:759.

Fay, P., 1965, Heterotrophy and nitrogen fixation in *Chlorogloca futschli, J. Gen. Microbiol.* **39**:11.

Fox, J. A., Booth, S. J., and Martin, E. L., 1976, Cyanophage SM-2: A new blue-green algal virus, *Virology* **73**:557.

Franca, S., 1976, On the presence of virus-like particles in the dinoflagellate *Gyrodinium resplendens* (Hulburt), *Protistologica* **12**:425.

Gibbs, A., Skotnicki, A. H., Gardiner, J. E., and Walker, E. S., 1975, A tobamovirus of a green alga, *Virology* **64**:571.

Ginzberg, D., 1973, The interaction between cyanophage LPP-1G and the photosynthetic system of the blue-green alga *Plectonema boryanum*, Ph.D. thesis, Hebrew University, Jerusalem.

Ginzberg, D., Padan, E., and Shilo, M., 1968, Effect of cyanophage infection on CO_2 photoassimilation in *Plectonema boryanum, J. Virol.* **2**:695.

Ginzberg, D., Padan, E., and Shilo, M., 1976, Metabolic aspects of LPP cyanophage replication in the Cyanobacterium *Plectonema boryanum, Biochim. Biophys. Acta* **423**:440.

Goldstein, D. A., and Bendet, I. J., 1967, Physical properties of the DNA from the BGAV LPP-1, *Virology* **32**:614.

Goldstein, D. A., and Bendet, I. J., Lauffer, M. A., and Smith, K. M., 1967, Some biological and physicochemical properties of blue-green algal virus LPP-1, *Virology* **32**:601.

Good, N. E., and Izawa, S., 1964, Selective inhibitors of photosynthesis, in: *Biochemical Dimensions of Photosynthesis* (D. W. Krogmann and W. H. Powers, eds.), pp. 62–73, Wayne State University Press, Detroit.

Goryushin, V. A., and Chaplinskaya, S. M., 1968, *The Discovery of Viruses Lysing Blue-Green Algae in the Dneprovsk Reservoirs*, Tsvetenie Vody, Dumka, Kiev.

Govindjee, 1975, *Bioenergetics of Photosynthesis*, Academic Press, New York.

Granhall, V., and von Hofsten, A., 1969, The ultrastructure of a cyanophage attack on *Anabaena variabilis, Physiol. Plant.* **22**:713.

Heytler, P. G., and Prichard, W. W., 1962, A new class of uncoupling agents—carboryl cyanide phenylhydrazones, *Biochem. Biophys. Res. Commun.* **7**:272.

Hoare, D. S., Hoare, S. L., and Smith, A. J., 1969, Assimilation of organic compounds by blue-green algae and photosynthetic bacteria, in: *Progress in Photosynthetic Research* (H. Metzner, ed.), pp. 1540–1573.

Hoare, D. S., Ingram, L. O., Thurston, E. L., and Walkup, R., 1971, Dark heterotrophic growth of an endophytic blue-green alga, *Arch. Mikrobiol.* **78**:310.

Hoffman, L., 1967, Observations on the fine structure of *Oedogonium*. III. Microtubular elements in the chloroplasts of *Oe. cardiacum, J. Phycol.* **3**:212.

Hoffman, L. R., and Stanker, L. H., 1976, Virus-like particles in the green alga *Cylindrocapsa, Can. J. Bot.* **54**:2827.

Holm-Hansen, O., 1968, Ecology, physiology and biochemistry of blue-green algae, *Annu. Rev. Microbiol.* **22**:47.

Horvitz, H. R., 1973, Polypeptide bound to the host RNA polymerase is specified by T4 control gene 33, *Nature (London) New Biol.* **244**:137.

Jensen, T. E., and Bowen, C. C., 1970, Cytology of blue-green algae. II. Unusual inclusion in the cytoplasm, *Cytologia* **35**:132.

Johnson, E. J., 1966, Occurrence of the adenosine monophosphate inhibition of carbon dioxide fixation in photosynthetic and chemosynthetic autotrophs, *Arch. Biochem. Biophys.* **114**:178.

Jones, L. W., and Meyers, J., 1964, Enhancement in the blue-green alga, *Anacystis nidulans, Plant Physiol.* **39**:938.

Josslin, R., 1970, The lysis mechanism of phage T4: Mutants affecting lysis, *Virology* **40**:719.

Kaiser, D., Syvanen, M., and Masuda, T., 1975, DNA Packaging steps in bacteriophage lambda head assembly, *J. Mol. Biol.* **91**:175.

Kellenberger, E., and Edgar, R. S., 1971, Structure and assembly of phage particles, in: *The Bacteriophage Lambda* (A. D. Hershey, ed.), pp. 271–295, Cold Spring Harbor Laboratory, Cold Spring Harbor, N.Y.

Kellenberger, E., Eiserling, F. A., and Boy de la Tour, E., 1968, Studies on the morphopoiesis of the head of phage T-even. III. The cores of head-related structures, *J. Ultrastruct. Res.* **21**:335.

Kenyon, C. N., Rippka, R., and Stanier, R. Y., 1972, Fatty acid composition and physiological properties of some filamentous blue-green algae, *Arch. Mikrobiol.* **83**:216.

Khoja, T., and Whitton, B. A., 1971, Heterotrophic growth of blue-green algae, *Arch. Mikrobiol.* **79**:280.

Kitajima, E. W., and Lauritis, J. A., 1969, Plant virions in plasmodesmata, *Virology* **37**:681.

Klein, R. M., and Cronquist, A., 1967, A consideration of the evolutionary and taxonomic significance of some biochemical, micromorphological, and physiological characters in the thallophytes, *Q. Rev. Biol.* **42**:105.

Laemmli, U. K., 1970, Cleavage of structural proteins during the assembly of the head of bacteriophage T4, *Nature (London)* **227**:680.

Laemmli, U. K., and Favre, M., 1973, Maturation of the head of bacteriophage T4. I. DNA packaging events, *J. Mol. Biol.* **80**:575.

Laemmli, U. K., Teaff, N., and D'Ambrosia, J., 1974*a*, Maturation of the head of bacteriophage T4. III. DNA packaging into preformed heads, *J. Mol. Biol.* **88**:749.

Laemmli, U. K., Paulson, J. R., and Hitchins, V., 1974*b*, Maturation of the head of bacteriophage T4. V. A possible DNA packaging mechanism: *In Vitro* cleavage of the head proteins and the structure of the core of the polyhead, *J. Supramol. Struct.* **2**:276.

Lang, N. J., 1968, The fine structure of blue-green algae, *Annu. Rev. Microbiol.* **22**:15.

Lang, N. J., and Whitton, B. A., 1973, Arrangement and structure of thylakoids, in:

The Biology of the Blue-Green Algae (N. G. Carr and B. A. Whitton, eds.), pp. 66–79, University of California Press, Berkeley.

Leach, C. K., and Carr, N. G., 1969, Oxidative phosphorylation in an extract of *Anabaena variabilis, Biochem. J.* **112**:125.

Leach, C. K., and Carr, N. G., 1970, Electron transport and oxidative phosphorylation in the blue-green alga *Anabaena variabilis, J. Gen. Microbiol.* **64**:55.

Lee, R. E., 1971, Systemic viral material in the cells of the freshwater red alga *Sirodotia tenuissima* (Holden) Skuja, *J. Cell Sci.* **8**:6231.

Lemke, P. A., 1976, Viruses of eukaryotic microorganisms, *Annu. Rev. Microbiol.* **30**:105.

Levine, M., 1972, Replication and lysogeny with phage P22 in *Salmonella typhimurium, Curr. Top. Microbiol. Immunol.* **58**:135.

Luftig, R., 1967, An accurate measurement of the catalase crystal period and its use as an internal marker for electron microscopy, *J. Ultrastruct. Research* **20**:91.

Luftig, R., 1968, Further studies on the dimensions of viral and protein structures using the catalase crystal internal marker technique, *J. Ultrastruct. Res.* **23**:178.

Luftig, R., and Haselkorn, R., 1967, Morphology of a virus of blue-green algae and properties of its deoxyribonucleic acid, *J. Virol.* **1**:344.

Luftig, R., and Haselkorn, R., 1968a, Studies on the structure of blue-green algae virus LPP-1, *Virology* **34**:664.

Luftig, R., and Haselkorn, R., 1968b, Comparisons of blue-green algae virus LPP-1 and the morphologically related virus GIII and coliphage T7, *Virology* **34**:675.

Luria, S. E., and Darnell, J. E., Jr., 1967, *General Virology*, Wiley, New York.

MacKenzie, J. J., and Haselkorn, R., 1972a, Physical properties of blue-green algal virus SM-1 and its DNA, *Virology* **49**:497.

MacKenzie, J. J., and Haselkorn, R., 1972b, An electron microscope study of infection by the blue-green algal virus SM-1, *Virology* **49**:505.

MacKenzie, J. J., and Haselkorn, R., 1972c, Photosynthesis and the development of blue-green algal virus SM-1, *Virology* **49**:517.

Manton, I., and Leadbeater, B. S. C., 1974, Fine structural observations of six species of *Chrysochromulina* from wild Danish marine nanoplankton, including a description of *C. campanulifera* sp. nov. and a preliminary summary of the nanoplankton as a whole, *Biol. Skr. Dan. Vid. Selsk.* **20**:1.

Markey, D. R., 1974, A possible virus infection in the brown alga *Pylaiella littoralis, Protoplasma* **80**:223.

Mattox, K. R., Stewart, K. D., and Floyd, G. L., 1972, Probable virus infections in four genera of green algae, *Can. J. Microbiol.* **18**:1620.

Mattson, T., Richardson, J., and Gooden, D., 1974, Mutant of bacteriophage T4D affecting expression of many early genes, *Nature (London)* **250**:48.

Mendzhul, M. I., Bobrovnik, S. A., Lysenko, T. G., and Shved, A. D., 1975a, Effect of certain physicochemical factors on infecting of cyanophages A-1, LPP-1A and LPP-3, *Mikrobiologia* **37**:73.

Mendzhul, M. I., Lysenko, T. G., Bobrovnik, S. A., Sagun, T. S., and Nesterova, N. V., 1975b, Detection of blue-green algae virus with octahedral type of symmetry, *Mikrobiologia* **37**:713.

Moestrup, Ø., and Hoffman, L. R., 1973, Ultrastructure of the green alga *Dicotomosiphon tuberosus* with special reference to the occurrence of striated tubules in the chloroplast, *J. Phycol.* **9**:430.

Moestrup, Ø., and Thomsen, H. A., 1974, An ultrastructural study of the flagellate *Pyramimonas orientalis* with particular emphasis on Golgi apparatus activity and the flagellar apparatus, *Protoplasma* **81**:247.

Mukai, F., Streisinger, G., and Miller, B., 1967, The Mechanism of lysis in phage T4 infected cells, *Virology* **33**:398.

Padan, E., and Shilo, M., 1969, Short trichome mutant of *Plectonema boryanum, J. Bacteriol.* **97**:975.

Padan, E., and Shilo, M., 1973, Cyanophages—viruses attacking blue-green algae, *Bacteriol. Rev.* **37**:343.

Padan, E., Shilo, M., and Kislev, N., 1967, Isolation of Cyanophages from freshwater ponds and their interaction with *P. boryanum, Virology* **32**:234.

Padan, E., Ginzburg, D., and Shilo, M., 1970, The reproductive cycle of cyanophage LPP-1G in *Plectonema boryanum* and its dependence on photosynthetic and respiratory systems, *Virology* **40**:514.

Padan, E., Raboy, B., and Shilo, M., 1971a, Endogenous dark respiration of the blue-green alga, *Plectonema boryanum, J. Bacteriol.* **106**:45.

Padan, E., Rimon, A., Ginzberg, D., and Shilo, M., 1971b, A thermosensitive cyanophage (LPP-1G) attacking the blue-green alga *Plectonema boryanum, Virology* **45**:773.

Padan, E., Shilo, M., and Oppenheim, A. B., 1972, Lysogeny of the blue-green alga *Plectonema boryanum* by LPP2-SPI cyanophage, *Virology* **47**:525.

Pan, P., and Umbreit, W. W., 1972, Growth of mixed cultures of autotrophic and heterotrophic organisms, *Can. J. Microbiol.* **18**:153.

Pearce, J., and Carr, N. G., 1969, The incorporation and metabolism of glucose by *Anabaena variabilis, J. Gen. Microbiol.* **54**:451.

Pearson, B. R., and Norris, R. E., 1974, Intranuclear virus-like particles in the marine alga *Platymonas* sp. (Chlorophyta, Prasinophyceae), *Phycologia* **13**:5.

Pearson, N. A., Small, E. A., and Allen, M. M., 1975, Electron microscopic study of the infection of *Anacystis nidulans* by the cyanophage AS-1, *Virology* **65**:469.

Pickett-Heaps, J. D., 1972, A possible virus infection in the green alga *Oedogonium, J. Phycol.* **8**:44.

Pienaar, R. N., 1976, Virus-like particles in three species of phytoplankton from San Juan Island, Washington, *Phycologia* **15**:185.

Rimon, A., 1971, The source of the LPP-1G cyanophage proteins in the blue-green alga *Plectonema boryanum* and the study of the relations between events in the infected host and the multiplication of the phage, M.S. thesis, Hebrew University, Jerusalem.

Rimon, A., and Oppenheim, A. B., 1974, Isolation and genetic mapping of temperature-sensitive mutants of cyanophage LPP-2SPI, *Virology* **62**:567.

Rimon, A., and Oppenheim, A. B., 1975, Heat induction of the blue-green alga *Plectonema boryanum* lysogenic for the cyanophage SPIcts1, *Virology* **64**:454.

Rimon, A., and Oppenheim, A. B., 1976, Protein synthesis following infection of the blue-green alga *Plectonema boryanum* with the temperate virus SPI and its ts mutants, *Virology* **71**:444.

Safferman, R. S., 1973, Phycoviruses, in: *The Biology of the Blue-green Algae* (N. G. Carr and B. A. Whitton, eds.), pp. 214–237, University of California Press, Berkeley.

Safferman, R. S., and Morris, M. E., 1963, Algal virus: Isolation, *Science* **140**:679.

Safferman, R. S., and Morris, M. E., 1964a, Growth characteristics of the blue-green algal virus LPP-1, *J. Bacteriol.* **88**:771.

Safferman, R. S., and Morris, M. E., 1964*b*, Control of algae with viruses, *J. Am. Water Works Assoc.* **56**:1217.

Safferman, R. S., and Morris, M. E., 1967, Observations on the occurrence, distribution, and seasonal incidence of blue-green algal viruses, *Appl. Microbiol.* **15**:1219.

Safferman, R. S., Schneider, I. R., Steere, R. L., Morris, M. E., and Diener, T. O., 1969*a*, Phycovirus SM-1: A virus infecting unicellular blue-green algae, *Virology* **37**:386.

Safferman, R. S., Morris, M. E., Sherman, L. A., and Haselkorn, R., 1969*b*, Serological and electron microscopic characterization of a new group of blue-green algal viruses (LPP-2) *Virology* **39**:775.

Safferman, R. S., Diener, T. O., Desjardins, P. R., and Morris, M. E., 1972, Isolation and characterization of AS-1, a phycovirus infecting the blue-green algae, *Anacystis nidulans* and *Synechococcus cedrorum, Virology* **47**:105.

Schneider, I. R., Diener, T. O., and Safferman, R. S., 1964, Blue-green algal virus LPP-1: Purification and partial characterization, *Science* **144**:1127.

Scholes, P., Mitchell, P., and Moyle, J., 1969, The polarity of proton translocation in some photosynthetic microorganisms, *Eur. J. Biochem.* **8**:450.

Sherman, L., 1970, Structure and replication of the blue-green algae virus, LPP-1, Ph.D. thesis, University of Chicago, Chicago.

Sherman, L. A., 1976, Infection of *Synechococcus cedrorum* by the cyanophage AS-1M. III. Cellular metabolism and phage development, *Virology* **71**:199.

Sherman, L. A., and Connelly, M., 1976, Isolation and characterization of a cyanophage infecting the unicellular blue-green algae *A. nidulans* and *S. cedrorum, Virology* **72**:540.

Sherman, L. A., and Haselkorn, R., 1970*a*, LPP-1 infection of the blue-green alga *Plectonema boryanum.* I. Electron microscopy, *J. Virol.* **6**:820.

Sherman, L., and Haselkorn, R., 1970*b*, LPP-1 infection of the blue-green alga *Plectonema boryanum.* II. Viral DNA synthesis and host DNA breakdown, *J. Virol.* **6**:834.

Sherman, L., and Haselkorn, R., 1970*c*, LPP-1 infection of the blue-green alga *Plectonema boryanum.* III. Protein synthesis, *J. Virol.* **6**:841.

Sherman, L. A., and Haselkorn, R. 1971, Growth of the blue-green algae virus LPP-1 under conditions which impair photosynthesis, *Virology* **45**:739.

Sherman, L. A., and Pauw, P., 1976, Infection of *Synechococcus cedrorum* by the cyanophage AS-1M. II. Protein and DNA synthesis, *Virology* **71**:17.

Sherman, L. A., Connelly, M., and Sherman, D. M., 1976, Infection of *Synechococcus cedrorum* by the cyanophage AS-1M. I. Ultrastructure of infection and phage assembly, *Virology* **71**:1.

Shestakov, S. U., and Khyen, N. T., 1970, Evidence for genetic transformation in blue-green alga *Anacystis nidulans, Mol. Gen. Genet.* **107**:372.

Silverberg, J., Rimon, A., Kessel, M., and Oppenheim, A. B., 1977, Assembly site of cyanophage LPP-2SPI in *Plectonema boryanum, Virology* **77**:437.

Simon, L. D., 1972, Infection of *E. coli* by T2 and T4 Bacteriophages as seen in the electron microscope: T4 head morphogenesis, *Proc. Natl. Acad. Sci. U.S.A.* **69**:907.

Simon, R. D., 1971, Cyanophycin granules from the blue-green alga *Anabaena cylindrica*: A reserve material consisting of copolymers of aspartic acid and arginine, *Proc. Natl. Acad. Sci. U.S.A.* **68**:265.

Singh, P. K., 1974, Isolation and characterization of a new virus infecting the blue-green alga *Plectonema boryanum, Virology* **58**:586.

Singh, P. K., 1975, Lysogeny of blue-green alga *Plectonema boryanum* by long tailed virus, *Mol. Gen. Genet.* **137**:181.

Singh, R. N., and Singh, P. K., 1967, Isolation of cyanophages from India, *Nature* (*London*) **216**:1020.

Singh, R. N., and Singh, P. K., 1972, Transduction and lysogeny in blue-green algae, in: *Taxonomy and Biology of Blue-Green Algae* (T. V. Desikachary, ed.), pp. 258–261, University of Madras, Madras, India.

Singh, R. N., Singh, P. K., and Varanasi, P. K., 1969, Lysogeny and induction of lysis in blue-green algae and their viruses, *Proc. 56th Int. Sci. Congr.* **56**:272.

Skotnicki, A., Gibbs, A., and Wrigley, N. G., 1976, Further studies on *Chara corallina* virus, *Virology* **75**:457.

Smith, A. J., 1973, Synthesis of metabolic intermediates, in: *The Biology of the Blue-green Algae* (N. G. Carr and B. A. Whitton, eds.), pp. 1–38, University of California Press, Berkeley.

Smith, A. J., London, J., and Stanier, R. Y., 1967, Biochemical basis of obligate autotrophy in blue-green algae and thiobacilli, *J. Bacteriol.* **94**:972.

Smith, K. M., Brown, R. M., Jr., Goldstein, D. A., and Walne, P. L., 1966a, Culture methods for the blue-green alga *Plectonema boryanum* and its virus with an electron microscope study of virus-infected cells, *Virology* **28**:580.

Smith, K. M., Brown, R. M., Jr., Walne, P. L., and Goldstein, D. A., 1966 b, Electron microscopy of the infection process of the blue-green alga virus, *Virology* **30**:182.

Smith, K. M., Brown, R. M., Jr., Walne, P. L., 1967, Ultrastructural and time-lapse studies on the replication cycle of the blue-green algal virus LPP-1, *Virology* **31**:329.

Stanier, R. Y., 1973, Autotrophy and heterotrophy in unicellular blue-green algae in: *The Biology of Blue-Green Algae* (N. G. Carr and B. A. Whitton, eds.), pp. 501–518, University of California Press, Berkeley.

Stanier, R. Y., Kunisawa, R., Mandel, M., and Cohen-Bazire, G., 1971, Purification and properties of unicellular blue-green algae (order *Chroococales*), *Bacteriol. Rev.* **35**:171.

Stent, G. S., 1963, *Molecular Biology of Bacterial Viruses,* W. H. Freeman, San Francisco.

Stevens, A., 1974, DNA dependent RNA polymerases from two T4 phage-infected systems, *Biochemistry* **13**:493.

Studier, F. W., and Maizel, J. V., Jr., 1969, T7-directed protein synthesis, *Virology* **39**:575.

Summers, W. E., 1970, The regulation of RNA metabolism in *E. coli* infected with phage T7, in: *RNA-Polymerase and Transcription* (L. Silvestri, ed.), pp. 110–123, North-Holland, Amsterdam.

Susor, W. A., and Krogmann, D. W., 1964, Hill activity in cell-free preparations of a blue-green alga, *Biochim. Biophys. Acta* **88**:11.

Swale, E. M. F., and Belcher, J. H., 1973, A light and electron microscope study of the colourless flagellate *Aulacomonas* Skuja. *Arch. Mikrobiol.* **92**:91.

Szybalski, W., Bovre, K., Fiandt, M., Hayes, S., Hradecna, Z., Kumar, S., Lozeron, H. A., Nijkamp, H. J. J., and Stevens, W. F., 1970, Transcriptional units and their controls in *Escherichia coli* phage λ: Operons and scriptons, *Cold Spring Harbor Symp. Quant. Biol.* **35**:341.

Teichler-Zallen, D., and Hoch, G., 1967, Cyclic electron transport in algae, *Arch. Biochem. Biophys.* **120**:227.

Tikchonenko, A. S., 1970, *Ultrastructure of Bacterial Viruses*, Plenum, New York.

Tikchonenko, A. S., and Zavarzina, N. B., 1966, Morphology of the lytic agent of *Chlorella pyrenoidosa, Mikrobiologiya* **35**:850.

To, C. M., Kellenberger, E., and Eisenstark, A., 1969, Dissassembly of T-even bacteriophage into structural parts and subunits, *J. Mol. Biol.* **46**:493.

Tomas, R. N., Lindsay, G. C., Nairn, R. S., and Cox, E. R., 1973, The symbiosis of *Peridinium balticum* (Dinophyceae). II. Virus-like inclusions, *J. Phycol.* **9**:16s (abstr.).

Toth, R., and Wilce, R. T., 1972, Virus-like particles in the marine alga *Chorda tomentosa* Lyngbye (Phaeophyceae), *J. Phycol.* **8**:126.

Travers, A., 1970, RNA polymerase and T4 development, *Cold Spring Harbor Symp. Quant. Biol.* **35**:241.

Travers, A., 1971, Control of transcription in bacteria, *Nature (London)* New Biol. **229**:69.

Ueda, K., 1965, Virus-like structures in the cells of the blue-green alga, *Oscillatoria princeps, Exp. Cell Res.* **40**:671.

Van Baalen, C., 1973, Mutagenesis and genetic recombination, in: *The Biology of Blue-Green Algae* (N. G. Carr and B. A. Whitton, eds.), pp. 201–213, U. of California Press, Berkeley.

Van Baalen, C., Hoare, D. S., and Brandt, E., 1971, Heterotrophic growth of blue-green algae in dim light, *J. Bacteriol.* **105**:685.

White, A. W., and Shilo, M., 1975, Heterotrophic growth of the filamentous blue-green alga *Plectonema boryanum, Arch. Microbiol.* **102**:123.

Whitton, B. A., Carr, N. G., and Craig, I. W., 1971, A comparison of the fine structure and nucleic acid biochemistry of chloroplasts and blue-green algae, *Protoplasma* **72**:325.

Wolk, C. P., 1968, Movement of carbon from vegatative cells to heterocysts in *Anabaena cylindrica, J. Bacteriol.* **96**:2138.

Wolk, C. P., 1973, Physiology and cytological chemistry of blue-green algae, *Bacteriol. Rev.* **37**:32.

Wu, J. H., Lewin, R. A., and Werbin, H., 1967, Photoreactivation of uv-irradiated blue-green algal virus LPP-1, *Virology* **31**:657.

Zaitlin, M., and Jagendorf, A. T., 1960, Photosynthetic phosphorylation and Hill reaction activities of chloroplasts isolated from plants infected with tobacco mosaic virus, *Virology* **12**:477.

Viruses of Fungi Capable of Replication in Bacteria (PB Viruses)

T. I. Tikchonenko

D. I. Ivanovsky Institute of Virology of the USSR AMS
Chair of Virology of the Moscow State University
Moscow, USSR

1. DISCOVERY

The discovery of DNA-containing bacteriophages in species of *Penicillium* happened by mere chance and in a laboratory quite skeptical of the possibility that viruses might be common to both prokaryotes (bacteria) and eukaryotes (fungi). Nevertheless, we have repeatedly demonstrated in fungi the presence of phages capable of replication in bacteria. These *Penicillium*-derived bacterial viruses or PB viruses were discovered as a byproduct of our interest in obtaining large quantities of double-stranded RNA (dsRNA) from fungi shown earlier to contain dsRNA mycophages (for review, see Lemke and Nash, 1974; Lemke, 1976). This interest, however, proved to be the first stone of an avalanche in our Institute of Virology, an institute which traditionally had studied the biochemistry and biophysics of conventional host-specific bacteriophages and which was not initially prepared to cope with or to interpret the events that followed.

Our involvement with dsRNA mycophages initially centered about a culture of *Penicillium brevicompactum* isolated in the USSR and

235

provided by Dr. G. F. Gause of the Institute of Antibiotics. Indeed, as expected, we found dsRNA mycophages in this culture, although their properties differed somewhat from those reported by Wood *et al.* (1971). Several papers were published from this work, but we had little opportunity to benefit from these findings.

At this point, trouble developed in our institute. It became increasingly more difficult to experiment with the conventional bacteriophages, which before this time we had maintained without incident for many years. Much to our dismay and disbelief, we found unknown phages growing indiscriminately in our stock strains of the *Escherichia-Shigella* group instead of the traditional T2, Sd, and DDV-II phages. The unknown phages were established and retained in our laboratory environment despite rigorous control and sterility measures. A variant of DDV-II phage was even lost in the turmoil, which lasted for nearly 6 months.

During this period, however, a trend became apparent, namely that our virus problem emanated from places where the work with *Penicillium* had been in progress. Completely at a loss to explain this association, I nonetheless gave a senseless recommendation to my co-worker, A. F. Bobkova, to carry out titrations for phage using the sterile culture fluid of *P. brevicompactum* against all the bacteria we were working with. She shrugged but obeyed, and the rest of our story will be discussed below in the scientific part of this chapter.

The discovery of PB viruses in *P. brevicompactum* was soon followed by the discovery of bacteriophages in other species of *Penicillium* and, more recently, by the discovery of similar viruses in a species of the genus *Cephalosporium* (Egorov *et al.*, 1976). The latter phages have been designated CB viruses.

In all candor, we did not for a long time adopt the hypothesis that PB viruses had fungal as well as bacterial host relationships, and we can understand the skepticism that others might have in this regard. Our own skepticism for some time kept us from intensively investigating PB viruses. Moreover, in our strictly biochemical laboratory, this problem was viewed as a sideline, and our collaboration with the mycological laboratory of Dr. Yu. E. Bartoshevich of the Institute of Antibiotics has begun rather recently. It is my hope that something of what is described below will urge skeptics to overcome their doubts sufficiently to isolate some completely new member of the PB virus brotherhood and to contribute independently to the study of this new group of viruses.

2. DETECTION

2.1. Test Systems

The problem itself implies that the simplest and easiest way to isolate and identify these viruses would be titration of mycelial extracts on the appropriate sensitive bacteria. Of course, considering the known specificity of viruses capable of replication only in a narrow host range, it is very difficult to decide in advance which strains and species of bacteria will be sensitive and can be used as test systems. This complexity, however, was purely theoretical for us. As mentioned, we unwillingly established in a practical way that a number of bacterial cultures of the *Escherichia-Shigella* group used routinely in our laboratory are lysed by PB viruses. These cultures included widely known *E. coli* B, *E. coli* C, and *E. coli* SK, which were subsequently used extensively in our studies, both for titration of already known PB viruses or their propagation in bacteria, and for examination of other fungal species to detect new members of this virus group. Sometimes, other species of bacteria were used, some of which were found to be sensitive to PB viruses (Table 1).

In addition, Mother Nature, obviously as a small compensation for all the troubles given us by PB viruses, endowed at least some of them (e.g., PBV-1 and PBV-3) with an unusually wide host-range spectrum (Velikodvorskaya *et al.*, 1972a; Tikchonenko *et al.*, 1974). This circumstance, described at length below, considerably facilitates selection of a sensitive test system.

A question may naturally arise whether some other viruses of the PB type may be present in mold undetected by us because they are incapable of replication in those bacteria which are used as test systems. Unfortunately, this question cannot be answered definitely because the appropriate experimental material is lacking. The maximum we dared to do (Table 1) was inclusion of *Bacillus subtilis* and some other more exotic baterial species into the number of the test systems. They were all found to be insensitive to the viruses from *P. brevicompactum* (PBV-1, PBV-2, and PBV-3), but it was not established whether this resistance was typical of these species or only of the given strain of this species. The capacity of different strains of even one bacterial species to support replication of PB viruses varies greatly, as is well demonstrated by the data in Table 2 from the study of Bobkova *et al.* (1975). Thus it may be stated that for some unknown reason fungal viruses show a special preference for the group of enteric bacteria.

TABLE 1
Titration of Mycelium Extracts of Different Molds on Bacteria

Fungus species[b]	Escherichia coli										Bacillus		Proteus		Pseudomonas		Serratia marcescens	Paracoli sp.
	C	B	SK	K12	W119	RH9	C600	CR63	W3350	uATCC 9637	sub-tilis	my-coides	mira-bilis	vul-garis	fluo-rescens	pyocy-aneum		
P. brevicompactum	+	+	+	+	–	–	+	+	+	–	0	–	–	–	–	–	–	–
P. stoloniferum	+	–	+	–	–	–	–	–	–	–	–	–	–	–	–	–	–	–
P. chrysogenum	+	+	0	+	+	0	–	–	–	+	–	0	0	0	0	0	0	0
P. nigricans	+	+	0	+	+	0	–	–	–	+	–	0	0	0	0	0	0	0
P. cyclopium	+	+	0	+	+	0	–	–	–	+	–	0	0	0	0	0	0	0
Cephalosporium acremonium	+	+	0	+	+	0	–	–	–	+	–	0	0	0	0	0	0	0

[a] +, The appearance of negative colonies on bacteria; 0, the absence of such colonies; –, titration of mycelium extract not performed on this bacterial strain.

[b] The experimental conditions for preparing fungus extracts: The mycelium or the spores grown under sterile conditions (Velikodvorskaya et al., 1972a) were cooled and separated from the culture medium by filtration through paper filters or by low-speed centrifugation. Starting from this step, the temperature must be kept between 0°C and +4°C. The collected sediment was washed with 0.1 M NaCl + 0.05 M phosphate buffer, pH 7, the volume being equal to that of the initial culture medium. The disintegration of washed cell material was performed either by grinding with quartz powder or by ultrasonication. The disrupted mycelia or spores were extracted with 0.1 M NaCl + 0.05 M phosphate buffer, pH 7 (the volume of extracting buffer being equal to 1/10 of that of the initial cultural medium). The supernatant after low-speed centrifugation was shaken once for 5 min with chloroform (1:1, v/v) by hand in a separatory funnel. The clear supernatant after the layer separation was concentrated either by ammonium sulfate precipitation (30 g per 100 ml of solution in the case of PBV-1–PBV-3) or by precipitation with 8% polyethylene glycol (PBV-5). Sometimes membrane filters were used for concentration of PBV-5. For stability of PBV-5 the presence of Ca^{2+} is required; therefore, 0.0005 M $CaCl_2$ was added to all solutions and media including agar (Velikodvorskaya et al., 1972a; Tikchonenko et al. 1974; Amosenko et al., 1976; Egorov et al., 1976).

TABLE 2
**Efficiency of Plating of PBV-5 on Different *E. coli*
Strains**[a]

| Bacterial strains | Titer of virus | |
	Grown in *E. coli* C	From the mold
E. coli C	336×10^4	114×10^2
E. coli W	56×10^4	22×10^2
E. coli K12	1.5×10^4	0.74×10^2
E. coli RI19	0	0

[a] From Bobkova *et al.* (1975).

At the same time, both the question of the most suitable bacterial host for PB viruses and the problem of yet undiscovered PB viruses whose detection requires other bacterial test systems remain open for discussion.

It is noteworthy in Table 2 that there is a very close ratio between infectious titers of PBV-5 of fungal and bacterial origin. If the infectious titers of PBV-5 obtained by plating in K12 cells are taken as 1, the ratio of titers in the four indicated strains will be 223:37:10 (for a mycelial extract) and 153:30:10 (after growing in *E. coli* C). This means that the virus directly from mycelium has the same biological-potencies and sensitivity to restriction as the virus after a replication cycle in bacteria. Consequently, passages of PBV-5 in bacteria do not change its biological properties and the titration procedure actually quantitates PB viruses in fungi.

Of the above-mentioned standard set of bacterial cultures, two, *E. coli* B and *E. coli* SK, possess strong restrictive systems, while *E. coli* C represents zero type with respect to host specificity (Arber, 1974; Nikolskaya *et al*., 1976). Furthermore, all three strains differ greatly in their physiological and biochemical properties, especially *E. coli* SK, which does not share bacteriophages with the other two strains employed (Krivissky, 1955).

The foregoing suggests that we had at our disposal sufficiently representative test systems for detection in mold of PB viruses, at least those species which are active against the genus *Escherichia*.

From the theoretical point of view, it seems very tempting and even desirable to employ direct electron microscopy of sections of mold mycelium for visualization of these viruses, as has been done in studies on mycophages with double-stranded RNA (for references, see Lemke and Nash, 1974; Lemke, 1976; see also Chapter 2 of this volume). The situation is somewhat complicated by the fact that PB viruses and mycophages are present in fungi simultaneously (Velikodvorskaya *et al.*, 1972*a*; Bobkova *et al.*, 1975), the latter being absolutely dominant in quantity. But since PBV-1–PBV-5 differ considerably from known mycophages in size and morphology, their identification in sections would not have been very difficult. The presence in mycelial sections of typical PB virions would have been a convincing and simple proof of their mycotic origin, which would have saved us a lot of time and effort spent in obtaining indirect or more complicated evidence.

This way, however, was closed for us. The matter is that titers of PB viruses in mold are very low, usually within the range of $10-10^4$ particles/g of dry weight of mycelium. The negligible amount of virus found in mycelium seems to be particularly relevant in terms of the amount in one cell. As far as we know, there are no definite data on the absolute weight of individual *Penicillium* cells in the literature, although this value may be approximated from the size of the genome of *P. brevicompactum* which we had determined by the kinetics of fungal DNA reassociation (Surkov *et al.*, 1975; Amosenko *et al.*, 1976). The size of the genome of this fungus was found to be 7×10^9 daltons or 0.12×10^{-13} g. Unfortunately, no data by other authors are available in the literature concerning the size of the genome of this species of *Penicillium*. However, the order of magnitude of the value found by us appears reasonable, since for the related *P. chrysogenum* the values from 0.3 to 1.3×10^{-13} g per haploid genome were presented (Lhoas, 1975), and in *Aspergillus* cells the absolute content of DNA is estimated to be 0.9×10^{-13} g (Storck, 1974). Our findings of the absolute content of DNA in *P. brevicompactum* represent the lower limit of magnitude in this series. It may be mentioned for comparison that similar values for absolute DNA content and its molecular weight are given for *Saccharomyces cervisiae*, the most extensively studied of mycotic fungi. With the content of DNA of 0.24×10^{-13} g per haploid genome, its molecular weight had been estimated by the kinetics of reassociation to be 9.2×10^9 (for references see Storck, 1974).

In these circumstances, therefore, it seemed to us reasonable to use the obtained value of the *P. brevicompactum* genome without further corrections. In making this decision, we were governed by the consideration that the analogous technique and calculation method were also used in our determinations of the size of genomes of PB viruses and of the number of their copies in fungal DNA. Therefore, if the use of the kinetics of reassociation in our hands led to some absolute systematic error in determinations of the size of the genomes, this error should be reduced considerably in comparisons of the size of these DNAs (see below).

At this value of the fungal genome and 0.2% content of DNA in mycelium, the mass of an individual cell of *P. brevicompactum* must be 6×10^{-12} g. In other words, 1 g of dry mycelium must contain 1.7×10^{11} *Penicillium* cells. This calculation is tentative, but possible deviations from this value appear not to exceed 1.5- to 2-fold. Thus it is easy to see that a titer of PB viruses in mycelium of $10-10^4$ particles/g of mass should correspond to $10^{-10}-10^{-7}$ virions per cell of mold. To look for this viral "needle" in a mold "haystack" by means of electron microscopy would be absolutely hopeless. It is quite evident that for successful electron microscopic studies it would have been necessary to increase markedly the yields of PB viruses in their natural host, but this we are not yet able to achieve. True, we succeeded in increasing the titers of some PB viruses in mold by UV irradiation and treatment with different mutagens (Lebed' *et al.*, 1975), but this increase is on the average no more than 1.5–2 log (see below), which in the light of the data on the amount of PB viruses *in situ* is definitely insufficient for electron microscopic search for virus in sections.

2.2. Conditions for Detection of PB Viruses

Unfortunately, direct titration on bacteria of fungal mycelium extracts or culture fluid supernatant most frequently gives negative or very low results. There appear to be several causes for this. First, both mycelium and the culture fluid seem to have an inhibitor which prevents demonstration of potential infectivity of PB viruses for bacteria (Velikodvorskaya *et al.*, 1972a; Bobkova, 1972). Mild deproteinization of mycelium extract with chloroform is the most universal means for its elimination (see footnotes to Table 1). Without this procedure, either no viruses titrable on bacteria can be detected in the mycelium extract or culture fluid, or their titer is 1–2 log lower. This holds true both for viruses of *P. brevicompactum* and for those of *P. chrysogenum* and *Cephalosporium acremonium* (Bobkova *et al.*, 1975; Egorov *et al.*, 1976). Second, in the case of antibiotic-producing strains of mold, there is the added presence of high concentrations of antibiotic preventing direct titration on bacteria. When *P. chrysogenum* was used, penicillin was either destroyed with penicillinase or removed in the process of concentration of PB viruses on membrane filters or with polyethylene glycol (Bobkova *et al.*, 1975; Egorov *et al.*, 1976). Third, very frequently the content of PB viruses in mold is so low that they can be detected only after preliminary concentration. Therefore, in a typical case, titration of a mycelium extract or culture fluid is preceded by chloroform deproteinization followed by concentration either with ammonium sulfate in case of *P. brevicompactum* (Velikodvorskaya *et al.*, 1972*a*) or with membrane filters or polyethylene glycol in the case of *P. chrysogenum* or *C. acremonium* (Egorov *et al.*, 1976; Amosenko *et al.*, 1976). Usually we concentrated the material 10- to 100-fold.

Without this pretreatment of mycelium extracts, the amount of PB viruses determined by titrations is considerably lower, and with a low initial amount of virus the titration on bacteria may be negative (Table 3). It will be seen that the average difference between the results of direct titrations of extracts and repeated titration of concentrates is 1.0–1.5 log, even without consideration of the inevitable virus losses in the process of concentration. Thus the pretreatment of the extract increases the number of infectious centers 100- to 200-fold.

Another obstacle in detection and quantitation of some PB viruses is their thermolability. From this point of view, a significant threat to PBV-5 viability is the melted agar used in titration of the virus in dishes according to the method of Gratia (1936). Therefore, we had to change the standard conditions of work developed by Gratia (46°C) and use melted agar of as low temperature as possible, not above 38–40°C (Bobkova *et al.*, 1975). In order to determine the extent of inactivation of any PB virus by melted agar, it is useful to compare the results of titration in dishes with titers determined by the end-point dilution method (see Adams, 1959). PB viruses from *P. brevicompactum* are quite thermostable and may be titrated by Gratia's method under standard conditions.

TABLE 3
Influence of Concentration Factor on the Absolute Quantity of PB Viruses Discovered in the Fungus

Source of PBV	Quantity of virus[a]	
	In the initial extract	After concentration
P. brevicompactum[b]		
a	3×10^1	2×10^2
b	0	1.5×10^1

[a] Titer/ml × volume (ml).
[b] The results of typical experiments obtained with high PBV concentration in the extract (preparation "a") and with low PBV concentration in the extract (preparation "b").

Fig. 1. Kinetics of thermo-inactivation of PBV-5. PBV-5 in the mycelium (△), in the extract (●) and in the suspension after replication in *E. coli* C (○). In the latter two cases, 0.1 M NaCl + 0.05 M phosphate buffer, pH 7, was used as a solvent. From Egorov *et al.* (1976).

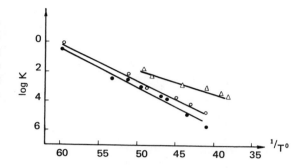

It must be mentioned that in a number of cases the sensitivity to thermal inactivation is determined by a combined effect of temperature and other environmental factors. In particular, the abovementioned considerable inactivation of PB viruses in titrations under standard conditions recommended by Gratia seem to be due not only to the high temperature itself but also to the effect of some compounds present in melted agar (Bobkova, 1972). Particularly sensitive to admixtures in agar is PBV-5, which was successfully titrated only on Noble agar (Difco) (A. Egorov, personal communication). We did not study this phenomenon in greater detail.

In a similar way, the thermal sensitivity of PB viruses may change greatly, depending on the environment of viral particles, as may be seen in Fig. 1, which shows the results of the experiments on thermal inactivation of PBV-5 directly in the mycelium, in mycelar extracts, and in a lysate of bacteria obtained after inoculation with this virus of *E. coli* C cells (Egorov *et al.*, 1976). From the inclination of the curves, it is easy to calculate the energy of activation for the three "conditions" of PBV-5 virions. For the virus in the intact mycelium, this value is minimal (27 kg-cal), while for the virus in the mycelar extract and bacterial lysate it is three- to fourfold higher (80 and 100 kg-cal, respectively). It should be noted that the energy of activation in these two cases on the whole corresponds to what is usually observed in free phages of the T series. The value of the activation energy registered upon heating of PBV-5 in mycelium should be considered to be abnormally low as compared not only to free forms of this virus but also to other bacteriophages. (Adams, 1959).

In the evaluation of these data it should be remembered that, in the experiments described, the thermostability of the virus was determined in extracts and lysates containing high concentrations of protein. Since the latter may exert a considerable thermostabilizing effect, these observations should be applied to preparations of purified virus with caution.

It should be remembered that higher temperatures may be used for preparation of virus-free variants of mold or variants with reduced capacity for production of PB viruses (Bobkova *et al.*, 1975). This effect may be achieved as a result of long-term cultivation of *P. chrysogenum* at a higher temperature (35°C) instead of the normal 30°C or by a short-term treatment of spores at 60°C. In both cases, however, virus-free preparations are obtained in which the capacity for virus production, although reduced, is restored after five to ten passages (see below).

The foregoing requires the observance of a strict temperature regimen in harvesting of the mycelium, its disruption, concentration of the extract, and further stages of purification. In our observations, the optimal temperature for these steps is within the range 0–4°C (in Tikchonenko *et al.*, 1974, there is an uncorrected erratum: the recommended temperature range is stated to be 0.4°C instead of 0–4°C).

The above conditions (the choice of the appropriate test system, preliminary concentration, and proper cooling) lend themselves readily to experimental control. Unfortunately, however, in the work with PB viruses, there exist a number of other additional difficulties which can become a stumbling block in successful isolation of viruses of this group. At our obviously insufficient level of knowledge on the relationship between the PB virus genome and its fungal host genome, these complexities appear nearly mystical and hardly lend themselves to experimental control.

First of all, isolation of PB viruses from one or another *Penicillium* surprisingly resembles an inverted calendar of agricultural work. Figure 2 presents a nomogram reflecting monthly distribution of positive results obtained in titrations of mycelar extracts of *P. chrysogenum* (PBV-5) on bacteria. Although it was not our purpose to check regularly the relationship between the number of our failures and the position of the constellations, the data in the laboratory records on which this nomogram is based indicate unequivocally seasonal variations in success with PB viruses. For instance, the gap in the summer months warns against attempts at PB virus isolation in summer. Our bitter experience permits a statement that the percentage of negative results from May through August is so high that fungal mycelium is practically sterile in the summer months at Moscow's latitude. To attempt isolations of PB viruses during the summer is as useless as seeding wheat in winter. The causes of this anomaly have their roots in

Fig. 2. Relationship among the seasons, presence of PBV-5, and biosynthesis of penicillin in *P. chrysogenum*. The left ordinate represents percentages of successful experiments on titration of PB viruses of fungi on bacteria with respect to the total number of titration experiments (solid line). The right ordinate represents the amounts of penicillin synthesized by the fungus as percentages with respect to the maximum synthesis taken as 100% (dotted line). The abscissa gives ordinal numbers of months starting from January (No. 1). The histogram is based on the data obtained during 2 years (1973–1974).

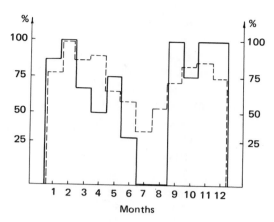

the fungus itself, as replication in bacteria of previously isolated PB viruses occurs normally throughout the year. It is impossible to explain this by a simple effect of higher temperatures, since laboratory refrigerators and thermostats are set at the same temperature at all seasons. So almost against one's will one has to think of such semimystical factors as solar activity, radiation, etc. Such a relationship between the amount of viruses titrable on bacteria on the one hand and the seasons of the year on the other is evidently due not to thermal inactivation of PBV-5 alone, since at the same time antibiotic production (the capacity of a highly active strain, No. 39, to synthesize penicillin) is also changed (Egorov *et al.*, 1976). Although we do not think that the capacity for penicillin synthesis is determined by the virus alone, nevertheless there exists some mediated relationship between the presence of PBV-5 virus and the production of antibiotic. No matter how vague these considerations are from the point of view of sound scientific pragmatism, the most practical advice to be given to those working with PB viruses is to use the summer for its best purpose—for holidays.

Another mysterious reef waiting for researchers in the stormy PB sea is their spontaneous elimination from fungal cultures upon long-term cultivation in artificial nutrient media. According to the observations of Maximova and Bobkova (to be published), *P. brevicompactum* mycelium after isolation from soil contains the largest amount of PBV-1–PBV-3, and this increased capacity for virus production persists about a year. Then, as a result of numerous passages in artificial nutrient media, virus titers in mycelium begin to decline to practically complete sterility. In such cases, we took the material again from the museum specimen stored in the soil which contributed to the restoration of maximally high (for a given object) virus titers. As a parallel, it should be mentioned that a similar relationship has been found by Wuest and Smith (1974) for mycophages with double-stranded RNA. As in our case, long-term passages of mycelium in artificial nutrient media gradually decreased its capacity to produce mycophage, which was restored after contact with soil.

It may be generally stated that mold mycelium, like the mythical Anteus, restores its "strength" after contact with Mother Earth. The scientific equivalent of this florid metaphor is that there is a certain cofactor in the soil which exerts a favorable influence on the capacity of the fungus to produce both PB viruses and mycophages. The explanation of the mechanism of action of this cofactor cannot, evidently, be based on the hypothesis of retention in the mycelium of a certain chemical compound picked up by the fungus in the soil. Understandably, under conditions of numerous passages in artificial nutrient media, this hypothetical compound would have been rapidly utilized or eliminated from the fungus. The explanation of this enigmatic phenomenon should be based on some more complicated mechanisms of which nothing can be said thus far.

In addition to the above factors, some still unknown features which may affect virus production probably exist in the interactions between PB viruses and molds. Most likely (see below), the relations between virus and mold genomes are of the type of lysogeny in bacteria (Tikchonenko et al., 1974; Lebed' et al., 1975). In this case, any factors affecting the frequency of spontaneous induction of provirus will naturally affect the virus yield. At the same time, it should be recognized that the extremely low amount of the infectious virus per cell (see above) attests to a very low level of spontaneous induction. A relatively small increase of the infectious phage in the mycelium treated with routine inducers of bacterial prophages indicates a very strong repressor.

In conclusion, a situation could be imagined when Dr. X from Moscow announces isolation of PB virus from a certain strain of *Penicillium*, while Dr. Y, let us say from Berkeley, examining a similar strain in his laboratory, finds no virus. In this situation, it must be admitted that if all the experiments have been done correctly, both scientists may be quite right. The correctness of Dr. Y does not mean the absence of PB virus in other populations of this strain of mold, whereas the correctness of Dr. X cannot force all the populations of this strain to contain PB virus in all circumstances. A virologist does not need to be told that not all *E. coli* populations will yield T2 or λ phage, and a similar conclusion should be applicable to PB viruses and their fungal hosts. It should be obvious that alongside with infected fungi which are natural carriers of the virus there exist also "healthy" fungal individuals. *Penicillium* may be sick in Moscow and healthy in Berkeley and the other way around. In any case, the loss by the fungus of PB viruses in the course of long-term cultivation or as a result of thermal treatment (see above) directly indicates such a possibility. From this point of view, another mandatory condition for detection of PB virus in mold is its undoubted presence in the form of either lytic virus or induced provirus. The other aspects of biology of PB viruses that are to some extent contrary to the latter view are discussed in other sections of this chapter.

3. PURIFICATION AND PROPERTIES OF PB VIRUSES

Because of the extremely small amount of virus in the initial mycelium, all the studies on concentration and purification were performed with PB virions grown in their second host, bacteria. In the sensitive bacteria under optimal conditions of cultivation, virus titers reached $1.0-1.5 \times 10^{11}$ particles/ml. Purification and concentration of PB viruses under study were not difficult and were done according to the methods routinely used with bacteriophages (Velikodvorskaya *et al.*, 1972a,b; Bobkova *et al.*, 1975; Tikchonenko *et al.*, 1974).

In order to extrapolate to PB viruses of fungal origin everything that had been established in studies on PB virions from bacteria, one must be sure that they are identical. We carried out identification by some key tests based on biological properties of viruses and requiring minimal amounts of materials of fungal origin. Table 4 presents such data for different viruses of the PB group.

According to the previously reported results, three different virus species were identified in *P. brevicompactum*; these were designated PBV-1, PBV-2, and PBV-3. It is very likely that in this object there exist additional species of virus because a diffuse peak of infectivity was also observed in the area of cesium sulfate gradients at a density of 1.32 g/ml and was overlapped by a peak in which dsRNA phages having no infectivity for bacteria were localized (Tikchonenko *et al.*, 1974). We failed to isolate a single virus from this gradient zone at one time. All

TABLE 4
Comparison of Some Properties of PBV Originating from Fungi and Bacteria

Source of virus	Virus	Virus from fungi				Virus grown in bacteria			
		Colonies on bacteria[a]		Constant of neutralization by antibody[b] (min^{-1})	Buoyant density in Cs_2SO_4 (g/ml)	Colonies on bacteria[a]		Constant of neutralization by antibody[b] (min^{-1})	Buoyant density in Cs_2SO_4 (g/ml)
		Type	Diameter (mm)			Type	Diameter (mm)		
P. brevicom-pactum	PBV-1	Clear without halo	1.5–2.0	14.5 ± 1.6	1.40	Clear without halo	1.5–2.0	15 ± 2.0	1.40
	PBV-2	Turbid	0.6–0.8	25 ± 2	1.38	Turbid	0.6–0.8	26 ± 1.5	1.38
	PBV-3	Clear with halo	1.5–2.0	20.8 ± 1	1.42	Clear with halo	1.5–2.0	12 ± 1.5	1.42
P. stoloniferum	PBV-4	Clear without halo	1.5	—	1.40	Clear without halo	1.5	—	1.40
P. chrysogenum[c]	PBV-5	Clear without halo	4.5–5.5	236	1.35	Clear without halo	4.5–5.5	236	1.35
P. cyclopium	PBV-6	Clear without halo	4.5–5.5	220	—	Clear without halo	4.5–5.5	220	—
Cephalosporium acremonium	CBV-1	Clear without halo	4.5–5.5	197	—	Clear without halo	4.5–5.5	197	—
P. nigricans	PBV-5	Clear without halo	4.5–5.5	230	—	Clear without halo	4.5–5.5	230	—

[a] Characteristics of colonies are given for E. coli C cells except PBV-2, which was tested on E. coli SK cells.
[b] Neutralization constants were measured with antisera against the viruses grown in bacteria. Neutralization constants for PBV-5, PBV-6, CBV-1, and PBV from P. nigricans were determined with antiserum against PBV-5.
[c] The description of PBV-5 colonies on bacteria given earlier is now revised as they were obtained with bacterial culture carrying prophage (Tikchonenko et al., 1974).

these viruses differ considerably in their host ranges, serological properties, structure, and morphology (Tikchonenko et al., 1974).

Another virus was discovered in *P. stoloniferum* and designated PBV-4, but has thus far not been studied in detail (Tikchonenko et al., 1974). A very peculiar small virus was identified in *P. chrysogenum* and designated PBV-5 (Bobkova et al., 1975). By the size of the colonies, its host range, and serological properties, it is close to the PB viruses present in *P. cyclopium, C. acremonium*, and *P. nigricans*, although these latter viruses have not been studied in detail either. At the same time, for the virus from *P. cyclopium* and *C. acremonium*, some data were obtained indicating their difference from PBV-5. Therefore, we have used designation PBV-6 for the virus from the former and CBV-1 for that of the latter. PB virus from *P. nigricans* has been left temporarily undesignated because the available evidence rather indicates its identity to PBV-5.

The data for PBV-1–PBV-3 and PBV-5 presented in Table 4 indicate unequivocally that the viruses isolated from the mold and those having replicated in bacteria are identical. In both cases, similar colonies are formed after plating on bacteria. Also, the same viruses of mold and bacterial origin have similar buoyant density and show similar neutralization constants on interaction with antiserum to bacterial virus. The treatment of mold extracts with antisera to PBV-1, PBV-2, and PBV-3 neutralized their infectivity for bacteria (Tikchonenko et al., 1974; Surkov et al., 1975). Similar results were obtained by Bobkova et al., (1975) for PBV-5. The foregoing results strongly suggest that the features of structure, composition, and morphology established in studies on PB viruses of bacterial origin hold true also for the virions formed in molds.

Table 5 summarizes the data characterizing the biochemical, biophysical, and morphological properties of some PB viruses. The PBV-1–PBV-3 group is similar to phages of moderate size containing double-stranded DNA; of these, PBV-2 has a short tail and PBV-1 and PBV-3 virions have long, noncontractile tails. The PBV-1 and PBV-3 particles are similar in size, and are antigenically related, but differ in buoyant density, neutralization constant, and host range.

The common antigen in PBV-1 and PBV-3 virions contributes to their reciprocal neutralization in reaction with heterologous sera. Thus the constants of inactivation rate of PBV-1 and PBV-3 in reaction with the antiserum against PBV-1 were 14.5 ± 1.6 and 6.8 ± 1.0, respectively, and in reaction with the antiserum against PBV-3 were 8.0 ± 1.2 and 10.8 ± 1.1, respectively (Chaplygina, 1974). However, judging by

TABLE 5
Structure and Composition of PB Viruses

Viruses and their sources	Morphology and dimensions[a]		Type of genome[a]	Buoyant density (g/ml)				$S_{20,w}$		Composition[b]					Molecular weight (megadaltons)	
	Head	Tail		Phage		DNA				Phage (%)			Mol % G+C			
				CsCl	Cs₂SO₄	CsCl	Cs₂SO₄	Phage	DNA	Protein	DNA	Da	T_{mel}	CsCl	Phage	DNA
P. brevicompactum																
PBV-1	I *D* = 45 nm	Long NC	DNA, DS, L	1.48	1.40	1.706	1.424	380	30	56	44	46.8	46.8	46.5	51	25
PBV-2	I *D* = 53 nm	Short	DNA, DS, L	1.48	1.38	1.703	1.427	610	30	57	43	44	—	—	67	25
PBV-3	I *D* = 45 nm	Long NC	DNA, DS, L	—	1.41	1.704	1.427	530	28	54	46	46.1	45.6	44.9	61	22
P. stolonife-rum																
PBV-4	—	—	—	—	1.39	—	—	—	—	—	—	—	—	—	—	—
P. chrysoge-num																
PBV-5	*D* = 27.5 ± 2.5 nm	None	DNA, SS, C	1.41	1.35	1.45	—	—	—	—	—	—	—	—	—	—

[a] Abbreviations: DS, double stranded; L, linear; I, icosahedron; NC, noncontractile; C, circular.
[b] Mol % of G+C was calculated in accordance with Da (direct analysis of the nucleotide composition), T_{mel} (melting temperature of DNA), and CsCl (buoyant density in CsCl).

the kinetics of reassociation, the genomes of these viruses show no marked homology (Amosenko *et al.*, 1976). It has already been mentioned that some fungal viruses have an extremely wide host range, and PBV-1 is a good example in this respect (Tikchonenko *et al.*, 1974). Besides, PBV-1 is not sensitive to B and K specific restriction (Table 6), limiting itself to SK restrictases, whereas PBV-3 is restricted in the usual way and modified in all the strains of *E. coli* tested (Nikolskaya *et al.*, 1976).

Virions of PBV-5 are very peculiar. The form of the tailless head and its size, typical of bullaviruses, indicate that this virus is similar to small phages belonging to this genus (Tikchonenko *et al.*, 1974; Bobkova *et al.*, 1975). To our great surprise, the antiserum to PBV-5 neutralized X174 phage, which is the typical species of this genus, the neutralization constants of both viruses coinciding within the limits of experimental error: 236 and 242 min^{-1}, respectively (Egorov *et al.*, 1976). Despite the presence of the common antigen, the surface properties of the protein capsid of PBV-5 and X174 phage differ greatly. In particular, the fungal virus typically shows an unusually high tendency to aggregation. Therefore, in concentrated viral suspensions there arise stable aggregates that upon titration cause the linear relationship between dose and virus dilution to be broken. The virus titer is higher the greater the dilution. This is manifested particularly well upon titration of concentrated virus suspensions (Table 7). The final yield of PBV-5 after concentration with polyethylene glycol is

TABLE 6
Efficiency of Plating of PBV-1 and PBV-3 on Different Strains of *E. coli*

Virus and its phenotype[b]	Efficiency of plating on strains[a]			
	C	B	SK	K
PBV-1·C	100	100	10^{-3}	—
PBV-1·B	100	100	10^{-3}	—
PBV-1·SK	93	100	100	—
PBV-1·K	100	—	—	100
PBV-3·C	100	10^{-2}	10^{-3}	—
PBV-3·B	100	90	10^{-3}	—
PBV-3·SK	100	10^{-2}	100	—

[a] Efficiency of plating is expressed as percentage of infectious titer of that on *E. coli* C strain devoid of any host restriction (Arber, 1974).
[b] The letters C, B, SK, and K designate the type of host specificity and represent the abbreviated designation of the last strain in which the phage was replicated.

TABLE 7
Anomalous Titration Behavior of PBV-5

Dilution of viral suspension[a]	Titer		Ratio of theoretically calculated and experimentally established titers
	Theoretically calculated	Experimentally established	
10^{-9}	8×10^2	7.5×10^1	10.6
10^{-10}	8×10^1	2.2×10^1	3.6
10^{-11}	8	17	0.47
Percentage of recovery	100%	213%	

[a] The figures given in the table are the results of a typical experiment with viral suspension concentrated by 8% polyethylene glycol. Theoretical values of infectious titers (T_c) in the concentrated suspension ($V_c = 10$ ml) were calculated from the amount of PBV-5 in initial suspension ($T_S = 8 \times 10^{10}$ particles, $V_s = 100$ ml):

$$T_c = \frac{V_s \times T_s}{V_c}$$

Percentage of recovery was calculated with respect to amount of PBV-5 in the initial suspension:

$$\frac{17 \times 10^{11} \text{ particles} \times 10 \text{ ml} \times 100\%}{8 \times 10^{10} \text{ particles} \times 100 \text{ ml}}$$

twice as high as that of the initial virus suspension, which is evidently due to elimination of the inhibitor (see above). Naturally, the inevitable losses of PBV-5 in the course of concentration were not taken into account.

We unsuccessfully tried to overcome this unpleasant tendency of PBV-5 to aggregate by the addition of ionic and nonionic detergents and mercaptoethanol, and by using low ionic strength and chelating agents, etc. This failure to prevent aggregation caused us to make seemingly unnecessary "extra" dilutions upon titration of PBV-5, for example, the dilution 10^{-11} in Table 7. Because of this tendency of the capsid to aggregate, destruction of PBV-5 is accompanied by formation of insoluble protein sediments (Tikchonenko *et al.*, 1977).

PBV-5 and X174 phage also differ greatly in their host-range spectrum. The effectiveness of infection of *E. coli* C22 and *E. coli* W3110 strains with X174 phage is only 0.02% and 0.004% as compared to the efficiency of plating on *E. coli* CR63. On the other hand, the efficiency of plating of PBV-5 on *E. coli* C22 and *E. coli* W3110 is 100%, whereas the effectiveness of infection of *E. coli* CR 63 strain is 1 order lower. Considerable differences between PBV-5 and X174 phage have also been found in the duration of the latent period (Egorov *et al.*, 1976). As expected, the nucleic acid of PBV-5 is a single-stranded DNA; however, detailed study of the genome of this virus has begun only recently (Tikchonenko *et al.*, 1976). All this shows that, despite remarkable similarities in morphology and antigenicity, these two viruses are distinct species, X174 phage being completely incapable of replication in the mold (see below).

Of doubtless interest in the determination of the extent of the relationship among PBV-5, PBV-6, and CBV-1 are the data on their sensitivity to UV irradiation (Egorov *et al.*, 1977). The pattern of the curves in Fig. 3 shows that, among this group of viruses, PBV-5 is the most radioresistant, whereas the level of inactivation of PBV-6 and CBV-1 is 4 log higher. Since all three fungal viruses are titrated in the same host (*E. coli* C22), the discovered differences in radioresistance should be ascribed exclusively to differences in the properties of the virus particles themselves. Unfortunately, the necessity of using another bacterial strain (*E. coli* CR) for studies on X174 phage deprived us of the possibility of comparing this group of viruses with X174 phage, whose level of inactivation is 6 log higher than that of PBV-5. In any case, the abovementioned high resistance of PBV-5 to UV irradiation is unusual for the group of spherical viruses with single-stranded DNA (Dityatkin *et al.*, 1967; Sinsheimer, 1968). Such abnormal radioresistance may possibly be associated with features of DNA–

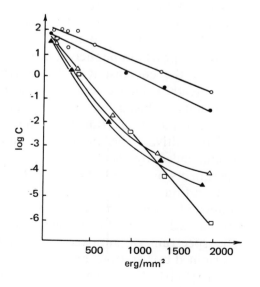

Fig. 3.   Sensitivity of some PB viruses to UV irradiation. PVB-5 from *P. chrysogenum* strain 39 (○), PBV-5 from *P. chrysogenum strain 194* (●), CBV-1 (▲), and X174 phage (□). Lysates of *E. coli* C were subjected to UV irradiation after the removal of the cell debris by low-speed centrifugation. Phage suspensions with titer $10^{10} - 10^{11}$ particles/ml were diluted 10^3- to 10^4-fold with 0.1 M NaCl + 0.05 M phosphate buffer, pH 7, before exposure to UV light of a wavelength of 259 nm. One milliliter of viral suspension was irradiated in 10-cm petri dishes.

protein interactions which may significantly affect virus sensitivity to irradiation (for references, see Tikchonenko, 1975). In this case, PBV-5 will be of undoubted interest to radiobiologists.

The general conclusion may be drawn that the PB viruses under study are similar to viruses of bacteria. Such similarity is not surprising if their "host" dualism and the known similarity of the cell structure of molds and bacteria (for references, see Kwapinski, 1974) are considered. Both types of cells have a rigid polysaccharide membrane, and therefore the mechanisms of infection and virus release from the mold and bacteria may be similar, leading to formation of similar structural elements of virions. Viruses of algae that also demonstrate certain similarities to bacterial viruses are a good analogy (see Chapter 3 of this volume).

4. THE FUNGAL HOST

4.1. The Possibility of Bacterial Contamination

The similar serological, biological, and physical properties of the virus extracted from the mold and grown in bacteria indeed establish the identity of these two viral entities, but cannot be considered to be serious arguments in favor of the fungal origin of PB viruses. If one assumes that in a fungal culture a trivial bacterial contamination is present with its own temperate or infectious bacteriophages, nothing will be left of our entire "biology-revolutionizing" concept of viruses common to prokaryotes and eukaryotes. Naturally, this particular possibility was subjected to the strictest experimental tests before we decided to proclaim publicly a new feature in biology. Subsequently, we came to the conclusion that the great effort spent to procure this experimental evidence was hardly worth it, because the explanation of our results as being due to bacterial contamination was ruled out by prime considerations. But this came later, and at the very beginning we tried to cultivate the mold under conditions almost as rigid as those recommended by the National Institutes of Health, U.S.A., for the experiments on the P3–P4 class with hybrid DNA, and in the course of fermentation took samples nearly every hour for sterility control. In a number of cases, fermentation of the mycelium of *P. brevicompactum* was carried out in the presence of neomycin, an antibiotic with a wide spectrum of antibacterial action. The amounts of PB viruses determined by titrations on bacteria with or without the antibiotic were similar (Velikodvorskaya *et al.*, 1972a; Tikchonenko *et al.*, 1974). We ended with the entire group of our laboratory "going into exile" to study PB viruses. We moved to the Institute of Antibiotics, where no research was done on either bacteria or bacteriophages. There, having broken all contacts with our phage past, we carried out the whole cycle of isolation and identification of PB viruses anew, and obtained the same results.

Table 8 presents the results of titrations of PBV-1 and PBV-3 in cultivation of *P. brevicompactum* and in the analysis of the fungus spores. Two circumstances are noteworthy here. First, PB viruses of *P. brevicompactum* at all periods of fermentation are found only in the mycelium and are absent from the culture fluid. The conditions of separation (filtration) into the sediment and the filtrate were such that the bacterial flora, together with bacteriophages, passed through the filter and were detectable in the culture fluid. PB viruses, however, went into solution only after preliminary grinding of the mycelium or its disruption by ultrasonication. In a similar way, we found no infectious viruses in an aqueous suspension of *P. brevicompactum* spores; however, disruption of the spores was accompanied by the appearance of infectivity toward bacteria. Combined with the abovementioned attempts at supersterility, these data rule out the hypothesis of bacterial contamination in *P. brevicompactum* cultures. The situation with *P. chrysogenum* is somewhat more complicated, since PBV-5 is present both in the culture fluid and in the sediment, even though also in this case the bulk of the virus is associated with the mycelium. Consequently, the negative conclusions concerning bacterial contamination of *P. brevicompactum* cultures hold true equally for *P. chrysogenum*.

4.2. The Presence of PBV Genomes in Fungal DNA

The above data rule out the hypothesis of trivial bacterial contamination as the source of PB viruses, but cannot deal with more sophisti-

TABLE 8
Distribution of PB Viruses in the Fungus Cultures and Spores

Fungus	Titer	
	Mycelium	Cultural fluid
P. brevicompactum[a]	1.2×10^4	0
P. chrysogenum[a]	8×10^3	7×10^1
Intact spores of *P. brevicompactum*	0	—
Disrupted spores of *P. brevicompactum*	1×10^3	—

[a] The data for the 6-day culture of fungi.

cated hypotheses concerning the presence of endosymbiotic bacteria in mycelium that might be the true hosts of our "orphan" PB viruses. The possibility of the existence of such prokaryotic endosymbionts in some fungi is indicated in reports reviewed by Lemke (1976). However, it is not clear to what extent the data for *Paramecium* and *Chlorella*, where such symbionts have been found, can be extended to *Penicillium*. Therefore, this possibility is thus far purely theoretical, based on the well-known triple system of *Paramecium*–endosymbiotic bacteriophage (see Preer *et al.*, 1974). And there are serious grounds for doubts concerning the validity of the "endosymbiotic hypothesis," the most important being our data on the presence of PB virus genomes in fungal DNA.

The hybridization method may be conveniently used for detection of the possible presence of copies of virus genomes in DNA isolated from mycelium. The presence of virus genomes in cellular DNA is known to increase the $C_0t_{1/2}$ of viral DNA in combined annealing with pre-denatured viral and cellular DNAs, the observed increase being proportional to the number of copies of viral DNA in the cell genome.

Fig. 4 presents reassociation curves of [³H]DNA of PBV-1 and DNA from *P. brevicompactum* and some other sources (Amosenko *et al.*, 1976). Similar data were obtained for PBV-3 DNA (Surkov *et al.*, 1975). When PBV-3 DNA and mold DNA were annealed, the $C_0t_{1/2}$ value was 4.5×10^{-3}, while in the control tests where this virus DNA and that of chick embryos were annealed this value was 6.0×10^{-2} (Surkov *et al.*, 1975). When DNAs from PBV-1 and fungal mycelium were annealed, the $C_0t_{1/2}$ value was 3.2–4.8×10^{-3}, while in the control test (PBV-1 DNA plus thymus DNA) it was 4.5×10^{-2} (Amosenko *et al.*, 1976). In comparing the reassociation kinetics of PBV-1 and PBV-3 DNAs from these two studies it should be remembered that the authors used different experimental conditions. The above data unequivocably attest to the presence in the DNA isolated from fungal mycelium of sites homologous to viral DNAs. The increase in reassociation rate for PBV-3 DNA was $6 \times 10^{-2}/4.5 \times 10^{-3} = 13.3$. Similarly, the increase in reassociation rate for PBV-1 DNA was $4.5 \times 10^{-2}/3.2 \times 10^{-3} = 14$. On the basis of the experimental data and the molecular weight of fungal and viral DNA, the number of virus genomes per haploid genome of *P. brevicompactum* can be calculated. The size of the fungal genome calculated from the reassociation kinetics is 7.1×10^9 daltons (Surkov *et al.*, 1975; Amosenko *et al.*, 1976), whereas the molecular weight of PBV-1 DNA is 25×10^6 and that of PBV-3 DNA is 22×10^6 (Chaplygina *et al.*, 1975; Amosenko *et al.*, 1976). Therefore, in the

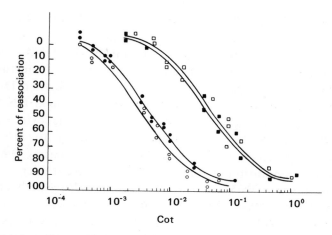

Cot

Fig. 4. Kinetics of reassociation of PBV-1 DNA in the presence of DNAs of different origin. A mixture of [³H]PVB-1 DNA (0.07 OD/ml with radioactivity of 70 × 10³ cpm/ml) and unlabeled *P. brevicompactum* DNA (7.0 OD/ml) (●); a mixture of [³H]PBV-1 DNA (0.07 OD/ml with radioactivity of 70 × 10³ cpm/ml) and unlabeled PBV-1 DNA (0.7 OD/ml) (○); a mixture of [³H]PBV-1 DNA (0.7 OD/ml with radioactivity of 70 × 10³ cpm/ml) and PBV-3 DNA (7.0 OD/ml) (■); mixture of [³H]PBV-1 DNA (0.7 OD/ml with radioactivity of 70 × 10³ cpm/ml) and unlabeled thymus DNA (7.0 OD/ml) (□). From Amosenko *et al.* (1976). OD values were measured at 260 nm. DNA in the incubation mixture was denatured by heating, and subsequent reassociation was performed at 68°C in 1 × SSC. The values of reassociation were measured after chromatography on the hydroxylapatite columns.

DNA extracted from *P. brevicompactum* there are 3.7 copies of the PBV-3 genome and 35.1 copies of the PBV-1 genome, altogether 38.8 viral genomes. No such data are thus far available for other viruses of the PB group.

For the proper evaluation of the experimental data, it should be remembered that a negligible amount of infectious virus particles is present in the fungal mycelium. For instance, PBV-1 and PBV-3 titers in the mycelium of *P. brevicompactum* used in the hybridization experiments were 1 × 10³ and 3 × 10³ particles/g of dry mycelium. At these titers, the number of genomes in the form of infectious virus will be 1 × 10⁻⁹ for PBV-1 and 3 × 10⁻⁹ for PBV-3 per fungal genome. Thus the total amount of virus genomes found in the mold is 10⁹–10¹⁰(!) that of virus genomes in the form of infectious virus. Despite the approximation of such calculations, the difference between the content of virus genome in the form of free DNA and in the form of infectious PBV particles is far greater than the possible error, which permits certain conclusions on the state of this DNA in fungal cells and, in particular, an evaluation of the likelihood of endosymbiotic bacterial origin of PB viruses.

First, with regard to "fungal" or "mycelial" DNA, we strictly speaking mean the total DNA preparation extracted from this source. Had endosymbiotic bacteria been present in the mold, their DNA could have got into the preparation together with truly fungal DNA. This is no more than a statement of possibility, since the existence in the fungus of bacterial endosymbionts and the conditions for DNA extraction from such bacteria remain unknown. Unfortunately, we failed to obtain a pure preparation of fungal nuclei or to use milder methods for isolation of fungal DNA (Amosenko *et al.*, 1976) which could have experimentally disproved the endosymbiotic hypothesis. In the light of the foregoing, therefore, it should be assumed that a preparation of DNA extracted from *P. brevicompactum* mycelium may, together with purely fungal DNA, also contain a hypothetical admixture of DNA of hypothetical endosymbiotic bacteria. In this case, the copies of virus genomes present in the preparation of mycellar DNA would be associated not with the fungal genome but with the bacterial genome, which represents part of the total DNA preparation. To date, we have failed to exclude this explanation experimentally, but some general considerations make it very unlikely.

According to the above calculations, 1 g of mycelium should contain 1.7×10^{11} cells of fungus and the appropriate amount of haploid fungal genomes. Since in one fungal genome there are 38.8 genomes of PBV-1–PBV-3, the total number of virus genomes will be 6×10^{12}. Now let us assume according to the "endosymbiotic" hypothesis that a bacterium lysogenized in the common way is the real carrier. For the sake of simplicity, let us say that each such endosymbiont is double lysogen and may, when necessary, carry both PBV-1 and PBV-3 prophages. Thus, ignoring the existence of the PBV-2 genome in the mycelium of *P. brevicompactum* (although like its odd-numbered brothers it should have a prophage state), one must postulate the presence of 1 g of fungal biomass of at least 3×10^{12} lysogenized bacterial cells. To fulfill this requirement, we should generously have to introduce into a fungal cell at least 15 bacteria (!). If it is assumed that a cell of *E. coli* type contains about 0.8% DNA (3×10^9 daltons or 5×10^{-15} g), the total mass of postulated endosymbionts will be 1.96 g, i.e., nearly twice the weight of the mycelium. Obviously, this is absurd. Of course, some inaccuracies may be found in the chain of reasonings and calculations. For instance, it is easy to imagine that in reality the number of copies of the virus genome is 2–3 times less or that the cell mass of mycelium is twice as large; however, these corrections will not change the essence of the matter. The hypothesis that endosymbiotic bacteria are possible carriers of PB proviruses in the fungus is conceivable only if the number of such endosymbionts does not exceed a quota of one or two bacteria per individual mycelium. Naturally, this condition only makes such an idea possible; it does not prove it. Thus far, our scientific evidence indicates that the more likely explanation lies in the lysogenic state of the fungus itself.

It seems, therefore, that our data on the marked difference in the total amount of virus genomes and the amount of virus genomes in the form of PB virus are best explained on the basis of analogy with integrational virus infections. In these cases, the main and sometimes the only form of the virus genome is provirus in the form of DNA integrated into the host chromosome or acquiring the status of plasmid (Zhdanov and Tikchonenko, 1974). The best analogy to describe all the accrued facts is, as has already been mentioned, lysogenized bacteria with the usual frequency of prophage induction of 10^{-6}–10^{-7} (Hayes, 1964). When suspensions of lysogenized cells are analyzed, usually one or two phage genomes per host genome are found in the form of prophage and 10^{-6}–10^{-7} phage genomes in the form of infectious virus particles. We assume, therefore, that the discovered system *Penicillium*–PB virus represents the known analogue of lysogenized bacteria, i.e., lysogenized fungus. From the fact that the amount of PBV-1 DNA in the form of infectious PB virus under standard conditions is 10^{9}–10^{10} times less than the amount of DNA existing in the form of provirus, it must be postulated that lysogenized provirus has a very strong repression system. Under certain conditions, infectious PB virus can practically disappear from mycelium (see above), which must also be associated with a very strict control of provirus transition into the lytic state. Actually, temperate phages may show great differences in the degree of control by the repressor, down to complete loss of the capacity for induction by defective lysogenic phages (Bertani and Bertani, 1971; Ravin, 1971).

4.3. Fungus as the Object of Infection by PB Viruses

The first question which the reader may ask the author of this chapter is whether it would not have been simpler to infect the fungus experimentally with PB viruses instead of making all the foregoing complicated postulations. We asked ourselves this basic question, but a lot of time elapsed before we could get a more or less definite answer to it.

The first stage of infection is known to be adsorption, and any self-respecting virus, before penetrating into the cell, must be specifically adsorbed on its surface. The specificity is the mandatory condition of this process, for had viruses been able to penetrate into all the cells they encountered, the kingdom of Vira would have long ceased to exist. The results presented in Table 9 answer these questions unequivocally. At least PBV-5 and PBV-6 adsorb onto mycellar cells specifically, and the degree of adsorption on the "parental" mold is much higher than on molds of other species. No such data for viruses of *P. brevicompactum* are now available.

It will be seen in Table 9 that, for example, PBV-5 adsorbs better on fungi than on bacteria, but the reverse is true for X174 phage. PBV-6 adsorbs best on the strain of fungus from which it originates and on susceptible bacteria, whereas on other fungal strains PBV-6 either does not adsorb or does so poorly (Egorov *et al.*, 1976). Thus the experimental data suggest that the surface structures of the fungal host and the virus somehow provide the conditions for their specific interaction. It is not clear yet whether there are special receptors on the mycelium surface with which PB virions interact.

It would seem that successful adsorption of PB viruses on the surface of mycelium creates conditions for the development of the

TABLE 9
Adsorption of PB Viruses by Different Host Cells[a]

| | Host cells | | | | |
| | Bacteria | | Fungi | | |
Virus	E. coli C	E. coli CR	P. chrysogenum	P. cyclopium	C. acremonium
PBV-5	36	—	53	0	0
PBV-6	18	—	0	20	11
φX174	—	80	6	0	0

[a] The decrease in virus titer after incubation with fungus or bacterium suspension has been considered as adsorption. Its value was calculated as percentage with respect to the initial amount of virus before the addition to the cell suspension. Adsorption experiments were performed with bacterial and fungal suspensions having equal values of turbidity, which corresponded to the concentration of *E. coli* C cells, 1×10^9/ml. Bacterial and fungal suspensions in 0.05 M phosphate buffer, pH 7, were incubated for 20 min at room temperature with the virus of known titer; thereafter, the cells were removed by low-speed centrifugation and washed once with the same volume of phosphate buffer. Both supernatants were pooled and titrated.

TABLE 10
Infectious Titers of PBV-5 and Penicillin-Synthesizing Activity in Strains of *P. chrysogenum*

Experiment No.	Strains	Penicillin-synthesizing activity[a] (%)	Titer of PBV-5/g of mycelium
1	39	100	2.0×10^4
	91	10	0
2	39	60	3.3×10^1
	91	6	0
3	39	92	1.7×10^2
	91	9	0
4	39	98	5×10^4
	91	12	0
5	39	98	1.4×10^3
	91	12	0
6	39	100	7×10^4
	91	9.8	0
7[b]	39	97	2×10^4
	91	12	7×10^1
8	39	100	2×10^3
	91	15	2×10^2
9	39	98	1.7×10^3
	91	12	1.0×10^1
10	39	102	2.5×10^4
	91	14	3.6×10^2

[a] Percentage of the activity of the industrial strain N194C taken as 100%.
[b] There was an interval of 7 months between the experiments 1 and 7.

infectious process. In this connection, attempts were made to artificially infect the mold with PB viruses. For this purpose, thermoresistant virus-free variants of *P. chrysogenum* were specially obtained (Bobkova *et al.*, 1975). It is of interest that the loss of virus was accompanied by a sharp reduction in the capacity for penicillin synthesis and by characteristic changes in the morphology and physiology of the mycelium (the so-called phenotype of sick mycelium).

If our assumption concerning lysogeny of fungi is correct, our selection of "virus-free" variants consisted of the elimination of inducible provirus, i.e., the lysogenic state itself, rather than in elimination of free PBV-5. As an alternative, it may also be assumed that we selected variants of cells with reduced capacity to provirus induction (compared with defective temperate viruses).

Later, it was found that within 8–9 months in such "virus-free" variants as strain 91 in Table 10 a virus infectious for bacteria appeared, possibly as a result of repeated laboratory contamination which, unfortunately, could not be reproduced in the controlled experimental conditions. It cannot be ruled out either that in a small portion of such virus-free variants of cells provirus remains. Replication of this cell population at normal temperature leads to the restoration of the lysogenic state which gives rise to the viruses infectious for bacteria, imitating a repeated infection. In any case, for the artificial infection experiments, virus-free variants had to be prepared anew, using strictly virus-free lines and checking spontaneous appearance of infectious PB viruses.

Noteworthy also in Table 10 is the occurrence of certain correlations between the capacity of a fungal strain for antibiotic production on the one hand and PB virus titers on the other, which we have already mentioned in connection with Fig. 2. Generally, in most active antibiotic producers of the type of strain 39 presented in Table 10, the titer of PBV-5 is higher than in the strains producing lower amounts of penicillin. Also, the strains actively producing the antibiotic and displaying highest PBV-5 titers have the typical altered morphology (the phenotype of sick mycelium) (Bobkova *et al.*, 1975). Virus-free strains of *P. chrysogenum* synthesizing small amounts of penicillin simultaneously recovered their normal morphology, the phenotype of healthy mycelium. No such relationship was found in *C. acremonium* (Egorov *et al.*, 1976). Although *P. brevicompactum* produces no penicillin, the strains which were active carriers of PB viruses also showed the altered morphology typical of the sick phenotype (Velikodvorskaya *et al.*, 1974).

Egorov *et al.*, (1976) made numerous attempts to infect with PBV-5 virions both intact cells of *P. chrysogenum* and fungal protoplasts obtained using snail enzyme (Bachman and Bonner, 1959; Lhoas, 1971; Tonino and Steyn-Parve, 1963). Incubation of the intact mycelium of virus-free strain 596 with PBV-5 preparations grown in bacteria, under different conditions and for various periods of time, produced no infection (18 independent attempts). The progeny of such cells after washing and transfer to the fresh medium were sterile and produced no PBV-5 (Egorov *et al.*, 1976). These authors observed that, during a long-term incubation of virus-free mycelium (strain 91 or 596) with the virus, the infectious titers declined steadily down to zero by the fifth day of incubation. It is not yet clear whether this is due to inactivation of PBV-5 because of irreversible adsorption on the mycelium surface or due to the effect of some inactivating factors in the medium.

Only the experiments in which *P. chrysogenum* protoplasts were infected with the virus were successful. Altogether, 148 variants obtained in several experiments and representing the progeny of the protoplasts giving rise to viable cells were tested. In 12 lines, i.e., in 8%, the progeny of infected protoplasts produced infectious PBV-5 in titers of $10^4 - 10^6$. The success of infection did not depend on the multiplicity of infection within the range of $10^{12} - 10^8$ particles/ml, but further reduction of the dose caused no infection of the protoplasts. The virus synthesized in the progeny of the infected protoplasts was typical PBV-5.

At the same time, this group of authors carried out experiments attempting infection of *P. chrysogenum* protoplasts with X174 phage (three experiments) and with deproteinized preparations of PBV-5 DNA (32 experiments) with negative results (Egorov *et al.*, 1976). As far as we know, no experiments with mold infection using other PB viruses have been done. In evaluating the significance of the experimental results, several possible explanations can be given: (1) in the experiments, under the influence of virus particle or viral nucleic acid penetrating into the protoplasts, the repression of provirus is weakened nonspecifically, causing an increase in the infectious titer of PBV-5; (2) actually, no infection but rather persistence of the infecting virus occurs; (3) protoplasts are actually infected with PBV-5 either in a lytic manner or with final production of provirus.

The first two explanations seem artificial and do not correspond to the facts. Against the first hypothesis is the fact that infection with X174 phage or PBV-5 DNA preparations as well as infection of the intact mycelium with PBV-5 virions does not produce the assumed biological effect (slackening of repression). Furthermore, the capacity to produce PBV-5 is retained in the progeny upon repeated transfers. The latter circumstance also excluded hypothesis (2). In our favor is the abovementioned fact that PBV-5 added to mycelar suspension is eliminated comparatively rapidly. In other words, the conditions of fungal cultivation do not provide for any long-term persistence of free virus in suspension.

It seems, therefore, that the contact of PBV-5 virions with *P. chrysogenum* protoplasts leads to infection of the mold and its lysogenization. The relatively low percentage of infected protoplasts (8%) may be due to the fact that the authors have not yet found the optimal conditions of infection and cultivation. It may also be due to a special biology of the mold and peculiar relationships of the mold and viral genomes, of which almost nothing is thus far known. The capacity of PBV-5 to infect only *P. chrysogenum* protoplasts resembles the behavior of RNA-containing mycophages, the successful infection with which was also possible only after removal of the rigid polysaccharide cell wall of the mycelium (for references, see Chapter 2, and Lemke and Nash, 1974)

4.4. Induction of PBV Proviruses in *Penicillium chrysogenum* Mycelium

The direct way to check the integrative nature of one or another virus infection is to determine the presence of virus genomes in the cellular DNA, which may be done either by physicochemical methods or biologically by provirus induction. The successful results obtained by the first approach and described above in Section 3.2 inspired us to carry out induction of PB provirus. Unfortunately, for a number of technical reasons, we had to begin this experiment not with PBV-1 or PBV-3, for which the presence of virus genome in the cellular DNA has been proved, but with PBV-5, for which we have no such evidence. This circumstance decreases the validity of the experimental data to some extent.

Beginning this series of experiments and preparing ourselves for a possible failure, we recalled that induction did not always accompany integrative infections. Oncogenic proviruses do not yield to induction and may only be sometimes "exhumed" from the transformed cell upon formation of symplasts or using biologically active DNA (for references, see Zilber *et al.*, 1975). Defective temperate phages are not induced (Bertani and Bertani, 1971; Ravin, 1971). There is no guarantee that the inducers active in lysogenic bacteria will be effective in molds.

Nevertheless, since one has to begin with something, we naturally began with traditional UV irradiation, proflavin, *N*-nitrosomethylbiuret, and acridine orange (Lebed' *et al.*, 1975). Different variations of the treatment to which the seed material had been subjected were tested. The treated spores or mycelia were subjected to additional fermentation for 48 hr in order to increase the cell mass, and then PBV-5 titres were determined in them and compared to those in controls not treated with these agents. The results are presented in Table 11.

It will be seen that the treatments usually producing effective induction of prophage in lysogenic bacteria exerted a similar effect on the mycelium of *P. chrysogenum* containing the PBV-5 genome. Unfortunately, thus far our studies have been limited to *P. chryso-*

TABLE 11
Influence of Possible Inducers on the Titer of PBV-5 in the Mold

		Titer			
		P. chrysogenum 39		*P. chrysogenum* 91	
Inducer	Conditions for treatment	Control	Experiment	Control	Experiment
UV light	2,000 erg/mm²	1.4×10^3	3.4×10^4	5×10^2	1×10^4
	12,000 erg/mm²	1.6×10^2	3.2×10^5	2×10^2	1×10^4
Nitrosomethyl-	0.05% 6 hr	8×10^2	2×10^3	1×10^2	5×10^1
biuret	0.05% 24 hr	2×10^3	2×10^4	5×10^2	5×10^1
Proflavine	0.005%	2×10^2	2×10^3	2×10^1	1×10^5

[a] *P. chrysogenum* 39 is highly active penicillin-producing industrial strain, *P. chrysogenum* 91 is its thermoresistant variant with low antibiotic-producing activity (see footnote to Table 10). The table represents results of a typical experiment.

genum and PBV-5. It appears to be too early to speak of the quantitative aspect of the observed phenomenon in molds, since at present the optimal conditions of induction are only groped for. Thus it will be seen in Table 11 that increased doses of UV irradiation, about $10-12 \times 10^3$ erg/mm², are optimal. Chemical inducers with one exception give a lower increase of the infectious titers, within the range of 1–1.5 log. The exception is the strain 91 with low antibiotic activity, in which treatment with proflavin produced an increase of the infectious titer by 4 log, a kind of record. It should also be noted that the conditions of induction of PBV-5 provirus in strains 39 and 91 differ considerably. On the whole, cells of strain 91 containing lower titers of the infectious virus are more sensitive to induction than the cells of strain 39. Another feature of this strain is that *N*-nitrosomethylbiuret causes not induction but rather elimination of the infectious PBV-5 in mycelium. Acridine orange in the concentrations tested exerts a toxic effect on mold cells and causes a loss in titer of PBV-5 (Lebed' *et al.*, 1975). Therefore, if the strains 39 and 91 of *P. chrysogenum* really have the inducible PBV-5 provirus, the conditions of its functioning and activation must differ considerably.

Despite a doubtless increase in the titer of the infectious virus, the level of induction even under optimal conditions (high doses of UV light and proflavin for strain 91) is quite low. Indeed, even under conditions of induction, the amount of PBV-5 per cell increases from 10^{-7}–10^{-10} to only 10^{-5}–10^{-6}. Consequently, in this case also, only a negligible proportion of cells in the population are capable of supporting replication of the infectious virus. Unfortunately, at the present time we have just started to determine the number of copies of the PBV-5 genome in mold mycelium, and therefore cannot properly explain the insignificant level of induction. It cannot be completely ruled out that, in contrast to PBV-1–PBV-3 and *P. brevicompactum*, only a small number of *P. chrysogenum* cells contain the PBV-5 genome in the form of provirus. Finally, another circumstance about which we thus far have kept delicate silence should be mentioned. This is that the PBV-5 genome is a single-stranded DNA. According to the classical theory of integration developed with λ phage, it is impossible to explain how a subtle genetic operation such as the "chaste conception" of provirus can be performed with such an imperfect form of the genome as a single-stranded DNA. True, such an outcome may be conceivable if one remembers the double-stranded ring replicative form which the DNA of bullaviruses acquires in the infected cell (for references, see Sinsheimer, 1968). There appears to be nothing seditious in the assumption that integration may occur also at the level of the replicative form. This possibility is supported by the fact that λ phage DNA also integrates in the replicative form, as a covalently closed circle, whereas the virion form of DNA is a linear duplex with cohesive ends (Yarmolinsky, 1971). At the same time, it may be recalled that the provirus of oncornaviruses is by no means single-stranded RNA (for references, see Zilber *et al.*, 1975). Even though the considerations presented here with regard to a possible mechanism permitting integration of the genome of a single-stranded DNA virus in a host DNA seem quite probable, more direct evidence is desirable. In particular, for PBV-5 it is urgently necessary to find out whether copies of the virus genome are present in cellular DNA, as mentioned above. With regard to the other viruses of this group (PBV-1–PBV-3), there appear to be no difficulties with the problem of provirus, since their DNA is double stranded, although it is still not quite clear whether it may produce replicative ring forms in the infected cell. In a similar way, we still do not know whether PBV-1 and PBV-3 proviruses can be induced; this must be the object of research in the nearest future.

5. CONCLUDING REMARKS

In presenting the experimental material from our studies on PB viruses, I deliberately avoided touching on two related but as yet unresolved problems. These problems should at least be mentioned so that silence does not imply deliberate avoidance. The problems are, first, a possible relationship between PB viruses and physiological-biochemical properties of the mold, including an important property such as the capacity for antibiotic synthesis, and, second, the relationship between DNA-containing PB viruses and dsRNA-containing mycophages.

In Fig. 2 and Table 10, the correlation between the capacity to produce penicillin and morphological variation in *P. chrysogenum*, on the one hand, and the titers of PB viruses, on the other, was described. In a number of cases, thermoresistant variants of the mold free of virus and synthesizing little penicillin exhibited characteristic alterations in morphology. A certain relationship between morphological features of the mycelium and the presence of PB viruses was observed also for *P. brevicompactum*. At the same time and in all cases, dsRNA mycophages were found in these mycelia, although we did not carry out any quantitative studies on the relationship between these two groups of viruses.

Thus, it cannot be ruled out that PB viruses and possibly mycophages are not merely parasites of viral nature causing one or another disorder of metabolism or damage to cell structure but are in some way integrated with the biology of *Penicillium* and influence in a positive way important physiological and biochemical features of the fungal cell. Without predetermining what is cause or effect and without drawing any ultimate conclusions, we can say with regard to the association of antibiotic production with the PBV-5 genome that it is logical to assume a certain influence of virus infection on the mold. It should be noted that such an assumption in a general form has been made before (see Lemke, 1976). Such an idea seems particularly complementary to the abovementioned concept of the lysogenic state of the mold. By analogy with temperate phages and lysogenic bacteria, it is easy to imagine that incorporation of provirus genomes into a fungal chromosome may directly affect the expression of one or another fungal gene, including genes responsible for antibiotic production.

The occurrence of viruses common to fungi and bacteria makes direct exchange of genetic information between prokaryotes and eukaryotes possible either in the form of defective transducing virions or in the form of pseudovirions. Such a "channel of genetic communication" between bacteria and fungi has definite evolutionary implications for the role of viruses as vectors in the evolution of life (Zhdanov and Tikchonenko, 1974).

6. REFERENCES

Adams, M. H., 1959, *Bacteriophages*, Interscience, New York.
Amosenko, F. A., Surkov, V. V., and Tikchonenko, T. I., 1976, On the presence of

PBV1 virus genomes in mycelium of *Penicillium brevicompactum, Vopr. Virusol.*
   3:361–366.
Arber, W., 1974, DNA modification and restriction. *Prog. Nucleic Acid Res. Mol.
   Biol.* **14**:1–37.
Bachman, B. J., and Bonner, D. M., 1959, Protoplasts from *Neurospora crassa, J. Bac-
   teriol.,* **78**:550–556.
Bertani, L. E., and Bertani, G., 1971, Genetics of P2 and related phages, *Adv. Genet.*
   16:200–237.
Bobkova, A. F., 1972, Isolation and biochemical properties of PBV2 page, Ph.D.
   thesis, Institute of Virology, Moscow.
Bobkova, A. F., Velikodvorskaya, G. A., Tikchonenko, T. I., Egorov, A. A., and
   Bartashevich, Y. E., 1975, Virus infective for bacteria in strains of *Penicillium
   chrysogenum* with various antibiotic activity, *Antibiotics* **7**:600–606.
Chaplygina, N. M., 1974, Studies of some new viruses isolated from the mold
   Penicillium brevicompactum, Ph.D. thesis, Moscow State University.
Chaplygina, N. M., Bobkova, A. F., Velikodvorskaya, G. A., Petrovsky, G. V., and
   Tikchonenko, T. I., 1975, Some biological and physicochemical properties of PBV1
   virus, *Vopr. virusol.* **1**:9–15.
Dityatkin, S. Y., Danieleytchenko, V. V., Zavilgelsky, G. V., and Ilyashenko, B. N.,
   1967, Comparison of UV sensitivity of $1\varphi 7$ and T7 phages and their infectious DNA,
   Genetika **11**:87–92.
Egorov, A. A., Lebed, E. S., Bartashevitch, Y. E., and Tikchonenko, T. I., 1977, Bio-
   logical properties of PB5 virus, submitted for publication.
Gause, G. F., Maximova, T. S., Dudnik, Y. V., Velikodvorskaya, G. A., Klimenko, S.
   M., and Tikchonenko, T. I., 1971, New strains of viruses isolated from mold fungi
   belonging to the *Penicillium* genus, *Microbiologia,* **40**(3):540–542.
Gratia, A., 1936, Des relations numeriques entre bacteries lysogenes et particles de bac-
   teriophage, *Ann. Inst. Pasteur* **57**:652–676.
Hayes, W., 1964, *The Genetics of Bacteria and Their Viruses*, Blackwell, Oxford.
Krivissky, A. S., 1955, Isolation and properties of S_d phage, in: *Problems of
   Pathogenesis and Immunology of Viral Infections* (A. S. Krivissky, ed.), Publishing
   House "Medgiz," Moscow.
Kwapinski, J. B. G., ed., 1974, *Molecular Microbiology*, Wiley, New York.
Lebed', E. S., Bartaschevich, Y. E., Egorov, A. A., Velikodvorskaya, G. A., and Tik-
   chonenko, T. I., 1975, Induction of PBV5 virus in mycelium of *Penicillium chryso-
   genum, Antibiotics* **7**:606–610.
Lemke, P. A., 1976, Viruses of eucaryotic microorganisms, *Annu. Rev. Microbiol.*
   30:105–145.
Lemke, P. A., and Nash, C., 1974, Fungal viruses, *Bacteriol. Rev.* **38**:1, 29–56.
Lhoas, P., 1971, Infection of protoplasts from *Penicillium stoloniferum* with double-
   stranded RNA viruses, *J. Gen. Virol.* **13**:365–367.
Lhoas, P., 1975, Process of cultivation of viruses in the molds, Patent C6F C12K 7/00
   No. 1332381, England.
Nikolsaya, I. I., Trushinskaya, G. N., and Tikchonenko, T. I., 1972, Some properties
   of mononucleotides from the DD VII phage DNA, *Dokl. Akad. Nauk USSR*
   205:241–243.
Nikolskaya, I. I., Lopatina, N. G., Chaplygina, N. M., and Debov, S. S., 1976, The
   host specificity system in *Escherichia coli* SK, *Mol. Cell. Biochem.* **13**:79–87.
Ohno, S., 1970, *Evolution by Gene Duplication*, Springer-Verlag, Berlin.

Preer, J. P., Jr., Preer, L. B., and Jurand, A., 1974. Kappa and other endosymbionts in *Paramecium aurelia, Bacteriol. Rev.* **38(2):** 113–163.

Ravin, V. K., 1971, *Lysogeny,* Publishing House "Nauka," Moscow.

Sinsheimer, R. L., 1968, Bacteriophage ϕX174 and related viruses, *Prog. Nucleic Acid Res. Mol. Biol.* **8:**115–150.

Storck, R., 1974, Molecular mycology, in: *Molecular Microbiology* (J. B. C. Kwapinsky, ed.), pp. 424–477, Wiley, New York.

Surkov, V. V., Chaplygina, N. M., Velikodvorskaya, G. A., and Tikchonenko, T. I., 1975, Fungal origin of PB viruses isolated from *Penicillium brevicompactum, Microbiologia* **44:**905–909.

Tikchonenko, T. I., 1975, Structure of viral nucleic acids *in situ,* in: *Comprehensive Virology,* Vol. 5 (H. Fraenkel-Conrat and R. Wagner, eds.), pp. 1–118, Plenum, New York.

Tikchonenko, T. I., Velikodvorskaya, G. A., Bobkova, A. F., Bartashevich, Y. E., Chaplygina, N. M., and Maksimova, T. S., 1974, New fungal viruses capable of reproducing in bacteria, *Nature (London)* **249:**454–456.

Tikchonenko, T. I., Surkov, V. I., Velikodvorskaya, G. A., and Bartashevich, Y. E., 1977, Biochemical characteristics of PB5 virus, submitted for publication.

Tonino, G. J., and Steyn-Parve, E. P., 1963, Localization of some phosphatase in yeast, *Biochim. Biophys. Acta* **67:**453–469.

Velikodvorskaya, G. A., Bobkova, A. F., Maximova, T. S., and Tikchonenko, T. I., 1972a, New viruses isolated from fungi of *Penicillium* genus, *Bull. Exp. Biol. Mec.* (USSR) **5:**90–93.

Velikodvorskaya, G. A., Bobkova, A. F., Petrovsky, G. V., and Tikchonenko, T. I., 1972b, Physical and chemical properties of PB2 phage, *Vopr. Virusol.* **3:**332–335.

Velikodvorskaya, G. A., Chaplygina, N. M., Petrovsky, G. V., Maximova, T. S., and Tikchonenko, T. I., 1974, Physicochemical properties of mycovirus isolated from *Penicillium brevicompactum, Vopr. Virusol.* **4:**442–445.

Wood, H. A., Bozarth, R. E., and Mislivec, P. B., 1971, Virus-like particles associated with an isolate of *Penicillium brevicompactum, Virology* **44:**592–598.

Wuest, P. J., and Smith, S. H., 1974, Use of sterilized soil as a storage medium for virus infected cultures of *Agaricus bisporus, Proc. Am. Phytopathol. Soc.* **1:**116–120.

Yarmolinsky, M., 1971, in: *The Bacteriophage Lambda* (A. D. Hershey, ed.), pp. 130–150, Cold Spring Harbor Laboratory, Cold Spring Harbor, N. Y.

Zhdanov, V. M., and Tikchonenko, T. I., 1974, Viruses as a factor of evolution: Exchange of genetic information in the biosphere, *Adv. Virus Res.* **19:**361–394.

Zilber, L. A., Irlin, I. S., and Kisselev, F. L., 1975, *Evolution of the Viral-Genetic Theory of Cancer Development,* Publishing house of "*Nauka,*" Moscow.

CHAPTER 5

Bacteriophages That Contain Lipid

Leonard Mindich

Department of Microbiology
The Public Health Research Institute of the City of New York, Inc.
New York, New York 10016

1. INTRODUCTION

The several types of lipid-containing bacteriophages that have been described so far do not constitute a natural grouping. They differ from each other in many ways including host range, nucleic acid type, mode of attachment to host cells, and location of the lipid in the virion. It is, however, useful to discuss them together since they are likely to prove of immense value in the elucidation of lipid–protein interactions and the biogenesis of structures containing lipid. The anticipated advantages of these viruses for the study of "membrane biology" are to be found in their rather simple and probably stoichiometric composition with respect to proteins and nucleic acids, which is characteristic of almost all viruses, as well as the ease of preparing large amounts of virus material for structural work and the possibility of isolating conditional-lethal mutants that will facilitate the investigation of the morphogenetic pathways employed by these viruses.

The structure of PM2 has been studied in great detail, and the virion has been reconstituted from its components. Bacteriophage $\phi 6$, which has not been studied in as great physical detail, has been studied genetically, and mutants have been isolated that are defective in proteins necessary for proper morphogenesis. Bacteriophages PR3, PR4, and PRD1 can infect *Escherichia coli* if certain plasmids are present in the host. This opens the study of phage lipid–protein structure and morphogenesis to manipulation with a large number of well-characterized mutations involving nucleic acid, protein, and lipid metabolism.

The working out of the developmental mechanisms of the bacteriophages that contain lipid should be of importance in the general understanding of membrane biogenesis and lipid–protein interactions. Among the questions that can be posed at the present are those of the asymmetry of the structure with respect to both lipids and proteins, how the proteins and lipids are assembled onto or into this structure, how they are transported from their sites of synthesis to the developing virion, and when in the time course of virus development the "membrane" is assembled. Is it at an early stage as in the case of poxviruses (Dales and Mosbach, 1968), or is it at a late stage as in the case of budding viruses such as the paramyxoviruses, herpesvirus, and togaviruses (Casjens and King, 1975)?

The first identification of a bacteriophage that contains lipid was the description of PM2 by Espejo and Canelo (1968a). Using information that Spencer (1963) had reported indicating that two bacteriophages active against marine pseudomonads were inactivated by organic solvents, they isolated several strains of marine pseudomonads from the Pacific ocean off the coast of Chile. A bacteriophage that grew on one of these strains was found in the same water and was named PM2. The host organism was later designated BAL-31 (Espejo and Canelo, 1968c). PM2 was found to be sensitive to inactivation by ether, chloroform, and 0.05% Sarkosyl. In cesium chloride, the buoyant density of PM2 was found to be 1.28 g/ml, and chemical analysis of the virus showed that the virion contained DNA phosphorus and an additional phosphorus-containing material that could be extracted with chloroform–methanol. This material proved to be phospholipid on the basis of thin-layer chromatography. The dry weight of the extracted lipids was found to be 10.5% of the dry weight of the purified virus.

Subsequent to the description of PM2 as a lipid-containing bacteriophage, there have been a number of claims of phages with lipid components. Many of these claims have been based on the low buoyant density of virus particles and their sensitivity to organic solvents. These two parameters, although indicative, are not sufficient to define a lipid component. For example, the filamentous single-stranded DNA phages of *E. coli* do not contain lipid, although they have a buoyant density of 1.29 and are sensitive to chloroform (Marvin and Hohn, 1969).

The lack of sensitivity to organic solvents is not a definite proof of the absence of structural lipid, either. The various lipid-containing bacteriophages differ markedly in their sensitivities to detergents and to organic solvents. Although the sensitivities have not been tested in a single experiment, it is of interest to compare the conditions and extents of inactivation with chloroform that have been reported for several of the lipid-containing bacteriophages. Bacteriophage ϕ6 is the most sensitive, being inactivated to the extent of 10^{10}-fold after several seconds of exposure (Vidaver et al., 1973); PM2 is inactivated 10^7-fold after 24 hr of exposure (Espejo and Canelo, 1968a); and PR3 is inactivated only ten- to a hundredfold after 4 hr in the presence of chloroform (Bradley and Rutherford, 1975). Bacteriophage fd, which does not contain lipid, is probably more sensitive to chloroform inactivation than is PR3, which does contain lipid. Even the determination of lipid in purified preparations of the virus particles, which is the only proper index of a lipid component, can be very difficult because of possible contamination of the virus preparation with host membrane fragments.

The buoyant density of the cytoplasmic membrane of gram-negative bacteria is 1.15–1.17, and that of the outer membrane is 1.21–1.23 in sucrose (Osborne et al., 1972). This material can easily contaminate preparations purified by isopycnic banding, since the density of lipid-containing phages is near 1.21 in sucrose, and membrane material distributes between the densities of the inner and outer membranes.

Some of the properties of lipid phages are listed in Table 1. Only those viruses that have been demonstrated to contain lipid by chemical analysis are listed. Although the list is small, it can be expected to expand, since most of the isolations are quite recent. Many bacteriophages have been described in the literature as containing lipid on the basis of buoyant density and organic solvent sensitivity (see Franklin, 1974); these are not listed since it is not yet demonstrated that they actually contain lipid. Most of the discussion in this chapter will deal with only two of the bacteriophages, PM2 and ϕ6. Although this choice is dictated by the fact that most of the work on lipid-containing phages has been done on these two viruses, it is fortuitous because the two are structurally quite different, and if there is any possibility at all of classifying lipid phages on the basis of structure, then ϕ6 will probably stand alone, whereas PM2 will be seen as a prototype for most of the other lipid-containing phages. The major difference between the two types is in the positioning of the lipids with respect to the exterior of the virion. Bacteriophage ϕ6 is composed of an icosahedral core that contains proteins and nucleic acid. The core is enveloped by a lipid and protein structure that can be removed by weak detergents, much as the envelope of herpesvirus (Spear and Roizman, 1972) or the togaviruses (Strauss et al., 1968) can be removed. The

TABLE 1
Some Characteristics of Lipid Phages

Phage	Nucleic acid	Lipids	Duoyant density in CsCl (g/ml)	Diameter	Attachment structure	Host	Reference
PM2	dsDNA	10%	1.28	60 nm	Spikes	*Pseudomonas* BAL-31	Espejo and Canelo (1968*a*)
φ6	dsRNA 3pc	12%	1.27	80 nm	?	*P. phaseolicola* HB10Y	Vidaver *et al.* (1973)
PR3, PR4, PRD1	dsDNA	12%	1.265	65 nm	?	*E. coli* and *P. aeruginosa* with plasmids of P, N, or W compatibility groups	Bradley and Rutherford (1975), Sands (1976)
AP50	RNA	14.3%	1.307	80 nm	?	*B. anthracis*	Nagy *et al.* (1976)
DS6A	n.d.	11.7%	n.d.	n.d.	n.d.	*M. tuberculosis* H37RV	Bowman *et al.* (1973)
φNS11	dsDNA	?	1.27	70 nm	Spikes	*B. acidocardarius* TA6	Sakaki and Oshima (1976)

lipids in $\phi6$ are accessible to phospholipase, indicating their proximity to the exterior of the particle (Vidaver et al., 1973). PM2 lipids are not at the outside of the virion. The PM2 structure is not sensitive to nonionic detergents and physical analysis, in particular, low-angle X-ray diffraction. Harrison et al., (1971) have shown that the lipid region of the virion is interior to the outer protein layer of the virion. In electron micrographs, $\phi6$ has the appearance of a core enveloped by a membranous structure (Fig. 7), whereas PM2, PR3, and the other lipid-containing bacteriophages show a double-layered compact structure (Figs. 1 and 5). A difference that does remain between PM2 and the other phages is that PM2 has spikes at the icosahedral vertices (Espejo and Canelo, 1968a), whereas PR3 may have a tail-like structure (Bradley and Rutherford, 1975) and PRD1 has neither spikes nor tails. Most of the lipid phages described so far reproduce in pseudomonads. It is not clear whether this reflects some aspect of the physiology or environment of these organisms, but the fact that phages such as PR4 and PRD1 which were originally isolated on pseudomonads can grow quite well on E. coli harboring the proper plasmids indicates that the host requirements of these phages are not particularly stringent. The finding of lipid phages that attack mycobacteria and the gram-positive bacilli shows that there is no specific requirement for host cells containing an outer membrane as found in gram-negative microorganisms.

The role of lipid in these bacteriophages is not clear. Perhaps it is simply a manifestation of the possible diversity in biological structures. Most organisms that are infected by lipid phages are also susceptible to the more conventional, lipid-free phages. One might expect that the lipids might protect the phage from deleterious agents such as acids and water-soluble reactive reagents. This possibility has not yet been established as valid for any of the lipid phages except ϕNS11 (Sakaki and Oshima, 1976), which is stable at low pH and at high temperatures.

The preparation of purified lipid phages follows current standard methods used for virus purification. Precipitation of virus material from lysates is usually accomplished by the addition of polyethylene glycol 6000 (Yamamoto and Alberts, 1970). In the case of PM2, the precipitate is collected, resuspended, and further purified in the ultracentrifuge on a cesium chloride step gradient, then another cesium chloride gradient, then a sucrose gradient (Hinnen et al., 1974). Bacteriophage $\phi6$ is precipitated with polyethylene glycol, resuspended, and applied to first a sucrose gradient for zone sedimentation and then isopycnic banding (Vidaver et al., 1973). Isopycnic banding of $\phi6$ in cesium chloride gradients has also been used successfully.

2. PM2

2.1. Virus Characterization

PM2 was the first lipid-containing bacteriophage to be described (Espejo and Canelo, 1968a). Its host is a marine pseudomonad designated BAL-31 (Espejo and Canelo, 1968c). This organism, which has not been further classified, has a requirement for magnesium as well as moderately high concentrations (0.5 M) of sodium chloride. BAL-31 grows at temperatures up to 35°C on a defined medium (Espejo and Canelo, 1968c). The strain behaves as if it contains a mutator gene in that the frequency of mutations to amino acid auxotrophy is much higher than found in other microorganisms (2% of the cells are amino acid auxotrophs in the absence of mutagenesis, Mindich, unpublished results). Another host for PM2 has been isolated from the waters of the East River in New York and is designated ER72. This strain has the growth properties of BAL-31 but does not seem to have a mutator gene and forms smaller plaques with PM2 than does BAL-31 (Mindich, unpublished results).

Electron micrographs of the virus show a particle of icosahedral shape, with a diameter of 60 nm (Fig. 1) and with spikes at the vertices of the icosahedral structure. Two-fold and three-fold axes of symmetry are observed. Two layers are seen in electron micrographs of negatively stained preparations (Espejo and Canelo, 1968a; Silbert et al., 1969), and sectioned specimens reveal a membranelike structure at the periphery of the virion (Espejo and Canelo, 1968b; Dahlberg and Franklin, 1970). The mode of attachment and penetration utilized by PM2 has not been studied to any great extent.

2.1.1. Viral DNA

The virus was originally reported to contain 14% DNA and 10% lipid by weight, with a particle weight of 5.1×10^7 (Espejo and Canelo, 1968a). The molecular weight of the DNA was found to be about 6×10^6, and the DNA was found to exist in the virion as a double-stranded circular molecule with supercoils. The DNA has about 90 superhelical turns per molecule (Franklin et al., 1976) and perhaps higher-order supercoiling as well (Brady et al., 1976). The base composition of the DNA was found to be 42% G+C on the basis of buoyant density and melting temperature (Espejo and Canelo, 1969; Espejo et al., 1969. There is no homology between PM2 DNA and that of the host BAL-31

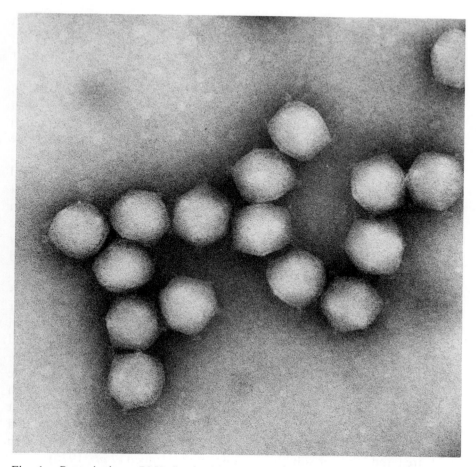

Fig. 1.   Bacteriophage PM2 fixed with glutaraldehyde and negatively stained with phosphotungstate (PTA). From Silbert *et al.* (1969).

(Franklin *et al.*, 1969). The DNA has a sedimentation coefficient of 26.2 S which decays to 21.2 S upon aging. This latter form is probably open-circle, double-stranded DNA. Contour length measurements of the open-circle form indicates a molecular weight of 5.9×10^6. A molecular weight of 6.4×10^6 was calculated on the basis of phage nonlipid phosphorus content (Camerini-Otero and Franklin, 1975).

2.1.2.   Protein Composition of PM2

Four proteins have been identified in purified virions (Datta *et al.*, 1971). Claims of additional proteins have so far not been corroborated

(Brewer and Singer, 1974). The four proteins have been designated in order of decreasing molecular weights as I, II, III, and IV (Datta *et al.*, 1971). Their molecular weights were originally determined by their mobilities in polyacrylamide gel electrophoresis (PAGE) relative to that of known standards (Brewer and Singer, 1974; Datta *et al.*, 1971). The molecular weights have varied between the two laboratories, but a consensus seems to have been reached. The proteins have been isolated and their amino acid contents determined recently. Table 2 lists the minimum weights according to the amino acid analyses (Hinnen *et al.*, 1976). The isoelectric points of the four proteins have been determined (Schäfer *et al.*, 1974a). They are also listed in Table 2. It is of interest that protein II is extremely basic and its isoelectric point varies from 9.0 in the absence to 12.3 in the presence of calcium ions; this will be commented on in more detail later. Earlier claims (Camerini-Otero *et al.*, 1972) that protein IV is a glycoprotein have not been substantiated (Brewer and Singer, 1974).

Two of the PM2 proteins, III and IV, have the solubility properties of proteolipids; they are soluble in organic solvents (Schäfer *et al.*, 1974a). Both of these proteins have calculated polarities (Capaldi and Vanderkooi, 1972) below 40% which correlates with integral membrane protein characteristics (Hinnen *et al.*, 1976); however, protein IV is probably not interacting with lipid. Protein II is not soluble in organic solvents but is rather insoluble in water. It has a high polarity (Table 2), and it is suggested that the polar groups are not accessible to the aqueous environment when the protein is isolated (Hinnen *et al.*, 1976). Protein I is easily extracted from the virus with 1 M urea and is water soluble when isolated (Hinnen *et al.*, 1976).

The relative proportions of the four proteins have been determined from the distribution of ^{14}C in the electrophoresis bands (Datta *et al.*,

TABLE 2
Characteristics of PM2 Proteins

Protein	Electrophoretic mol. wt.[a]	Compositional mol. wt.[b]	PI[c]	Polarity[d]	Mol/virion
I	43,000	43,580	6.2	49.4	~80
II	26,000–27,000	27,660	9–12.3	44.9	~820
III	12,500	13,040	5.8	40.1	~500
IV	4,700	6,640	5.5	36.2	~300

[a] SDS-PAGE (Schäfer *et al.*, 1974a).
[b] Minimal molecular weight by amino acid analysis (Hinnen *et al.*, 1976).
[c] Schäfer *et al.* (1974a).
[d] Hinnen *et al.* (1976).

1971). The proportions are I, 11%; II, 65%; III, 19%; and IV, 6%. The values for the number of moles of each protein per mole of virus in Table 2 are calculated from these figures and an estimate of a protein content of 33×10^6 g/mol of virus. The PM2 virion contains an RNA polymerase activity (Schäfer and Franklin, 1975a) that seems to be associated with protein IV. The activity is a polynucleotide-pyrophos-phorylase utilizing either ribo- or deoxyribonucleotide triphosphates in the presence of manganese ions. Although the reaction is stimulated by nucleic acids, it does not appear to copy sequences with any fidelity. Although sensitive to proteolysis, the activity is not diminished by heating or detergents. The role of this activity in phage development is not known, but it may be part of a polymerase that includes host proteins as well. Since isolated PM2 DNA is infectious for BAL-31 spheroplasts (Van der Schans et al., 1971), the polymerase activity is probably not important for infection.

An endonuclease activity was reported for PM2 (Laval, 1974), but there is evidence that it may be a contamination by a host protein (Franklin, 1976).

2.1.3. Lipid Composition of PM2

The lipid content of PM2 has been determined in three ways. In the first, the percent dry weight lipid was determined by extracting the lipids from the virion with chloroform–methanol (Espejo and Canelo, 1968a). A lipid content of 10.5% was found. In the second method, the fraction of total phosphorus extractable from the virion by lipid solvents was determined. In this case, it was found that phospholipid accounts for 118 μg/mg virus. Since neutral lipid constitutes 7% of the viral lipid, a lipid content of 12.6% was calculated (Camerini-Otero and Franklin, 1972). Extracting virus uniformly labeled with [^{14}C]acetate yielded a figure of 14% lipid by weight.

The lipid composition of the virus is of interest in that it is predominantly phosphatidylglycerol (PG), with the remainder primarily phosphatidylethanolamine (PE). The composition is 64% PG, 27% PE, 7% neutral lipid, and about 1% acyl PG. This is in contrast to the composition of the host lipids, which are 75% PE and 23% PG (Braunstein and Franklin, 1971; Tsukagoshi et al., 1976a). In gram-negative organisms, the inner membrane has been found to be richer in PG than is the outer membrane. The inner membrane of BAL-31 has 71.5% PE and 23.5% PG, whereas the outer membrane has 79% PE and 16% PG (Diedrich and Cota-Robles, 1974). The content of PG in BAL-31 inner membrane is not as high as that found in the PM2 lipids; therefore, it is not likely that the PM2 lipids simply reflect the lipid composition of a particular membrane of the host.

The fatty acid composition of the PM2 lipids is rather similar to that found in the host BAL-31. The predominant fatty acids are $C_{16:1}$, 56.6%; $C_{16:0}$, 12.0%; $C_{18:1}$, 13.9%; and $C_{17:cy}$, 7.3% (Camerini-Otero and Franklin, 1972).

A mutant of BAL-31 that requires unsaturated fatty acids for growth has been isolated, and this organism will grow when supplemented with a variety of normal and abnormal fatty acids (Tsukagoshi *et al.*, 1975). In general, the fatty acid compositions of both the host and PM2 reflect the characteristics of the supplemented fatty acids; however, the phospholipid composition of the virus was found to vary with the nature of the supplemented fatty acid. The ratio of PG/PE in PM2 derived from, wild-type BAL-31 was found to be 1.51; in PM2 derived from the mutant grown on *cis*-16:1-palmitoleate or *cis*-18:1-oleate the ratio was 1.24 and 1.21, respectively; however, PM2 derived from mutant cells grown on *trans*-16:1-palmitelaidate had a ratio of 0.81, indicating that the virion contained more PE than PG. It is therefore clear that although the phospholipid composition of the virion is different from that of the host and is rather specific, it is not defined completely by charge or by the chemical composition of the polar groups.

Analyzed by electron spin resonance, membranes derived from cells grown on different fatty acids show phase transitions at temperatures similar to those found for the extracted lipids (Tsukagoshi *et al.*, 1976*b*). PM2 development takes place in *trans*-16:1-palmitelaidate-grown cells at 10°C, which is below the 12.4°C transition temperature found for membranes and lipids of these cells.

2.2. Structure of PM2

The virion appears in electron micrographs as an icosahedral structure with brushlike spikes at the vertices (Fig. 1). The diameter of the structure was originally reported as 60 nm (Espejo and Canelo, 1968*a*). Measurement of the particle weight has been done by several methods. This work has been very well covered in a review by Franklin (1974). Table 3 gives the dimensions and particle weight determinations obtained by several methods in Franklin's laboratory. There is very good agreement between the determinations performed with the different techniques. It is interesting that the physical methods also give a

TABLE 3
Size and Particle Weight of Bacteriophage PM2[a]

Method	Particle diameter (nm)	Particle weight × 10^{-6}
Electron microscopy	60–61	—
Low-angle X-ray scattering	60	—
Neutron scattering	63	
Turbidity		
At λ = 436 nm	—	43.5 ± 2.1
At λ = 546 nm	—	43.8 ± 1.6
Extrapolation to infinite wavelength	77	45.4 ± 1.5
Light scattering	67	46.0 ± 1.5
Sedimentation-diffusion	—	47.9 ± 1.7
Equilibrium sedimentation	—	44.1 ± 1.2
Average particle weight		45.1 ± 1.8

[a] From Franklin (1974).

particle dimension very similar to that found with electron microscopy. Camerini-Otero and Franklin (1975) also determined the hydration of the PM2 virion and found that water constituted 62% of the particle volume, and 42% of the water was located inside the particle.

PM2 has been studied with low-angle X-ray diffraction to determine the position of the lipids in the virion. A region of low electron density, 30 electrons/nm³, is found at 22 nm from the center of the particle. This probably represents the lipids and indicates that the presumptive bilayer is also occupied by proteins since the electron density of a pure lipid bilayer would be expected to be about 26 electrons/nm³ at the center. The particle seems to extend to about 30 nm, but there are regions of increased electron density just inside and outside the region of the minimum at 22 nm and these are interpreted as being due to the polar groups on the phospholipids. There is also a region of high electron density exterior to the area of the presumed bilayer. This region may well correspond to the outer layer of material seen in electron micrographs (Harrison et al., 1971), which is the major capsid protein (II).

Although the X-ray diffraction data do not prove the existence of a true bilayer, they are consistent with it. Certainly the lipid material seems to be concentrated at a definite distance from the center of the particle. The amount of lipid in the particle is not enough to construct a continuous bilayer at that distance, but, as mentioned above, there is evidence that the lipid region does contain protein. Camerini-Otero and

Franklin (1972) calculated that between 39% and 52% of the bilayer volume is occupied by lipid on the basis of the virion lipid content and the bilayer volume as found by Harrison *et al.* (1971).

It is not possible to establish the bilayer nature of the PM2 lipids by electron spin resonance studies; however, assuming that such a structure exists, the data obtained by Scandella *et al.* (1974), showing a strong temperature dependence of order parameter S_3, can be interpreted to indicate that much of the bilayer is occupied by proteins that interact strongly with the lipids, resulting in a high orientational energy for the fatty acids, a lower rotational correlation time, and a larger bond orientation potential (Franklin, 1974).

The structure of PM2 should be rather uncomplicated since the composition is so simple, containing DNA, primarily two kinds of phospholipid, and four proteins. We know from the X-ray diffraction studies that the lipid is located in a lipid–protein bilayer structure approximately 20 nm from the center of the particle. Two methods of structural analysis have been applied to PM2. In the first, the virion was treated with agents such as proteolytic enzymes or chemical reagents that do not penetrate the lipid bilayer, and the consequences of this treatment were ascertained. Molecules that react are considered to be more exterior than those that do not react. Experiments with proteolytic enzymes have been rather clear with PM2; however, those involving the labeling of viral proteins by iodination in the presence of lactoperoxidase or the labeling with [^{14}C]glycine ethyl ester in the presence of guinea pig transglutaminase (Brewer and Singer, 1974) are difficult to interpret because the intrinsic extent of labeling for the various proteins has not been shown in detail. Comparisons of the degree of labeling of proteins must be evaluated in light of the number of protein molecules of each species per virion and the maximum extent to which each protein can be labeled in the intact as well as completely disrupted virions. Labeling the virion with [^{35}S]diazonium salts of sulfanilic acid (DSA) (Hinnen *et al.*, 1974) results in the accumulation of radioactivity in the peak for protein II, indicating an exterior location for this protein; however, the same qualifications must be kept in mind in this case as for the enzymatic labeling studies.

In the second approach, the particle is disassembled with reagents such as urea or treatments such as freezing and thawing, and the resulting particles are analyzed for the presence of particular components. A third approach is to treat the virus with cross-linking reagents and then determine the association of components after dissociating the structure of the particle. Using these techniques, a model for the structure of

PM2 has been arrived at and is shown in Fig. 2. The DNA is at the center of the particle in association with protein IV. Protein III is exterior to protein IV and the DNA. The lipid bilayer is outside this core, with protein II constituting the major outside protein and protein I constituting the spikes at the vertices of the icosahedral structure. The lipid bilayer is asymmetrical, with most of the PG on the outside and most of the PE on the inside. The evidence for this model is as follows.

2.2.1. Protein I as the Spike Protein

Treatment of the virion with proteolytic enzymes, and particularly with bromelin at 1 mg/ml, results in a particle lacking only protein I, unable to adsorb to host cells, and lacking the spikes in the electron micrographs (Schäfer *et al.*, 1974*b*; Hinnen *et al.*, 1974).

2.2.2. Protein II as the Major Exterior Protein

Labeling of the virion with radioactive iodine, glycine ethyl ester, or DSA indicated a surface location for protein II; however, this assignment suffered from the deficiencies remarked upon earlier. Protein II is the most abundant of the viral proteins, and on the basis of packing considerations it was felt that this protein might constitute the layer exterior to the lipids (Harrison *et al.*, 1971).

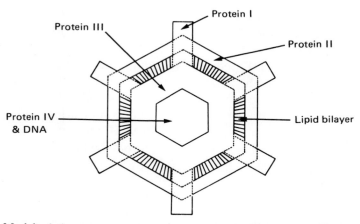

Fig. 2.   Model of the structure of bacteriophage PM2. The lipid bilayer consists of phosphatidylglycerol outside and phosphatidylenthanolamine inside. From Schäfer *et al.* (1974*a*).

A more satisfactory analysis of the distribution of proteins was effected by the use of dissociation techniques. In the presence of 1 M urea, the particle was degraded to several classes of subviral structures, including a major one containing DNA, lipid, and proteins III and IV, and a small amount of protein II, indicating that proteins I and II were exterior to the lipid in the virion (Hinnen *et al.*, 1974). PM2 is also disrupted by cycles of freezing and thawing. After such a treatment, particles containing DNA, lipid, and proteins III and IV and a small amount of protein II are obtained (Brewer and Singer, 1974), again indicating that proteins I and II are at the exterior of the particle. The "nucleocapsid" particles isolated after urea treatment are smaller than untreated virus, having a diameter of 40–50 nm, and they are irregular in shape (Hinnen *et al.*, 1974).

2.2.3. Proteins III and IV as Core Proteins

The results listed above show that proteins III and IV are located interior to proteins I and II. Extraction of PM2 with 4.5 M urea resulted in a particle containing DNA and proteins III and IV (Hinnen *et al.*, 1974). These particles, although lacking lipid, retained their icosahedral shape and were of similar size to those produced in 1 M urea. It would seem that the lipid layer is exterior to proteins III and IV and the DNA. Protein IV seems to be closely associated with the DNA since particles treated with sodium bisulfite result in the association of over 70% of protein IV with DNA isolated after complete disruption of the particles (Hinnen *et al.*, 1974). Other cross-linking studies have shown the association of protein III with phosphatidylethanolamine (Schäfer *et al.*, 1975). On the basis of the relative amounts of proteins III and IV, it would seem reasonable that IV would be associated with the DNA and III would form the exterior of the "nucleocapsid" core (Franklin, 1974). The solubility properties of protein IV and the data on its amino acid composition and sequence can be reconciled with its binding to DNA (Hinnen *et al.*, 1976). Both the *N*- and *C*-terminal regions of IV are polar, while the center of the molecule is hydrophobic; an electrostatic interaction with DNA is therefore reasonable for this rather hydrophobic protein.

2.2.4. Asymmetry of the Bilayer Lipid Distribution

The phospholipids of the erythrocyte plasma membrane are distributed in an asymmetrical manner with respect to the bilayer

(Bretscher, 1973). The phospholipids of PM2 also appear to be distributed asymmetrically with respect to the bilayer (Schäfer *et al.*, 1974*a*). Radioactive sulfanilic acid diazonium salt (DSA) reacts with PE to form a derivative that can be isolated chromatographically. PG is oxidized by DSA to an aldehyde. This compound can be labeled by incubating in the presence of tritium-labeled sodium borohydride. None of the lipids of intact PM2 is very accessible to DSA, but they can be made accessible by treating with increasing concentrations of lithium chloride. At 0.75 M LiCl, viral infectivity is conserved but most of the phospholipid becomes accessible to DSA. The labeling of proteins I and II, which appear to be located outside of the bilayer, is also maximal at this concentration. Above 0.75 M LiCl, the core proteins become accessible to the reagent, indicating a disruption of the structure. In the presence of 0.05% Triton X-100, viral infectivity is also preserved, but core proteins are labeled by DAS, indicating an opening up of the structure.

In the presence of 0.75 M LiCl, there was little modification of PE but extensive modification of PG; the extent of labeling of PE increased tenfold in the presence of 0.05% Triton X-100, while that of PG increased only slightly (Table 4). These results are consistent with a model in which PG is found primarily in the outer sheet of the lipid bilayer and PE is found almost exclusively in the inner sheet. There is a possibility that the distribution of reactivity reflects the accessibility of the polar groups to the surface and to the reagent rather than a true asymmetry, but the concept of asymmetrical phospholipid distribution appears to be finding support in many membrane studies and is probably true in this case (Bretscher, 1973). Franklin (1974) has proposed that protein II, which is in itself basic and even more so when naturally complexed with calcium ions, is interacting electrostatically

TABLE 4
Asymmetry of the PM2 Bilayer[a]

PM2 medium	Phosphatidyl-ethanolamine[b] (counts/min)	Phosphatidyl-glycerol[c] (counts/min)
0.75 M LiCl	945	86,000
0.05% Triton X-100	10,650	117,000

[a] From (Schäfer *et al.* (1974*a*).
[b] Phosphatidylethanolamine was labeled with [^{35}S]sulfanilic acid diazonium salt.
[c] Phosphatidylglycerol with ^{3}H after treatment with NaB^{3}H$_4$. The values represent the total radioactivity of the modified lipids found after separation on thin-layer plates.

with the negatively charged PG and that this interaction plays a vital role in the assembly and stabilization of PM2.

The ability of protein II to interact with phospholipids was tested by studying the distribution of phospholipids PE and PG in two phase systems containing chloroform and an aqueous phase containing proteins I and II or III and IV in the presence of urea and mercaptoethanol (Schäfer and Franklin, 1975b). In the presence of proteins I and II, at pH values less than 11.5, PG was drawn into the aqueous phase. The mixture of proteins III and IV also pulled some of the phospholipid into the aqueous phase but in proportions reflecting the total lipid composition. The mixture of proteins I and II appeared to react specifically with PG. This interaction did not change appreciably in the presence of added EDTA or calcium ions and did not change when phospholipids containing *trans* fatty acids were used instead of those composed of *cis* fatty acids. Schäfer and Franklin (1975b) assumed that the observed interaction was reflecting the properties of protein II in the mixture.

Membrane vesicles of dimensions similar to those of PM2 appear to have a natural asymmetrical distribution of phospholipids. Twice as much PG as PC (phosphatidylcholine) is in the outer layer of vesicles prepared from equimolar amounts of the two (Michaelson *et al.*, 1973), and it was proposed that this distribution is favored energetically because the negatively charged polar groups of PG are farther apart from each other in the outer layer (Israelachvili, 1973). The asymmetrical distribution of PG and PE may therefore be due to both the electrostatic interactions between PG molecules themselves and their interactions with protein II (Franklin, 1974).

2.3.   Reconstitution of PM2

Schäfer and Franklin (1975b) succeeded in reconstituting the core of PM2 from DNA and proteins III and IV (Fig. 3). They were also able to use nucleocapsids isolated from virus particles through dissociation with 4.5 M urea to reconstitute infectious particles in the presence of phospholipids and proteins I and II.

The nucleocapsid was reconstituted from proteins III and IV, which are isolated by disrupting the virion in 8 M urea. Treatment of the virion with 8 M urea resulted in complete dissociation; dialysis vs. a buffer containing 8 M urea at pH 4.5 resulted in the precipitation of all components except proteins III and IV. These were then combined with

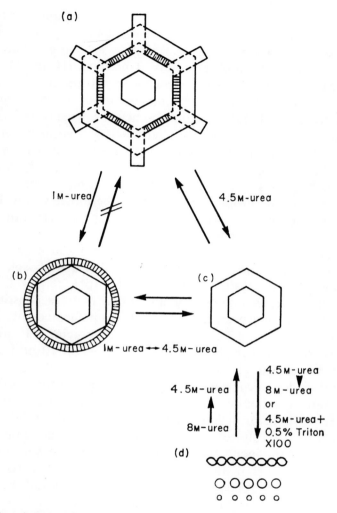

Fig. 3. Dissociation and association reactions of bacteriophage PM2 *in vitro*. Intact virus (a), lipid-containing nucleocapsid (b), nucleocapsid (c), solubilized protein III, protein IV, and PM2 DNA (d). From Schäfer and Franklin (1975b).

purified PM2 DNA in the presence of 8 M urea. The mixture, which was in the ratios found in the intact virion, was dialyzed for 12 hr vs. 4.5 M urea, 0.5 M NaCl, 10 mM $CaCl_2$, 20 mM β-mercaptoethanol, and 25 mM tris-HCl (pH 7.5 at 4°C). The conditions of pH, temperature, and ionic composition were quite specific. The particles could be isolated with a yield of 2.3% by sucrose gradient sedimentation. If the reconstitution was carried out in the presence of 10% bovine serum albumin, the yield could be raised to 33%. The appearance of the reconstituted particles was similar to that found for particles prepared directly from virions by treatment with 4.5 M urea. The buoyant density of the particles and the ratios of proteins III and IV to DNA were the same as for isolated nucleocapsids. More recent experiments (Franklin, 1976) indicate that the nucleocapsid can be reconstituted with pure proteins III and IV in the absence of bovine serum albumin.

If nucleocapsids isolated from virions by treatment with 4.5 M urea were incubated with PE and PG, suspended by sonication in 4.5 M urea, and dialyzed vs. 1 M urea, 1 M NaCl, 10 mM $CaCl_2$, and 25 mM tris-HCl (pH 7.5 at 4°C), particles resulted that had incorporated lipid in approximately the proportion found for the intact virion. However, the ratios of the phospholipids were dependent on their ratios in solution; there did not seem to be selectivity.

If nucleocapsid-containing lipid as described above was incubated with stoichiometric amounts of proteins I and II and phospholipid in 1 M urea and dialyzed against 25 mM tris-HCl (pH 7.5 at 4°C), 10 mM $CaCl_2$, 20 mM β-mercaptoethanol, and 0.5 M NaCl, reconstitution did not take place. If, however, nucleocapsid was incubated with phospholipid in the presence of 4.5 M urea along with proteins I, II, and bovine serum albumin and then dialyzed vs. 0.5 M NaCl, 10 mM $CaCl_2$, 10 mM β-mercaptoethanol, and 25 mM tris-HCl (pH 7.5 at 4°C), then reconstitution did occur and infectious particles were obtained. The concentration of albumin was critical, as reconstitution did not take place in its absence or at high concentrations but did take place with 30–50 mg/ml. The ratio of PG to PE was the same as found in native particles even if the incubation mixture contained the phospholipids in the proportion found in the host cell membranes. The reconstituted particles had a normal buoyant density and proportion of the four proteins and absorbance ratio at 260 nm and 280 nm (Table 5). Physical reconstitution occurs with phospholipids containing *trans*-16:1 fatty acids as well as *cis*-16:1 fatty acids, but no infectivity is recovered with the former (Franklin, 1976). Acyl PG does not substitute for PG to yield infectious particles in the reconstitution reaction (Tsukagoshi *et al.*, 1976a).

It is not known how the reconstitution results in the proper selection of the phospholipids for the particle. However, it was found by Schäfer and Franklin (1975b) that a mixture of proteins I and II had a selective affinity for PG and could remove it from the organic phase in a partitioning experiment. It is not clear whether the asymmetry of the phospholipid distribution on the particles is due to the specific interactions with specific virion proteins or whether it is a property of

TABLE 5
Reconstitution: Nucleocapsid → Intact Virus[a]

A. Lipid mixture	Lipid in the reconstituted virus	Yield	Biological activity (plaque-forming units, total)
a. 70% PE 30% PG	31% PE 69% PE	21 μg(0.5%)	8×10^3
b. 30% PE 70% PG	28% PE 71% PG	121 μg(2.5%)	6×10^4

B. Physical properties	Reconstituted virus	Native virus
a. ρCsCl	1.28	1.29
b. A_{260}/A_{280}	1.40	1.42
c. Ratio of proteins I/II/III/IV	1:3.86:1.8:0.89	1:4.2:1.9:0.95

[a] Comparison of the lipids present in the reconstitution assay with the lipid content, yield, and biological activity of the bacteriophage after reconstitution (A). Physical properties of reconstituted virus and native virus (B). From Schäfer and Franklin (1975b).

membranes with a high degree of curvature. As pointed out by Schäfer and Franklin (1975b), vesicles prepared from phosphatidylcholine (PC) and PG have most of the PG on the outside (Michaelson et al., 1973); this may be due to the repulsion of the negatively charged PG molecules. The observation that the ratio of PG to PE in particles formed in vivo changes dramatically when the phospholipids contain trans-16:1 fatty acids indicates that charge is probably not the sole determinant of the phospholipid distribution (Tsukagoshi et al., 1975).

2.4. Biogenesis of PM2

Although the structural analysis of PM2 has reached a high level of sophistication, the analysis of biogenesis is only beginning. Very little has been done on the isolation of viral mutants, and those that have been isolated have not been fully characterized (Brewer, 1976). Analysis of the production of virus-specific proteins in infected cells has also not progressed very far, probably due to the experimental difficulty in shutting off host protein synthesis in infected cells to allow the identification of virus-specific proteins. Most of the information relevant to the biogenesis of the virus has come from studies utilizing the electron microscope (Cota-Robles et al., 1968; Dahlberg and Franklin, 1970; Brewer, 1976).

The mode of attachment of PM2 to the host cell has not been studied in detail and has not been resolved. Protein I, which forms the spikes at the vertices of the particle, appears to be necessary for adsorption to the host (Hinnen *et al.*, 1974). The adsorption rate constant for PM2 is 3.4×10^{-9} ml/min (Sands, personal communication), which is rather fast (the constant for $\phi 6$ is $3.3–3.8 \times 10^{-10}$ ml/min, Vidaver *et al.*, 1973). The latent period is 40–45 min, and mature particles can be isolated from infected cells at about 36 min post-infection (Franklin *et al.*, 1969).

Soon after infection of BAL-31 by PM2, there is an increase in the rate of DNA synthesis. This increase is due to the synthesis of PM2 DNA. Host DNA synthesis stops early in infection (Franklin *et al.*, 1969). The host DNA is stable and continues to serve as a template for the synthesis of host-specific mRNA (Franklin *et al.*, 1969). Viral mRNA is synthesized throughout the infection. Synthesis appears to begin at 4 min after infection at a low rate which increases at 10 min after infection to the maximal rate which is maintained until lysis (Espejo *et al.*, 1971*a*). Radioactive thymidine was incorporated into three species of PM2 DNA, corresponding to supercoiled PM2 DNA of $s_{20,w}$ of 28.1 S, circular nicked PM2 DNA of 21.2 S, and a 24 S replicative intermediate. The supercoiled DNA isolated during infection has a superhelix density that is slightly lower than that found in the DNA of the mature virion (Espejo *et al.*, 1971*b*). Pulse-chase experiments indicate that the nicked circular DNA is a precursor to the supercoiled form and that the 24 S form is a precursor of the nicked one. The 24 S structure is visualized as closed circles of double-stranded DNA with single branches of variable length up to the size of the PM2 genome. The 24 S structure yields single-stranded DNA on denaturation of lengths both longer and shorter than PM2 DNA. It is proposed that the replication of PM2 proceeds according to the rolling circle model of Gilbert and Dressler (1968).

The virus appears to mature at the periphery of the cell. Mature particles are seen at approximately 40 min after infection (Dahlberg and Franklin, 1970; Cota-Robles *et al.*, 1968). The particles appear to line up at the membrane, and as infection proceeds additional layers of particles are seen in layers interior to the membrane. Early in infection, many particles have the appearance of empty vesicles; the proportion of these particles relative to those having a dense core decreases with time, and it is possible that the empty vesicles are precursors to the mature virus. Whereas 85% of the particles are "empty" at 40 min after infection, only 25% are empty at 70 min (Dahlberg and Franklin, 1970). The empty particles are distributed in a similar manner to the mature particles, although there seems to be somewhat more of them in interior layers than at the membrane in cells containing many virus particles (Fig 4).

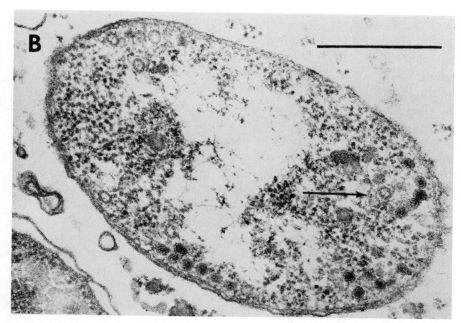

Fig. 4. BAL-31 close to lysis after infection with PM2, showing empty and filled particles. Arrow points to empty vesicle. Bar equals 0.5 μm. From Dahlberg and Franklin (1970).

Brewer (1976) has studied two interesting temperature-sensitive mutants of PM2. Cells infected with mutant *ts*1 show an abundance of the vesicles at the restrictive temperature; cells infected with *ts*5 show an abundance of these forms at both the restrictive and permissive temperatures. It may be that these forms are truly precursors of the mature virion, and the unspecified mutations leading to temperature sensitivity of development cause their accumulation by blocking some late step in the morphogenetic pathway. The empty vesicles may be analogous to the prohead structures of the large DNA bacteriophages (Casjens and King, 1975). There is no evidence at this time that budding of the inner membrane of the host cell is involved in viral development.

2.5. Origin of Viral Lipids

The differences in the lipid composition between PM2 and the host BAL-31 are very dramatic. PM2 has an abundance of PG and a small amount of PE, whereas the opposite is true for the host (Braunstein and Franklin, 1971). Phospholipids are presumed to be synthesized on the inner membrane of gram-negative bacteria, since the enzymes of phospholipid synthesis are located there (Bell *et al.*, 1971). It is possible that there is contact between the developing virions and the inner membrane of the host, but this has not been established. How, then, do the lipids move from their site of synthesis to the developing virion? Are the viral lipids newly synthesized, and how are they selected for placement on the virion? And how is the asymmetry of the distribution of the membrane lipids with respect to the bilayer surfaces established? These are the major questions relevant to the incorporation of the lipids into PM2. The only one of these questions that can be answered by experimental findings with PM2 is that of the relationship of synthesis to incorporation into the virion. The lipids that are incorporated into the virion include molecules synthesized before and after infection (Tsukagoshi and Franklin, 1974; Snipes *et al.*, 1974; Diedrich and Cota-Robles, 1976).

Because the lipid composition of the virus is markedly different from that of the host, it was of interest to determine whether this difference is reflected in the metabolism of the infected cells. Tsukagoshi and Franklin (1974) proposed that the lipid metabolism of infected cells changes after infection. At 90–120 min after infection, at a time when lysis has begun, prelabeled PE is broken down, whereas PG is not. In uninfected cells, both compounds are stable. Synthesis of PE also slows down in infected cells at this time, whereas the rate of synthesis of PG increases. These changes are rather late in the infectious cycle, and Diedrich and Cota-Robles (1976) point out that they are occurring after the maturation of the bacteriophage and probably reflect events related to lysis. A similar criticism was made of the work of Snipes *et al.* (1974). Diedrich and Cota-Robles (1976) found that the synthesis of PE or PG, as measured by pulses of [^{32}P]phosphate, showed no change during the first 60 min of infection. They found that the only changes attributable to the infection were in the turnover rates for the glycerol moieties of PG. Although the net radioactivity in glycerol of PG does not change in infection, the amount in the unacylated glycerol increases at the expense of the diacylated glycerol. It is not clear how or whether this change in turnover pattern is associated with phage development.

Most of the lipid is incorporated into the particles by 45 min after infection (Tsukagoshi and Franklin, 1974). Labeling with [^{32}P]phosphate at various times after infection shows a dramatic decrease in labeling of viral PE and PG at 45 min. An experiment to determine the fraction of viral phospholipids synthesized before and after infection involved the prelabeling of cells with tritiated glycerol and infecting in the presence of [^{32}P]phosphate. Virus was harvested at 210 min after infection and the ratio between ^{3}H and ^{32}P was determined for PE and PG. Infected cells were harvested at intervals during infection, and the ratio of the labels was determined for PE and PG as well. Using the assumption that the assembly of the particles is taking place at 45 min after infection and that the lipids will be derived from new synthesis and previously synthesized lipids of the proportions present in the bulk host lipids at 45 min, then one can calculate that approximately 30% of the PE and PG of the virion is derived from material synthesized before the assembly of the viral membrane. This calculation rests on several untested assumptions, including that of the rapid equilibration of lipids of different ages in the outer and inner membranes of the host cell, and the lack of exchange between lipids of mature virions with those of the host cell. In any case, it seems clear that the viral lipids contain molecules that were synthesized before and during assembly of the virion. Diedrich and Cota-Robles (1976) reported that the specific activities of phage produced in cells prelabeled with [^{32}P]phosphate or labeled after infection indicate that the viral lipids are taken from the host membrane at some time late in infection with no preference with respect to the time of synthesis.

3. PR3, PR4, AND PRD1

Bacteriophages PR3, PR4, and PRD1 can be grouped together since they all infect bacteria harboring drug-resistance plasmids of either the P, N, or W compatibility group (Stanisich, 1974; Bradley and Rutherford, 1975; Olsen *et al.*, 1974). The phages of this group have an icosahedral structure with a diameter of about 65 nm and have the variable appearance of a tail-like structure (Fig. 5). Negatively stained preparations show a double-layered structure similar to that found for PM2. Although the phages are seen attached to the surface of susceptible bacteria, they have not been observed attached to pili of the P or N type, although they have been observed attached to the tips of W-type pili (Bradley, 1976). These phages are of particular interest since they will infect strains of *E. coli* harboring the above drug-resistance factors. This should facilitate the development of genetic and physiological studies of these viruses.

PR3 and PR4 have a low degree of sensitivity to chloroform and a buoyant density of 1.267 and 1.264 g/ml in cesium chloride. The titer of PR3 is reduced less than a hundredfold by prolonged mixing with

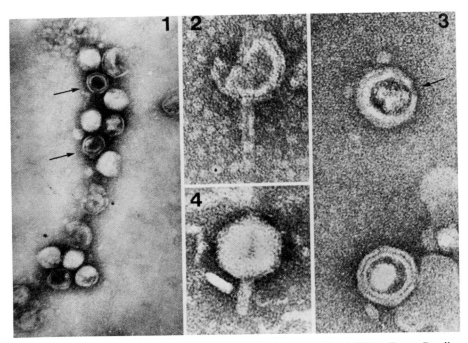

Fig. 5. Bacteriophage PR3 negatively stained with neutralized PTA. From Bradley and Rutherford (1975).

chloroform (Bradley and Rutherford, 1975). The onset of lysis occurs at 130 min for PR3 and 95 min for PR4. PRD1 DNA is linear, contains 49.2% G+C, and has a molecular weight of 24×10^6 (Olsen *et al.*, 1974). The phospholipids of PR4 have been investigated, but the lipid contribution to the phage particle weight has not been determined (Sands and Cadden, 1975). Virus prepared from *E. coli* growing in [^{32}P]phosphate was purified and the phospholipids were analyzed. It was found that PR4 phospholipids contain 21% PE, 20% PG, 6% cardiolipin, 20% PS, and 33% of an unidentified lipid. The composition changes somewhat with different media (Sands and Cadden, 1975). Further studies on the composition of PR4 (Sands, 1976) showed 22% PE, 32% PG+DPG, 15% PS, and 31% unidentified phospholipids. The nature of the unidentified phospholipids is not known; however, their chromatographic behavior is consistent with that of lysophospholipids, and they may be degradation products similar to those found in PM2 lipids in earlier studies (Braunstein and Franklin, 1971). The relative rates of synthesis of the cellular phospholipids do not change during infection, although there is a slight overall decrease in rate of synthesis (Sands, 1976).

An experiment was performed to determine the time of synthesis of PR4 phospholipids (Sands, 1976). In this study, cells were labeled with [^{32}P]phosphate either for several generations up to the time of infection, or up to and including the time of infection, or only after infection. Virions were purified and analyzed for the distribution of phospholipid radioactivity. A comparison of the amount of label in virions prepared from cells labeled at various times indicated that approximately 25–40% of the viral lipids had been synthesized before infection, and that relatively more of the PE was synthesized before infection and more of the PG was synthesized after infection (Sands, 1976). Since the formation of viral membrane takes place sometime after infection, the most significant distinction would be the relative amounts of viral lipid that are incorporated from preexisting lipids and lipids synthesized at the time of assembly. The findings with PR4 and PM2 both indicate that a substantial amount of the viral lipid is synthesized before assembly. In the case of PM2, the amount of previously synthesized lipid was calculated as about 30% (Tsukagoshi and Franklin, 1974). In the experiment of Sands (1976), the calculation of 25–40% should be a minimum since the measurement is of synthesis before and after infection rather than relative to the time of assembly. It appears that there is probably not a requirement for the direct transfer of lipids from their sites of synthesis to the developing virions but rather that lipids can be taken from the host plasma membrane for virus assembly. Nascent phospholipids may be preferentially incorporated but are probably not required.

The proteins of PR4 have been analyzed and six peptides have been observed in autoradiograms of electropherograms performed on SDS-polyacrylamide systems. The proteins were found to have molecular weights of (1) 52,000, (2) 38,000, (3) 31,500, (4) 28,300, (5) 20,300, and (6) 16,000. Protein 2 constitutes about 58% of the viral proteins by weight (Cadden and Sands, 1976) and appears to be synthesized late in infection. This information is preliminary and undoubtedly will be extended.

4. BACTERIOPHAGE φ6

4.1. Virus Characterization

Bacteriophage φ6 is a truly unique bacteriophage in that it has double-stranded RNA as its genetic material and a lipid-containing membrane surrounding an icosahedral nucleocapsid. The bacteriophage's natural host is *Pseudomonas phaseolicola* HB10Y (Vidaver *et al.*, 1973). This organism is a prototrophic phytopathogen that does not have particular growth requirements and can be cultured in simple defined media. Alternate hosts for the bacteriophage have been found in *P. pseudoalcaligenes* (Mindich *et al.*, 1976a). This organism is quite different from the natural host in that it grows at temperatures up to 42°C and is unable to grow on many of the carbohydrates, including glucose, that support the growth of *P. phaseolicola*. Attachment of the virus to host cells has not been studied in great detail except that the adsorption constant of $3.3–3.8 \times 10^{-10}$ m-min^{-1} has been determined and phage has been visualized in electron micrographs attached to long pili on the host cells (Vidaver *et al.*, 1973) (Fig. 6). The latent period in infection is about 90 min, and approximately 300–500 particles are released upon lysis (Vidaver *et al.*, 1973); Productive infection takes place at temperatures below 28°C, and even in the *P. pseudoalcaligenes* hosts, which grow at 42°C, infection is not productive at higher temperatures (Mindich, unpublished results). The nature of the heat-sensitive process in $\phi6$ development has not been studied in detail, but some experiments indicate that it may be a late step (Sands *et al.*, 1974).

Electron micrographs of negatively stained virions show a somewhat amorphous particle with a diameter of 60–70 nm and the variable appearance of an amorphous tail-like appendage (Vidaver *et al.*, 1973). Treatment of the virus with chloroform (Ellis and Schlegel,

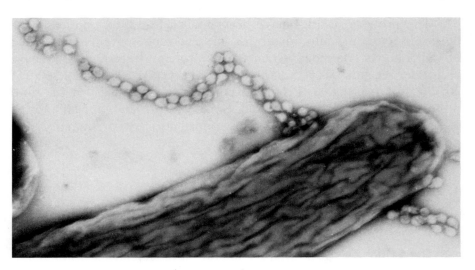

Fig. 6. Bacteriophage $\phi6$ adsorbed to pili of *P. phaseolicola* HB10Y. From Gonzalez, Langenbert, Van Etten, and Vidaver (unpublished).

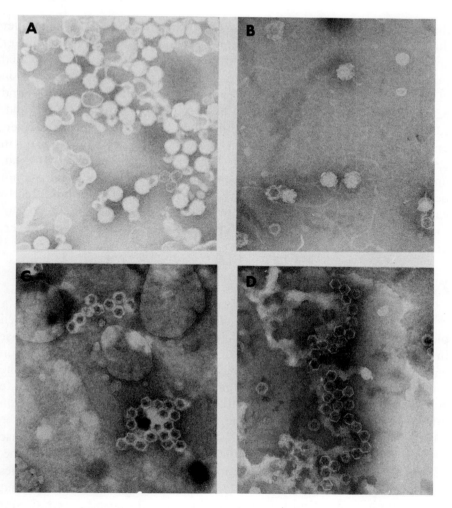

Fig. 7. Bacteriophage φ6 negatively stained with neutralized PTA. Samples were treated with various concentrations of Triton X-100. (A) Untreated; (B) 0.5%; (C) 1%; (D) 2%. From Sinclair *et al.* (1975).

1974) or with detergents results in a particle of hexagonal outline that is probably the icosahedral core capsid (Sinclair *et al.*, 1975). This particle has a diameter of 50 nm (Ellis and Schlegel, 1974) (Fig. 7). Thin sections of virus show an electron-dense inner structure of 30-nm diameter that appears ringlike and probably represents the nucleic acid of the virus (Bamford *et al.*, 1976). Around this material is a less dense region bounded by an additional layer at a diameter of about 50 nm; this is probably the core capsid (Ellis and Schlegel, 1974; Bamford *et al.*, 1976). Exterior to this layer is another that probably corresponds to the outer membrane of the virion (Bamford *et al.*, 1976). This layer does not usually display a definite structure; however, a bilayer appearance has been reported preliminarily for preparations embedded in water-soluble epoxy plastics (Gonzalez *et al.*, 1976). The composition of the virus which was originally reported as 25% lipid, 13% RNA, and 62% protein (Vidaver *et al.*, 1973) has been redetermined as 12% lipid, 23% RNA, and 65% protein (Partridge, Van Etten, and Vidaver, unpublished results). The virus has a buoyant density of 1.27 g/ml in cesium chloride (Vidaver *et al.*, 1973), and 1.21 in sucrose (Mindich, unpublished results). The physical characterization of the virion has not been done in great detail; however, equilibrium centrifugation has led to values of 45.6×10^6 for the particle weight of the virus and 20.3×10^6 for that of the nucleocapsid (Partridge, Van Etten, and Vidaver, unpublished results).

4.1.1.   Viral RNA

The RNA of the virion is composed of three double-stranded pieces. The RNA has T_m of 91°C and is resistant to pancreatic RNase in $2 \times$ SSC but is sensitive in $0.01 \times$ SSC. It is sensitive to digestion with KOH and has a buoyant density of 1.605 g/ml in Cs_2SO_4. The molecular weights of the individual pieces were determined by rate zonal centrifugation, polyacrylamide gel electrophoresis, and electron microscopy, and were found to be approximately 2.3, 3.1, and 5.0×10^6 (Semancik *et al.*, 1973; Van Etten *et al.*, 1974). The molar base composition of the RNA was found to be 27% cytosine, 29% guanine, 22% uracil, and 22% adenine. The individual dsRNA molecular species have been isolated (Van Etten *et al.*, 1974), and it has been found that there is no hybridization between the three species despite their similar base compositions. Single-stranded RNA could be prepared from the double-stranded material by heating in the presence of formaldehyde (Van Etten *et al.*, 1974). In the absence of formaldehyde, the small- and medium-sized species reannealed very efficiently upon cooling. The three pieces of dsRNA appear to be present in equimolar amounts in the virion.

4.1.2.   Lipid Composition of ϕ6

The infectivity of ϕ6 is extremely sensitive to organic solvents (Vidaver *et al.*, 1973). Whereas the infectivity of PM2 or PR3 is reduced 100- to 10,000-fold by extended treatment with chloroform, that of $\phi6$ is reduced over 10^{10}-fold by exposure to chloroform for several seconds. The sensitivity to ethyl ether and detergents is also much greater for $\phi6$ than for the other lipid-containing bacteriophages. Bacteriophage $\phi6$ is also sensitive to phospholipase A (Vidaver *et al.*, 1973). These results are consistent with a structure in which the lipids are closer to the surface of the particle than is the case for the other lipid-containing phages.

The lipid composition of $\phi6$ has been determined by labeling with [^{32}P]phosphate (Sands, 1973). The molar composition is 57% phosphatidylglycerol, 35% phosphatidylethanolamine, and 8% diphosphatidylglycerol. The identification of the lipids was in terms of chromatographic behavior; it is possible that the material identified as DPG is acyl PG as in the case of PM2 lipids (Tsukagoshi *et al.*, 1976a). The percent lipid in the particle was determined by dry weight measurements and is now reported to be 12%. The fatty acid composition of the viral lipids is similar to that of the host, with 61% of the fatty acids being unsaturated 16:1 and 18:1, and 33% being saturated 16:0 (Vidaver *et al.*, 1973).

4.1.3.   Protein Composition of $\phi6$

The purified virion contains 11 peptides when analyzed on SDS-polyacrylamide gels (Sinclair *et al.*, 1975). It was originally reported that only ten peptides were present in the virion, but protein P11, which seems to be derived from the same peptide as protein P5, is found in mature virions to a variable extent (Mindich, unpublished observation). The virus directs the synthesis of a twelfth protein, P12, that has a role in viral morphogenesis but is not found in the virion (Sinclair *et al.*, 1975; Mindich *et al.*, 1976c) (Fig. 8). The molecular weights have been determined by analysis of electrophoretic mobility in SDS-containing gels (Sinclair *et al.*, 1975; Van Etten *et al.*, 1976), and in the case of protein P9 by the minimum molecular weight calculated from the amino acid composition (Sinclair *et al.*, 1975). Table 6 lists the molecular weights obtained for the $\phi6$ proteins in several studies. The protein analysis done by Sinclair *et al.* (1975) utilized. radioactively labeled virus, whereas that of Van Etten *et al.* (1976) utilized staining to identify the various species. Recent work indicates that the correct molecular weights for P1, P2, and P3 are between those of Sinclair *et al.* (1975) and those of Van Etten *et al.* (1976). It is of additional

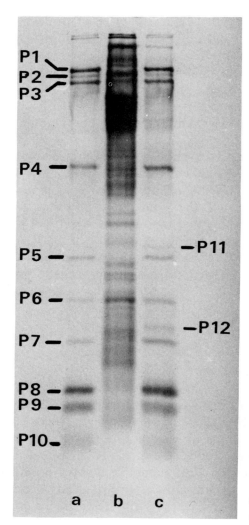

Fig. 8. Autoradiogram of ϕ6 proteins in virions and infected cells. Electrophoresis was in a 15% discontinuous polyacrylamide slab gel. The autoradiogram was exposed for 6 days; the origin is at the top of the pattern. Cells were infected with ϕ6 at m.o.i. 50. At 60 min after infection, rifampin was added to a final concentration of 50μg/ml. Incubation was continued for 10 min, followed by the addition of [^{14}C]leucine, 2 μCi/ml. (a) Purified ϕ6; (b) uninfected cells; (c) ϕ6-infected cell lysate. From Sinclair et al. (1975).

interest that phage proteins P5, P11, P9, and P10 are not labeled by radioactive methionine. The amino acid analysis of P9 also indicates a lack of methionine (Sinclair et al., 1975).

Treatment of ϕ6 with chloroform or detergents reduces the infectivity markedly (Vidaver et al., 1973). These treatments also result in the loss of the membranelike structure enveloping the core (Ellis and Schlegel, 1974; Sinclair et al., 1975; Van Etten et al., 1976) (Fig. 7). Extraction with the nonionic detergent Triton X-100 results in the loss of the viral lipids (Van Etten et al., 1976). In addition, several of the viral proteins are lost. Proteins P3, P6, P9, and P10 are extracted with

TABLE 6
Characteristics of φ6 Proteins

| | Mol. wt.[a] | Present in virion | Extracted with Triton X-100, buffered with | | | Total pure virus[b] (%) | Total infected cells[b] (%) | Mol/virion[d] |
			Tris[b]	Phosphate[c]	Methionine[b]			
P1	82,000	+	−	−	+	15.6	13.3	56
P2	80,000	+	−	−	+	6	3.7	22
P3	75,000	+	+	+	+	9.8	12.3	39
P4	37,000	+	−	−	+	6.8	9.1	54
P5	24,000	+	−	+	−	3	2.7	37
P6	21,000	+	+	+	+	2	3.5	28
P7	19,900	+	−	−	+	2.9	6.9	43
P8	10,500	+	−	−	+	26.3	26.7	742
P9	8,700	+	+	+	−	18	12	613
P10	<6,000	+	+	+	−	9.5	6.8	470
P11	25,000	+			−	e	0.6	
P12	20,000	−			+		2.0	

[a] SDS-PAGE (Sinclair et al., 1975; Van Etten et al., 1976; Mindich, unpublished).
[b] Sinclair et al. (1975).
[c] Van Etten et al. (1976).
[d] Calculated on the basis of a virion of 45.6 × 10⁶ daltons with 65% protein (Partridge, Van Etten, and Vidaver, unpublished results) and a uniform distribution of leucine in all proteins.
[e] The content of P11 is low and variable.

1% Triton X-100 in 20 mM tris buffer (Sinclair *et al.*, 1975). Protein P5 and probably P11 in addition to proteins P3, P6, P9, and P10 are extracted by Triton X-100 in phosphate buffer (Van Etten *et al.*, 1976). These proteins and the lipids are therefore likely to be components of the "membrane" of the virus. Other evidence will be presented at a later point to support the designation of proteins P3, P5, P9, P10, and P11 as components of the "membrane" structure.

Proteins P8 and P9 have been purified (Sinclair *et al.*, 1975), and the amino acid composition of P9 has been determined. The amino acid composition does not have any unique aspects with respect to membrane structure. The degree of hydrophobicity is not exceptional, although in the range of other membrane proteins. The polarity as calculated by the method of Capaldi and Vanderkooi (1972) is 38.8%. These authors found that only 2% of soluble proteins analyzed showed polarities less than 40% and that membrane proteins with polarities less than 40% can be separated from their membranes only by detergent treatment.

The core particles have a higher buoyant density than the complete virions. In cesium sulfate gradients the density of the core is 1.33 g/ml, whereas that of the untreated virus is 1.22 (Van Etten *et al.*, 1976). The core structure is disrupted in 10 mM EDTA. It also contains the viral RNA and a polymerase activity (Van Etten *et al.*, 1976). This polymerase activity incorporates tritiated nucleoside triphosphates into high-molecular-weight material that has the characteristics of double-stranded RNA (Van Etten *et al.*, 1973). Incorporation using complete particles requires a brief heat shock; however, incorporation mediated by cores does not require this treatment (Van Etten *et al.*, 1976). The products of the incorporation exhibited the electrophoretic mobilities of the three viral double-stranded RNA species, but the extent of incorporation of the various labeled nucleotide triphosphates varied with the RNA species. UMP was preferentially incorporated into the small- and medium-sized pieces, and GMP was preferentially incorporated into the large piece. CMP and AMP were distributed equally (Van Etten *et al.*, 1973). The incorporation is dependent on the presence of all four nucleoside triphosphates and manganese ions. The extent of incorporation is dependent on the amount of phage material used and proceeds only to a limited extent, which may indicate the filling in of single-stranded gaps or ends in the dsRNA (Van Etten *et al.*, 1973). Preliminary work (Coplin *et al.*, 1976) has suggested that the polymerase reaction may result in the displacement of one of the strands of the dsRNA by the new strand synthesis. Since the core is composed of proteins P1, P2, P4, P7, and P8 (Sinclair *et al.*, 1975; Van Etten *et al.*, 1976), it can be assumed at this point that the polymerase activity will involve one or more of these proteins. We will show later that at least P1 and P2 are involved in some type of RNA polymerase activity.

The placement of the viral proteins has not been determined in any great detail beyond the assignment to the nucleocapsid or the membrane by detergent treatment. The assignment of P8 as the major component of the nucleocapsid envelope is primarily on the basis of its abundance in the core (Table 6). A study of the accessibility of $\phi6$ proteins to iodination with ^{125}I in the presence of lactoperoxidase indicated that P3 is at the exterior of the particle and that P8 and other core proteins become accessible when the membrane is removed (Van Etten *et al.*, 1976) (Table 7). Of particular interest is the inaccessibility in intact virions of proteins P9 and P10 which are the major components of the "membrane." This may indicate that they are normally located interior to the lipid bilayer.

4.2. Genetics of $\phi6$

As stated earlier, the primary usefulness of the lipid-containing bacteriophages is for the study of the biogenesis of lipid–protein structures and the interactions necessary for their maintenance. One of the

TABLE 7
Enzymatic Labeling of $\phi6$ and $\phi6$ Core Proteins with ^{125}I

Protein	$\phi6$ Intact[a] (%)	$\phi6$ Disrupted[b] (%)	$\phi6$ core Intact[a] (%)	$\phi6$ core Disrupted[b] (%)
P1	4.2	11.8	19.0	12.5
P2	3.4	14.4	4.5	14.6
P3	73.9	13.5	—	—
P4	3.1	4.3	7.9	11.0
P5	2.9	n.d.	—	—
P6	0.01	4.4	—	—
P8	2.1	9.9	2.6	12.4
P8	8.8	31.2	65.9	50.0
P9+P10	1.7	13.9		
Totals	100.1	103.4	99.9	100.5

[a] Van Etten *et al.* (1976).
[b] Partridge, Van Etten, Lane, Gonzalez, and Vidaver (unpublished).

major tools for investigations of this sort is the development of conditional-lethal mutants. Two types of conditional mutants which have been used extensively in the study of bacteriophage morphogenesis are temperature-sensitive and nonsense mutants. Temperature-sensitive mutants have been used in the study of animal virus development as well; however, nonsense mutants have been profitably exploited only in bacteriophages. Nonsense mutants can be grown only on bacterial hosts that harbor nonsense suppressor mutations. They are of special value in phage morphogenesis studies in general and for $\phi6$ in particular because they produce fragments of the protein products of the mutated gene when grown in wild-type infected cells. This enables one to identify the mutant allele by analysis of gel electrophoresis patterns wherein the missing or shortened peptide can be identified.

The isolation of nonsense mutants of $\phi6$ depended on the isolation of suppressor mutants of the host organism. The procedure for accomplishing this was to introduce a plasmid designated pLM2 into the host organism (Mindich et al., 1976a). This plasmid is derived from RP1, a plasmid of wide host range among gram-negative bacteria, and carries genes for resistance to kanamycin (kan), tetracycline (tet), and ampicillin (amp), as well as the determinant for sensitivity to bacteriophage PRD1 (Olsen et al., 1974); however, the tet and amp genes have nonsense (amber) mutations that prevent the expression of drug resistance. Bacteria containing this plasmid that acquire resistance simultaneously to ampicillin and tetracycline are usually nonsense suppressor mutants (Mindich et al., 1976a). Although this procedure suffices for the isolation of suppressor mutants in many species of bacteria, it did not work for the natural host of $\phi6$, which is P. phaseolicola HB10Y. An alternate host was found for $\phi6$ and was designated P. pseudoalcaligenes ERA. Using the plasmid pLM2, it was possible to isolate suppressor strains with this organism, and these strains were tested further for their ability to support the growth of nonsense mutants of bacteriophage PRD1. The most stable suppressor mutant was designated P. pseudoalcaligenes ERA (pLM2)S4 and will be called S4 in further discussions.

Wild-type $\phi6$ has a very low efficiency of plating (e.o.p.) on strain ERA. It was found that the plaques that did form were due to host-range mutants that retained their ability to plate on ERA after passage on P. phaseolicola HB10Y (called HB in further discussions). The host-range mutation is designated hl. In addition, it was found that $\phi6hl$ had a lower e.o.p. on the suppressor strain S4 than it had on the nonsuppressor host ERA. An additional mutation for high e.o.p. on S4 was selected for, and the strain bearing this mutation is designated $\phi6hls$.

Mutations were induced in $\phi 6$ or $\phi 6hls$ by several different treatments. Virus was exposed to nitrosoguanidine for long periods of time at high concentrations, with a yield of approximately 5% temperature-sensitive mutants and 0.5% nonsense mutants (Mindich et al., 1976b). Temperature-sensitive mutants formed plaques at 17°C but not at 27°C, whereas nonsense mutants formed plaques on S4 but not on HB or ERA. Mutants were also isolated by treating infected cells with fluorouracil. The reversion frequency for the mutants ranged from 0.2% to about 0.001%.

Genetic crosses between the temperature-sensitive mutants resulted in the establishment of three sets of mutants, the members of each set not yielding wild-type progeny in crosses with each other but yielding wild-type progeny in crosses with members of the other two sets. No inconsistencies were found in the genetic behavior of the members of the three sets. The assignments are consistent with the demonstration by Semancik et al. (1973) that $\phi 6$ contains three pieces of double-stranded RNA in approximately equimolar amounts. The observed recombination is similar to that found for other RNA viruses with segmented genomes, i.e., influenza (Simpson and Hirst, 1968) and reovirus (Fields and Joklik, 1969), in which reassortment of segments takes place but in which intramolecular recombination has not been demonstrated.

The nonsense mutants could also be placed in the same three sets. A simple plate test modified from that of Edgar et al. (1964) was used for the rapid mapping of phage mutants. This technique requires both complementation and recombination. True complementation analysis was not effective with $\phi 6$ mutants. Yields were very low and unreliable. This situation is reminiscent of the situation found for other RNA viruses, including the double-stranded segmented RNA reovirus (Fields and Joklik, 1960).

The host-range marker hl has also been mapped and found to reside in set B (Mindich et al., 1976b).

4.3.   Synthesis of $\phi 6$ Proteins

The general procedure for following the synthesis of bacteriophage proteins is to irradiate the host cells with ultraviolet light in order to shut off the host protein synthesis. The cells are then infected and radioactive amino acids are added for a given time period. The cells are then heated with SDS and mercaptoethanol and subjected to poly-acrylamide gel electrophoresis in SDS; the proteins are resolved and an autoradiogram is prepared from the gel. In the case of $\phi 6$ infecting strain HB, UV treatment drastically reduces the rate of phage protein synthesis (Sinclair *et al.*, 1975). The rate of phage protein synthesis decreases exponentially with increasing time between irradiation of the cells and infection. Not all the phage proteins are made after irradiation; the four proteins observed were P1, P2, P4, and P7 (Sinclair *et al.*, 1975). If the incubation was allowed to extend for 170 min, which is much longer than the normal latent period of the infection, then small amounts of other proteins could be visualized in autoradiograms (Mindich and Sinclair, unpublished).

Rifampin is known to inhibit the initiation reactions of bacterial RNA polymerase (Wehrli and Staehelin, 1971). This drug inhibits $\phi 6$ production in rifampin-sensitive host cells but does not inhibit production in rifampin-resistant mutants (Mindich, unpublished observation). However, the synthesis of phage proteins can be followed in cells treated with rifampin shortly before the addition of radioactive amino acids. The extent of phage protein synthesis observed falls exponentially with the length of time of exposure to rifampin, with a half-life of 7 min; however, the differential in effect of rifampin on host and viral protein synthesis is enough that the synthesis of the viral proteins can be followed. Although the reason for the effects of UV irradiation and rifampin on phage protein synthesis has not been elucidated, it seems reasonable that some host protein synthesis is probably necessary for the initiation and continuation of phage protein synthesis. Host factors have been found necessary in the development of many bacteriophages and in particular for RNA phages where several host factors have been found necessary for RNA synthesis (Happe and Jockusch, 1975; Van Dieijen *et al.*, 1976).

Pulse labeling of infected cells at various times during infection demonstrated that some of the phage proteins are synthesized earlier than others (Sinclair *et al.*, 1975) (Fig. 9). Proteins P1, P2, P4, and P7 are synthesized throughout the infection, whereas the remainder of the proteins begin their synthesis at approximately 45 min after infection. The effects of UV irradiation are therefore seen to be a slowing down of the time course of phage protein synthesis, with the consequent predominant synthesis of early proteins.

Pulse-chase experiments show little indication of cleavage or processing reactions involved in phage development. Only in the case of protein P8 is there any indication of a higher molecular weight precursor (Sinclair *et al.*, 1975). A cleavage may be involved in the production of proteins P5 and P11; however, this would have to take

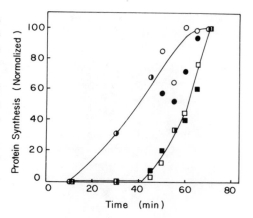

Fig. 9. Appearance of viral proteins during infection. The extent of synthesis of four viral proteins at various times during infection was determined by analyzing band densities in an autoradiogram of an SDS-polyacrylamide gel of cells infected with $\phi6$ and labeled in the presence of rifampin with radioactive leucine at various times after infection. The results are presented for P4 (●), P7 (○), P8 (■), and P9 (□). The extent of synthesis at various times is normalized to that at 70 min. From Sinclair et al. (1975).

place at the time of synthesis since no transfer of radioactivity from P11 to P5 takes place in pulse-chase experiments.

4.4. Nonsense Mutants of $\phi6$

The nonsense mutants isolated using the *P. pseudoalcaligenes* suppressor strain S4 were analyzed with respect to the protein synthesis that they directed in the nonsuppressor host. Seventy-four mutants were tested in this manner, and 62 mutants showed patterns of protein synthesis that are illustrated in Fig. 10 and Table 8. The eight classes, with the exception of 1 and 2, are designated according to the proteins missing during infection of a nonsuppressor host. The proteins involved in the eight classes include all but three of the 12 whose synthesis is directed by the virus (P4, P7, and P10). These proteins include membrane proteins, the nonstructural protein, the major core capsid protein, and two presumptive polymerase proteins (Sinclair et al., 1976). Five of the mutants synthesized all viral proteins after infection, indicating either that they are extremely leaky, that the nonsense fragments might be indistinguishable from the normal protein, or that they are nonsense mutants in an additional undetected viral protein. In seven mutants, no viral proteins were synthesized after infection. These seven could be nonsense mutants in an early protein whose synthesis is necessary for the formation of all other proteins, or they could be mutants that adsorb or infect at low efficiency.

The proportion of each mutant type obtained differed with the method of mutagenesis used. Since the larger genes within a chromosome have a greater target size and therefore an increased

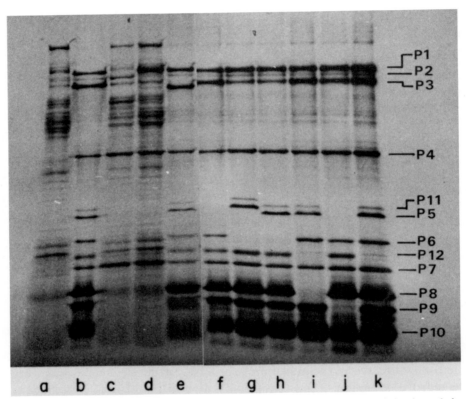

Fig. 10. Autoradiogram showing the pattern of protein synthesis after infection of the nonpermissive host with nonsense mutants of ϕh1s. The infected cells at 60 min after infection were treated with 200 μg rifampin/ml for 20 min, then labeled with [¹⁴C]leucine for 15 min. Electrophoresis was in a 15–25% gradient gel. (a) Uninfected cells; (b) infected with ϕh1s; (c) mutant susF132 of class 1; (d) susF108 of class 2; (e) sus45 of class 3; (f) sus32 of class 511; (g) sus55 of class 6(511m); (h) sus28 of class 6; (i) sus59 of class 812; (j) sus97 of class 9511; (k) sus95 of class 12. From Sinclair et al. (1976).

probability of being affected by the mutagen, one would expect a greater number of mutants for them; however, this expectation was not borne out for nitrosoguanidine mutagenesis (Table 8). There seems to be a greater number of mutants in certain genes, specifically those coding for P9, P5, P11, and P3, a pattern which is inconsistent with the target size of the genes involved. The results in Table 8 suggest that the frequency of different classes of mutants obtained with fluorouracil mutagenesis coincides more closely with the target size of the genes involved. P10 is a small protein (molecular weight <6000) (Sinclair et al., 1975), and would be expected to constitute only 1.6% of the total

TABLE 8
Nonsense Mutant Classes Obtained by Different Methods of Mutagenesis[a]

Class	Missing proteins										Linkage set	Example in Fig. 10	Number and (fraction) mutants isolated		Fraction expected[a]	Mutant designation
													FU[b]	NTG[c]		
1	**P1**[e]		P3	P11	P5	P6	P12	P8	P9	P10	A	c	1 (0.1)	—	0.21	susF132
2		**P2**	P3	P11	P5	P6	P12	P8	P9	P10	A	d	4 (0.4)	3 (0.058)	0.21	sus9, sus17, susF99, susF122, susF124
3			**P3**								B	e	4 (0.4)	16 (0.31)	0.20	sus45, sus60, sus64, sus74
511				**P11**	P5						C	f	—	14 (0.27)	0.065	sus22, sus32, sus48, sus49, sus72
6						**P6**					B	g,h	—	3 (0.057)	0.055	sus28, sus50, sus55
812							P12	**P8**			C	i	—	3 (0.057)	0.027	sus30, sus59
9511				P11	P5				**P9**		C	j	—	10 (0.192)	0.023	sus26, sus65
12							**P12**				C	k	1 (0.1)	3 (0.057)	0.052	sus87, sus97, susF123

[a] From Sinclair et al. (1976).
[b] FU, mutagenesis with fluorouracil.
[c] NTG, mutagenesis with nitrosoguanidine.
[d] Calculated on the basis of the relative sizes of the proteins.
[e] Boldface protein is presumed product of mutant gene.

nonsense mutants on the basis of target size. Since fewer than 100 mutants have been tested, it is not surprising that a mutant for P10 has not been found. However, that is not the case for mutants lacking P4 and P7, whose molecular weights are 36,800 and 19,900, respectively, and would be expected to compose 9.6% and 5.2% of the mutant population. Both P4 and P7 may be proteins that are not essential for viral development, a situation that has been found with several early bacteriophage T7 proteins (Studier, 1973a). It is also possible that the particular amino acid substitution for the nonsense codon by S4 renders both P4 and P7 inactive.

4.5. Identification of the Nonsense Classes

Four of the mutants make detectable polypeptide fragments as a result of the nonsense mutations. In the case of class 1 (Fig. 10c), the fragment is located below what is normally the position of P3 and is evident as a dark band among some lighter host bands. The class 2 mutant shown (Fig. 10d) makes a fragment that migrates just below the position of P3 and is very light in intensity, as is P2 in cells infected by wild-type virus (Fig. 10b). The class 3 mutant in Fig. 10 has a large fragment that migrates directly below the normal position for P3. A mutant in class 12 makes a fragment which migrates below the position of P9. There is a host band which also appears in this region; however, the fragment is evident as a darker band in the same region (Fig. 10k). Two additional mutants that may produce polypeptide fragments are 812 and 9511. In the pattern of protein synthesized by a class 812 mutant (Fig. 10i), the band at P10 is considerably thicker than that observed in the wild-type pattern. In addition, the upper region of the band appears to be darker than the lower, suggesting that it may be a fragment of P8. In the mutant of class 9511 (Fig. 10j), there appears to be a region below P10 that is heavier than that observed in the wild-type pattern. If this is a true viral band, it may correspond to a polypeptide fragment of the mutant protein, but it is not clear whether this protein is P9 or the P5,P11 precursor.

There are a number of mutants which are lacking more than one viral protein, as seen in Fig. 10. The revertant frequencies of these mutants correspond to that expected for a single mutation. Revertants isolated from the pleiotropic class synthesize all viral proteins during infection of the nonsuppressor host.

Nonsense mutants which lack either P1 (Fig. 10c) or P2 (Fig. 10d) are also missing the "late" class of viral proteins (P3, P5, P11, P6, P8,

P9, P10, and P12) (Sinclair *et al.*, 1975) indicating that P1 and P2 are both involved in turning on late viral protein synthesis after infection. It has been found that P1 and P2 are responsible for the synthesis of viral dsRNA. Perhaps they are both involved in transcription of late functions as well. Reduction of host protein synthesis occurs in all mutants except those in class 1 or class 2, as can be seen in Fig. 10. The inhibition detected may be indirect in that viral mRNA may have a greater affinity for ribosomes than host mRNA, a possibility that has been suggested for various animal RNA viruses (Nuss *et al.*, 1975). Since more viral mRNA is made late in infection than early, the reduction of host protein synthesis would not be observed until a later time. Another possibility is that the host polymerase or some of its components interact with P1 and P2 for late transcription and/or replication, thereby reducing the amount of polymerase available for transcription of host DNA. Since no mutant in P10 has been isolated as yet, its possible involvement in the inhibition of host protein synthesis must be considered.

Polar relationships have been found in several of the "late" genes of $\phi6$. Class 12 mutants lack P12, yet synthesize normal amounts of P8, while another class has been found, called 812, that lacks P8 and P12 (Fig. 10i). A similar relationship is true for proteins 9, 5, and 11. Mutants lacking P5 and P11 make normal amounts of P9, whereas when P9 is missing, so are P5 and P11. The determinants of all five proteins, P5, P8, P9, P11, and P12, have been shown by genetic mapping to be on the same chromosomal segment. There are three double-stranded RNA segments in the viral genome (Semancik *et al.*, 1973). Single-stranded RNAs that correspond to complete transcripts of the three segments have been found in infected cells (Coplin *et al.*, 1975). If those single-stranded RNAs observed are polycistronic messengers, then the effects mentioned above are likely to be translational rather than transcriptional regulation. The effects in classes 812 and 9511 are independent of each other and may reflect secondary structure in the mRNA, as has been found with the single-stranded RNA bacteriophages (Fiers, 1975), since the genes for these proteins are all in linkage set C (Table 8). In such a scheme, the translational initiators for proteins 8 and 9 would be exposed for binding of ribosomes, while those for P12, P11, and P5 would be located in a double-stranded region of the message. The cistrons for proteins 11 and 5 would then be distal to that for P9, and the region coding for protein 12 would be distal to that for P8.

There seems to be a unique polar relationship between proteins 3 and 6. The cistrons for both proteins are located in the same chromosomal segment. Nonsense mutants in class 3 that produce a large fragment of protein 3 show a slight reduction in the relative amount of P6 made (Table 9). If a mutant in class 3 produces no detectable fragment of protein 3, the decrease in P6 is even more marked. These results seem to correspond to a classic gradient of polarity, as has been observed with β-galactosidase (Zipser et al., 1970; Newton, 1969) and Qβ (Ball and Kaesberg, 1973). However, mutants in class 6, which lack P6, also show a considerable reduction in the relative amount of P3 made (Table 9). This reduction is greater than that observed for P6 in the class 3 mutants. A similar phenomenon has been reported to occur in the trp operon of E. coli (Yanofsky and Ito, 1966; Yanofsky et al., 1971) and has been named antipolarity. Whereas polarity can diminish the synthesis of distal proteins up to 85% or 95%, antipolarity diminishes the synthesis of proximal proteins only about 50%. This is similar to the finding with P3 and P6, if we consider P6 to be closer to the 5′ end of the mRNA. One possible model for this reciprocal effect between P3 and P6 is the following: The cistron for P6 could be proximal to that for P3 on a polycistronic mRNA that contains secondary structure. The structure would be such that the internal initiation site for P3 is located in a double-stranded region of the message. The terminal region of the cistron for P3 would contain some bases that are complementary to a region near or in the translational initiation site for P6, as illustrated in Fig. 11. The complementary bases at the terminal region of P3 would provide some competition with the ribosomes for binding at the translational initiation site. However, it also is possible

TABLE 9
Polar Effects in Class 3 and Class 6 Mutants[a]

Class	Figure	Relative amounts of proteins[b]		
		P3/P5	P6/P5	P4/P5
Wild type	10b	(1.0)	(1.0)	(1.0)
812	10c	1.01	1.25	1.3
6	10h	0.19	—	1.0
6(511m) altered P5,P11	10g	0.13	—	0.93
3, with fragment	10e	—	0.74	1.74
Wild type	—	—	(1.0)	(1.0)
3	—	—	0.42	0.97

[a] From Sinclair et al. (1976).
[b] The relative amounts of the individual proteins were estimated from a Joyce-Loebl microdensitometer tracing of the autoradiograms. The ratios are given relative to wild-type levels.

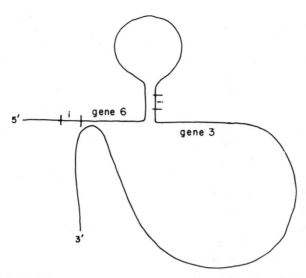

Fig. 11.   Model of reciprocal polarity between P6 and P3. Single-stranded mRNA has double-stranded regions that interfere with initiation of translation at initiation sites *i*, as suggested by the structure of MS2 RNA (Fiers, 1975).

that proteins 3 and 6 can act as positive regulators of each other. In such an arrangement, P6 would increase the synthesis of P3, and *vice versa*, while both proteins can be made at low levels without any regulatory effect. It has been found that both P3 and P6 are required for adsorption to the host bacterium, and each is required for the addition of the other to a viral particle *in vivo*. Thus it is feasible that they exert a positive regulatory effect on each other. Another possibility to explain the reciprocal effect would be that either P3 or P6 is unstable in the cell unless associated with the other.

As stated earlier, P5 and P11 both lack methionine, and no mutants missing one protein and not the other have been found. Several mutants have also been found in which the electrophoretic migration of both P5 and P11 changes to the same extent. In one case, a mutant *sus*55 was found to be a nonsense mutation in gene 6, and at the same time P5 and P11 migrated as if larger than the normal proteins. This mutant has been designated as class 6(511*m*) (see Fig. 10g). Revertants that can grow on nonsuppressor strains still retain the altered electrophoretic mobility of P5 and P11, although they are now capable of synthesizing P6. The change in the apparent size of P5 and P11 does not, therefore, affect function. In another case, a revertant of a class 9511 mutant was found to synthesize all three proteins in a nonsuppressor host, but P5 and P11 migrated as if smaller than normal. In all cases, the pair of proteins differ in molecular weight by about 1000 and are able to function despite having three different molecular weights. The pair P5,P11 is not dispensible as shown by the lack of growth of class 511 nonsense mutants on nonsuppressor hosts. In addition, the relative amounts of P5 and P11 are found to vary with conditions of infection. At low temperatures, one finds that $\phi 6$ produces more P5 than P11. At higher temperatures, relatively more P11 is synthesized, and in some cases one finds greater amounts of P11 than P5. $\phi 6hls$ does not produce large amounts of P11 relative to P5 at high temperature in HB, and infection of S4 with $\phi 6hls$ leads to the synthesis of almost no P11, while P5 is found in normal amounts.

Since there is a constant difference in the molecular weights of P5 and P11 and since the average molecular weight of the proteins can vary as well, it appears that these proteins are derived from a common gene and that both the N- and C-terminal ends of the molecule are variable. The difference in molecular weight between P11 and P5 could be explained by a cleavage of P11 to produce P5. Pulse-chase experiments have not demonstrated a precursor–product relationship between P11 and P5; however, it may be that most P11 is cleaved directly on formation, while a certain proportion of molecules are never cleaved. Another possibility is that there are two initiation sites for this gene as suggested for gene A of $\phi X174$ (Linney and Hayashi, 1974), or P11 may result from read through of gene 5 as found in the case of the coat protein of $Q\beta$ (Weiner and Weber, 1971). In this case, a working model would involve either two initiation sites or readthrough and a variable cleavage site near the other end of the protein. The cleavage site would be subject to variation due to missense mutations.

It is clear from the analysis of the defects in the nonsense mutants (Sinclair *et al.*, 1976) that the number of genes is greater than the number of linkage groups. This indicates that the genome segments are polycistronic. The polarity relationships are also consistent with the polycistronic nature of the messenger molecules produced by the virus. Most RNA viruses that have segmented genomes have only one cistron per segment. This is true of single-stranded viruses such as influenza (Pons, 1976) and double-stranded viruses such as reovirus (McDowell *et al.*, 1972), but segmented polycistronic RNAs are also common.

The assignment of a genetic group to a physical structure can be done for the case of set A. This set contains nonsense mutants of class 1 and class 2, both of which have defects in polymerase factors (Sinclair and Mindich, 1976; Sinclair *et al.*, 1976). Set A also contains $ts10$, a pleiotropic mutant that synthesizes only early proteins and no dsRNA at 27°C (Sinclair and Mindich, 1976). The molecular weights of proteins P1 and P2 are 82,000 and 80,000, respectively. Using a coding ratio of 19:1, based on triplet coding for amino acid residues, results in a calculated minimum molecular weight of 3.1×10^6 for the dsRNA segment of set A. Bacteriophage $\phi6$ contains three dsRNA segments of molecular weights 5.0×10^6, 3.1×10^6, and 2.3×10^6 (Van Etten et al., 1974). Since there must be some nonstructural information in the RNA, only the largest piece has the capacity to contain the genetic information for proteins P1 and P2. Proteins P4 and P7 are related to P1 and P2 in that all four are synthesized early in infection and all are synthesized by mutants that cannot synthesize late proteins and cannot synthesize the medium and small single-stranded viral RNA species. It seems highly probable that the genes for P4 and P7 are also located on the same segment as P1 and P2, i.e., the largest segment. If segment A does contain the genes for P1, P2, P4, and P7, it would then mean that the early proteins, all of which are found in the core of the virion and two of which are certainly involved in RNA synthesis, are clustered genetically.

Set B contains the genes for P3 and P6. These proteins are part of the membrane of the virion and are extractable with Triton X-100 (Sinclair et al., 1975). Proteins P3 and P6 are also involved in adsorption of the virus to host cells (Table 10) in that particles formed by nonsense mutants defective in the synthesis of either do not adsorb to host cells. In addition, the host-range character hl also maps in set B. This gene would be expected to affect the structure of the attachment apparatus of the virion. Set B, therefore, seems to be constituted of two genes that deal with one function.

TABLE 10
Composition of Subviral Particles[a]

Mutant	Mutant class	Missing protein in lysate	Missing proteins in particles	RNA	Lipid	Adsorbs to HB	Column in Fig. 15
sus45	3	P3	P3, P6	+	+	−	g
sus28	6	P6	P3, P6	+	+	−	d
sus55	6(511m)	P6	P3, P6	+	+	−	h
sus32	511	P5, P11	P5, P11	+	+	+	f
sus97	9511	P9, P5, P11	P3, P6, P9, P5, P11, P10	+	−	−	c
sus95	12	P12	P3, P6, P9, P5, P11, P10	+	−	−	e
sus59	812	P8, P12	All but P1	−	−	nd	i
wt	wt	None	None	+	+	+	k
wt	wt(p)	None	All but P1 and some P4	−	−	nd	j

[a] From Mindich et al. (1975c).

Set C contains the genes for the remainder of the viral proteins. These include proteins P9, P5, and P11, which are membrane proteins (Sinclair *et al.*, 1975; Van Etten *et al.*, 1976). P12, which is a nonstructural protein involved in membrane formation (Sinclair *et al.*, 1975; Mindich *et al.*, 1976*c*), and P8, which is the major structural protein of the core of the virion (Sinclair *et al.*, 1975; Van Etten *et al.*, 1976). Thus set C is a clustering of the genes involved in the formation of the exterior of the particle. Mutants involving membrane protein P10 have been found recently and they map in set B (Lehman and Mindich, unpublished).

The finding of three linkage groups does not prove that they constitute distinct molecules of dsRNA, although it is certainly consistent with this possibility. The polarity relationships found with nonsense mutants (Sinclair *et al.*, 1976) are also consistent with the mapping in that all proteins involved in polar relationships are found to map in the same set, with the exception of the class 1 and class 2 mutants which prevent the formation of all late proteins.

4.6.   Viral RNA Synthesis

RNA synthesized during the infectious cycle of $\phi 6$ has been shown to include three segments of dsRNA corresponding in size to the viral genome, three species of ssRNA that appear to be complete transcripts of the viral genome, and large replicative-intermediate-like structures (Coplin *et al.*, 1975) (Fig. 12). It has not yet been established whether the three ssRNA transcripts are messenger RNA, replicative intermediates, or both. There is an early and late pattern of RNA synthesis in infected cells.

Relatively more of the large (*l* ssRNA) piece is made early in infection, and chloramphenicol added early in infection increases the proportion of the large ssRNA (Coplin *et al.*, 1975). The effect of chloramphenicol on ssRNA synthesis is shown in Fig. 13. The proportions of ssRNA change markedly at later times during infection in that there is a great preponderance of the medium and small ssRNA transcripts relative to the large ones. At 60 min after infection, the large ss pieces constitute only 4% of the total ssRNA, whereas 48% would be expected if the pieces were synthesized in equimolar amounts. Despite the relative differences in the amounts of ssRNA, the proportions of the three pieces of dsRNA are almost equal late in infection (Coplin *et al.*, 1976). Early in infection there is a relative paucity of

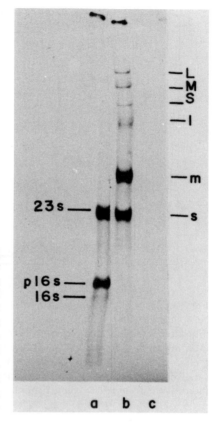

Fig. 12. Autoradiogram of RNA synthesized during φ6h1s infection of S4. At 70 min after infection, cells were treated with 200 μg rifampin/ml and 100 μg chloramphenicol/ml for 10 min, then pulsed with [³H]uracil for 5 min. (a) Uninfected S4 with no antibiotic treatment, showing the positions of host ribosomal RNA; (b) φ6h1s-infected S4, treated with chloramphenicol and rifampin; (c) uninfected S4 treated with chloramphenicol and rifampin. From Sinclair and Mindich (1976).

large dsRNA. Replication of φ6 may proceed through a single-stranded intermediate (Coplin et al., 1975), a process that has been demonstrated in reovirus (Acs et al., 1971; Schonberg et al., 1971), an animal virus which also contains a genome consisting of segmented double-stranded RNA. If so, there must be a mechanism for regulating the extent of synthesis of each species of dsRNA independently of the amount of precursor material. This mechanism might also be responsible for packaging one copy of each segment in the mature virion.

Viral proteins P1 and P2 are both involved in the synthesis of double-stranded RNA. All nonsense mutants isolated except those in classes 1 and 2 synthesize large amounts of double-stranded RNA at later stages of the infectious process. Since mutants that are defective in the synthesis of early proteins P4 and P7 or late protein P10 have not been isolated, the role of these proteins in RNA synthesis is unknown. It is unlikely that P10 is involved since it is a late protein and is part of

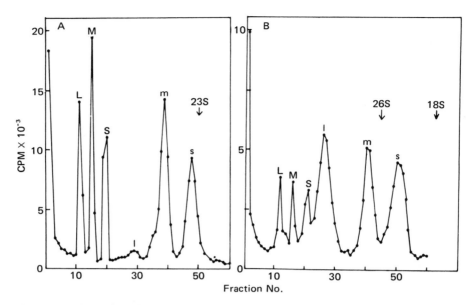

Fig. 13.   Polyacrylamide gel electrophoretic analysis of viral RNA extracted from (A) cells treated with rifampin at 40 min after infection and pulsed with [³H]uracil 50–53 min after infection, and (B) cells as in A but treated with chloramphenicol at 30 min after infection. Arrows show positions of *P. phaseolicola* 23 S rRNA and *Rhizopus stolonifer* 18 S and 26 S rRNA markers. L, M, and S refer to the large, medium, and small dsRNA segments; *l*, *m*, and *s* are the corresponding ssRNAs. The direction of migration is from left to right. From Coplin *et al.* (1975).

the membrane structure of the virion; however, P4 and P7 are found in the core of the virion (Sinclair *et al.*, 1975) and are synthesized early. They may be involved in RNA synthesis.

There is a small amount of double-stranded RNA synthesis that occurs early during infection. In the wild-type virus, this early dsRNA synthesis is inhibited approximately 90% if chloramphenicol is added before infection. This RNA synthesis apparently requires the synthesis of new proteins. In the case of class 1 or 2 mutants, where protein P1 or P2 is missing, one finds that the small amount of early dsRNA synthesis still occurs but to an extent much less than in the case of wild-type virus. This synthesis is reduced only 50% if infection takes place in the presence of chloramphenicol, even though overall protein synthesis is reduced by 96% by the antibiotic treatment. This indicates that the dsRNA synthesis observed in this case is probably not due to newly synthesized proteins. The residual activity may be due to functional proteins P1 and P2 that are associated with the genome of the infecting virion. This view is supported by the results obtained in infection by a temperature-sensitive mutant, $ts10$ (Sinclair and Mindich, 1976). This mutant is similar to nonsense mutants of classes 1 and 2 both genetically and in its pattern of protein synthesis at the nonpermissive temperature. Class 1 and class 2 mutants, as well as $ts10$, map in linkage set A (Mindich et al., 1976b). Only early viral proteins are made by this mutant at high temperature, and there is no double-stranded RNA synthesis. There is no residual dsRNA synthesis at high temperature with $ts10$, a result that might be expected if the polymerase itself were temperature sensitive. A polymerase activity has been found in purified virions (Van Etten et al., 1973); however, its relationship to in vivo synthesis is not clear. Since proteins P1 and P2 are present in the virion (Sinclair et al., 1975), they may be responsible for this activity. The early dsRNA synthesis and the synthesis catalyzed by virions in vitro have not been completely characterized, although it has been proposed that it is the finishing up of single-stranded regions in the genome (Van Etten et al., 1973). Isolated dsRNA does behave as if it has about 1–2.5% single stranded RNA (Van Etten et al., 1974). This may be contaminating material, but if dsRNA synthesis occurs in the particle as part of maturation there may be pieces that have not been completed, and these may be the templates for the in vitro synthesis and the early in vivo dsRNA synthesis.

Single-stranded viral RNAs, corresponding in size to complete transcripts of the medium and small dsRNA, are made by all nonsense mutants except those of classes 1 and 2. These results suggest that proteins P1 and P2 are also necessary for the synthesis of medium and small single-stranded viral RNAs.

There is no viral ssRNA species observable during infection by the nonsense mutants of classes 1 and 2 and the temperature-sensitive mutant $ts10$, at the nonpermissive temperature. Since early proteins are synthesized, viral messenger RNA must be made. The large viral ssRNA is difficult to detect during infection, especially after infection of HB by mutant and wild-type $\phi6hls$. Therefore, it is probable that the large viral ssRNA is made but is not detectable in infections by class 1 and 2 nonsense mutants and $ts10$.

If the large dsRNA segment is transcribed early in wild-type infection and during infection with class 1 and class 2 nonsense mutants and at high temperature in $ts10$, the question remains open as to what proteins are involved in this transcription. One possibility is that the host DNA-dependent RNA polymerase, in conjunction with other host proteins, may be capable of transcribing the large dsRNA segment. This

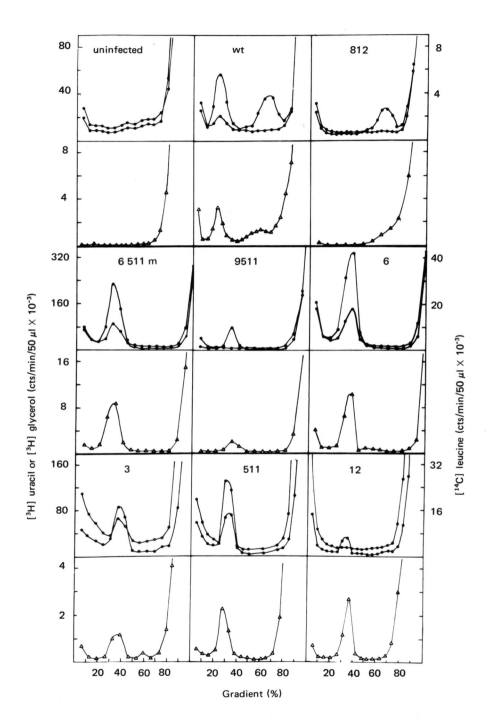

transcript would then act as messenger RNA for the synthesis of P1 and P2 so as to allow transcription of the *m* and *s* segments and consequently the synthesis of late proteins. Such a model would be somewhat analogous to the case of T7, where a virus-specific RNA polymerase is among the early proteins (Studier, 1973a,b). Alternatively, some combination of virion proteins and host proteins may be responsible for the early transcription. There are not as yet mutants affecting early proteins P4 and P7, and these might be involved in early RNA synthesis.

Another open question is the role of P1 and P2 in the transcription and the formation of dsRNA. The effects of mutations in P1 and P2 appear to be similar. Possibly the two proteins act together or form part of the same enzyme. This enzyme might be both a transcriptase and a dsRNA polymerase.

The $\phi6$ core contains approximately 60 molecules of P1 (Table 6). Experiments described in the next section show that a structure containing predominantly P1 is made during infection. It is possible that this is a capsidlike structure inside of which the dsRNA is synthesized. The major capsid protein P8 may form around this structure.

4.7. Particles Formed by Nonsense Mutants

The particles formed by wild-type $\phi6hls$ and six classes of nonsense mutants in a nonsuppressor host were isolated after centrifugation in sucrose gradients, as illustrated in Fig. 14. Particles doubly labeled with tritiated glycerol and [^{14}C]leucine were run on separate gradients from those labeled with tritiated uracil. Labeled particles are formed in all cases, and, with the exception of those formed by class 812, they all sediment with rates similar to those of intact infectious virions. Uninfected cells do not form observable particles, whereas

Fig. 14. Sucrose gradient centrifugation of phagelike particles accumulating in mutant lysates. Samples of 0.3 ml were layered on 5–20% sucrose gradients and centrifuged at 23,000 rev/min for 45 min at 21°C. The gradients are displayed in pairs, the upper of which contains the lysate prepared from cells labeled with 2-[^{3}H]glycerol (O) and with [^{14}C]leucine (●), and the lower containing lysate labeled with 5-[^{3}H]uracil (△). The final lysate supernatants contained approximately 1.5×10^7 counts/min/ml for glycerol, 3×10^6 counts/min/ml for leucine, and 5×10^6 counts/min/ml for uracil. Infectious virus was found at the position of the peak located at 30% of the distance from the bottom of the gradient. From Mindich *et al.* (1976c).

those formed by the class 812 particles lack RNA. These particles also lack lipid and all proteins but P1 (Table 10).

Particles formed by wild-type virus contain RNA, lipid (Fig. 14), and all the virion proteins (Fig. 15k). The extent of labeling of the virion proteins in the particles is different from that found in the analysis of whole, infected cells (Sinclair *et al.*, 1975). In particular, the

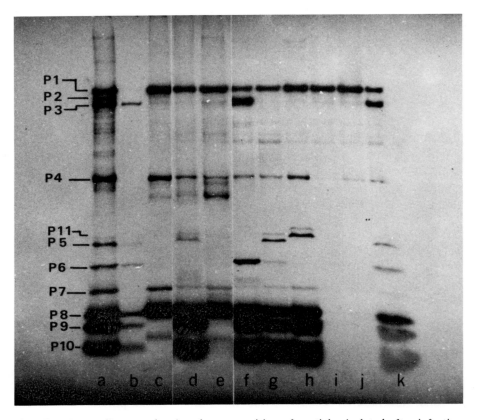

Fig. 15. Autoradiogram showing the composition of particles isolated after infection of HB with nonsense mutants. Aliquots of the peak fractions shown in Fig. 14 were heated with SDS-mercaptoethanol and applied to a 15–25% gradient gel. Samples are (a) purified wild-type virus; (b) wild-type particles; (c) class 9511 particles; (d) class 6 particles; (e) class 12 particles; (f) class 511 particles; (g) class 3 particles; (h) class 6(511m) particles; (i) class 812 particles; (j) wild-type particles from the slower-sedimenting peak of Fig. 14; (k) wild-type particles from the faster-sedimenting peak of Fig. 14. Sample (a) is wild-type virus labeled during the entire infection and purified by zone sedimentation and equilibrium isopycnic banding in sucrose gradients. Sample (b) is wild-type virus labeled late in infection and isolated from a gradient similar to that shown in Fig. 14.

TABLE 11
Labeling of Viral Lipids with 2-[³H]Glycerol[a]

Source of lipids	Fraction of counts in lipid class[b]		
	PE	PG	CL
HB lipids[c]	0.84	0.14	0.02
Infected cells[d]	0.64	0.15	0.21
Particles[e]			
wt	0.22	0.66	0.12
Class 3	0.16	0.64	0.19
Class 6	0.34	0.62	0.04
Class 6(511m)	0.25	0.66	0.08
Class 511	0.21	0.68	0.11

[a] From Mindich et al. (1976c).
[b] Since PE, Pg, and CL contain 1, 2, and 3 molecules of glycerol, respectively, it would be necessary to divide the incorporated respective counts by these factors to obtain the extent of incorporation on a molar basis.
[c] A culture of HB was labeled with 2-[³H]glycerol and lipids were extracted and analyzed.
[d] A culture of HB infected with φ6h1s was labeled during infection with 2-[³H]glycerol and lipids were extracted and analyzed.
[e] Viral particles were isolated and lipids were extracted and analyzed.

early proteins, P1, P2, P4, and P7, are poorly labeled, probably because a large pool of unlabeled early proteins exists in the infected cells since the label was added at only 60 min after infection. In Fig. 15b, one can see this in an extreme case where the early proteins are barely detectable. The lipid composition of the particles formed by wild-type particles is close to that described for φ6 (Sands, 1973) in that PG is more abundant than PE (Table 11). The values obtained are not meant to be exact reflections of the lipid composition of the virions, since the label is present only during infection and the virion incorporates lipids formed before infection as well (Sands and Lowlicht, 1976); however, it can be seen that the lipid composition of the particles is in all cases similar to that of the wild-type particles and none of the mutants forms a particle with only one type of lipid.

The wild-type virus also forms a particle similar to that formed by class 812, lacking RNA and lipid and composed primarily of P1 (Fig. 15j, Table 10). This may be a precursor particle to the mature virion and appears to have more proteins than the 812 particle. At least, it seems to contain some P4. It is not usually seen in such high relative amounts compared to the complete particle, and its presence in extracts is often not apparent, because of its slow rate of sedimentation.

Class 511 Particles. Particles formed by mutants missing protein P5 and P11 contain RNA, lipid, and all virion protein but P5 and P11

(Fig. 15f, Table 10). These particles adsorb quite well to host cells, indicating that proteins P5 and P11 do not play a role in the formation of particles or in attachment. Recent unpublished experiments from this laboratory indicate that protein P5 has the activity of a cell wall lysin.

Class 3 and Class 6 Particles. Particles formed by mutants lacking protein P6 contain RNA, lipid, and all virion proteins except P6 and P3 (Fig. 15d,g, Table 10). This is of special interest since P3 and P6 have been identified as possible membrane proteins by Triton X-100 extraction (Sinclair *et al.*, 1975). Mutants lacking P3 form particles that contain RNA, lipid, and all proteins except P3 and a small amount of P6. There seems to be a reciprocal relationship between P3 and P6 in that when one is missing from the particle the other is also missing, although in both cases the second missing protein is made by the cell (Sinclair *et al.*, 1976). The particles lacking P3 and P6 do not adsorb to host cells, indicating a role for these proteins in attachment. This is of special interest because the host-range mutation *hl* maps in set B, which contains the genes for P3 and P6 (Mindich *et al.*, 1976*b*). The high reactivity of P3 relative to P6 to iodination of intact particles (Table 7) indicates that P3 may be attached to the membrane through P6. The double mutant lacking P6 and showing altered migration for P5 and P11 forms a particle lacking P3 and P6 (Fig. 15h), but the altered P5 and P11 are incorporated into the particles.

Class 9511 Particles. Particles formed by class 9511 mutants contain RNA but no lipid (Fig. 14). The early proteins, P1, P2, P4 and P7, are present, but only P8 of the late proteins is seen in the particle (Fig. 15c, Table 10). Proteins P10, P3, and P6 are missing, indicating that P9 plays a vital role in the formation of the outer membrane of $\phi6$. It is of interest to compare class 9511 with class 511 since the latter forms a complete particle lacking only the proteins P5 and P11. The effects of the absence of P9 can therefore be identified, and they appear to be very marked. A band is seen in Fig. 15c at a position intermediate between the normal positions of P9 and P10. It is not known whether this material is of host or viral origin.

Class 12 Particles. Particles formed by class 12 mutants are identical to those formed by class 9511 particles. RNA is present but lipid is absent, as well as P9, P10, P6, P3, P5, and P11 (Fig. 15e, Table 10). These results indicate that P12, which is the only nonvirion protein formed by $\phi6$, plays an essential role in the formation of the viral membranelike structure. It is also of interest to note that P5 and P11 are missing from these particles. These proteins were not found to be extractable with 1% Triton X-100, as were P9, P10, P6, and P3.

However, their absence in these particles indicates a location near the exterior of the virion, as least exterior to P8. Van Etten *et al.* (1976) found that P5 is extractable from particles with 2.5% Triton X-100.

The particles formed by class 12 and class 9511 mutants do not adsorb to host cells but do adsorb to glass and plastics. It is of interest that the particles formed by class 12 and class 9511 sediment at the same rate as those containing lipid and membrane proteins (Fig. 14). This is probably due to a compensating increase in density and decrease in mass. Van Etten *et al.*, (1976) have found that Triton-extracted particles (cores) can sediment at the same rate as intact virions in high concentrations of sucrose. These authors found that the density of cores was greater than that of intact particles. We have found class 12 and class 9511 particles to have a higher buoyant density than complete particles.

The particles formed by the various classes of mutants allow us to develop some tentative notions as to the pathway leading to the formation of the complete virion of $\phi 6$ (Fig. 16). Protein P8 is the most abundant in the virion. Since it is not extracted with detergent (Sinclair

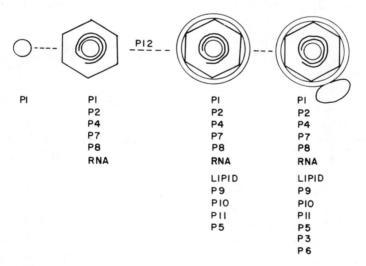

PI	PI	PI	PI
	P2	P2	P2
	P4	P4	P4
	P7	P7	P7
	P8	P8	P8
	RNA	RNA	RNA
		LIPID	LIPID
		P9	P9
		PIO	PIO
		PII	PII
		P5	P5
			P3
			P6

Fig. 16. Model for the structure and formation of bacteriophage $\phi 6$. A particle is formed consisting primarily of P1. This develops into a core particle that contains the early proteins P1, P2, P4, and P7, as well as RNA and the major virion protein P8, which probably constitutes the icosahedral core capsid. In a step dependent on protein P12, lipid and proteins P9, P10, P5, and P11 are incorporated into a membranelike structure exterior to the core. Finally, proteins P3 and P6 form the attachment apparatus, which may be a separate structure as depicted or distributed over the surface of the particle. From Mindich *et al.* (1976c).

et al., 1975), and since it is not affected by the class 12 or 9511 mutations, it seems reasonable that it would be the major component of the icosahedral inner capsid of the virion.

Proteins P5 and P11 seem to be exterior to P8, since they are missing, along with the other membrane proteins in the class 12 particles. Therefore, it seems that the early proteins, P1, P2, P4, and P7, form an inner core structure and the "late" proteins, with the exception of P8, are involved in the formation of the outer membrane.

Protein P1 forms a particle without RNA in class 812 mutants. A similar particle is seen in wild-type cells as well, but the latter seem to have other proteins associated. Whether these are part of the structure is not clear. Protein P1 has been identified as a component of the RNA polymerase of the virus (Sinclair and Mindich, 1976), and it may be that an early step in virion development is the formation of a particle of P1 molecules. The virion is estimated to have about 56 molecules of P1. This could be a particle of about 4.6×10^6 daltons, which might be expected to have a sedimentation rate similar to that found for the "slow" particles. This particle might form a core around which the dsRNA might be wound or a capsidlike structure inside of which the RNA would be packed. Other double-stranded viruses such as reovirus (Morgan and Zweerink, 1975) seem to form complexes of polymerase and newly formed double-stranded RNA. The double-stranded RNA appears to be synthesized inside a particle. The results of experiments on particle morphogenesis are summarized in Fig. 16. An inner core particle is formed with RNA, P8, and the early proteins. P8 is not necessary for the formation of dsRNA. It is not possible, from the experiments described, to determine whether the formation of the structure containing P8 precedes the association of RNA with the inner core of the particle. However, if dsRNA must form inside a capsid structure in analogy with reovirus, then this structure is likely to be composed of P1. If P9 is missing, the outer structure does not form, but P5, P11, P3, and P6 are not essential. A model wherein a lipid-containing membrane would form before the packaging of RNA would not seem to be satisfactory for $\phi 6$. It is apparent that the lipid-containing structure forms later with the help of P12. The function of P12 is unknown, but it is possible that this protein is a phospholipid exchange protein similar to those described in eukaryotic systems (Wirtz, 1974).

Phospholipid exchange proteins have not been found in bacteria, although this is not unexpected since the current idea of the role of these proteins is to transfer phospholipids from their sites of synthesis to other developing membrane structures, for example, from the endoplasmic reticulum to mitochondrial membranes (Wirtz, 1974). Since bacteria have a single, continuous membrane structure that contains the enzymes of phospholipid synthesis, it would seem that lateral diffusion of lipids in the plane of the lipid bilayer would suffice to move them to any site on the membrane. Only in the eukaryotic systems where membranes are physically isolated from each other would a diffusible, water-soluble exchange protein be necessary. However, in the case of $\phi 6$, the formation of the viral membrane seems to take place in isolation from the cell membrane, which is the site of the lipid synthesis. A device of some sort may be necessary to carry the lipids to the developing virion, and this may be the role of P12. A role for the phospholipid exchange protein in the development of vaccinia has been proposed (Stern and Dales, 1974); however, in that case, the host lipid exchange protein has been shown to be able to exchange viral lipids, and there may be no need for the virus to specify the synthesis of a new protein.

The formation of the outer envelope of the virion after the formation of the core would appear to proceed first through the formation of the membranelike structure containing lipid, P9, P10, P5, and P11. P5 and P11 are not necessary for the formation of this structure, but P9 is. A nonsense mutant lacking P10 has not yet been isolated; therefore, we do not know whether its presence is essential for membrane formation. P12, however, is essential for the construction of the envelope. P6 and P3 form a structure that appears necessary for attachment. An amorphous tail-like structure is seen in electron micrographs of the virus. It is sometimes linear and sometimes ringlike (Vidaver *et al.*, 1973; Sinclair *et al.*, 1975). This may be the attachment device; however, it is possible that the attachment structures are not the tail-like form but are spread over the surface of the virion.

The source of the viral lipids is probably the host cell membrane. The viral lipids are derived from lipids synthesized primarily before infection (Sands and Lowlicht, 1976). It has been reported that the phospholipid metabolism of the infected cell changes so that there is an increased rate of PG synthesis, although the rate of total phospholipid synthesis remains constant (Sands and Lowlicht, 1976). This change may be a result of the incorporation of predominantly PG into virion or could play a role in the determination of the virus lipid composition It would be of interest to study turnover patterns of the phospholipids, since some of the apparent biosynthetic characteristics may be a reflection of turnover rather than net synthesis.

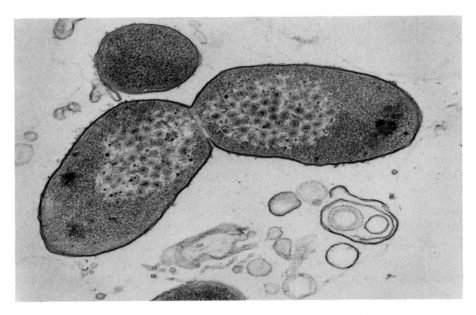

Fig. 17. *P. phaseolicola* HB10Y, 120 min after infection with $\phi 6$. Virions are concentrated at the interior of the cell away from the membrane. From Dales and Mindich (unpublished).

The $\phi 6$ virions have been shown to concentrate in the interior of the cell, away from the periphery (Ellis and Schlegel, 1974; Bamford *et al.*, 1976) (Fig. 17). Core particles of 50-nm diameter can be seen in electron micrographs at about 45 min after infection. At this time, 65–75 nm particles corresponding to mature virions begin to appear (Bamford *et al.*, 1976). This development is consistent with the pathway proposed on the basis of the mutant particle studies.

5. SUMMARY AND CONCLUSIONS

The lipid-containing bacteriophages have been studied for only a short period of time. Much of their structure and morphogenesis remains to be elucidated. Whereas PM2 has been well characterized structurally, its biology is poorly understood. Conversely, very little structural work has been done on $\phi 6$, whereas a fair amount of progress has been made on its genetics and biogenesis. As the progress in this area has been rapid of late, it can be expected that a full picture for some of the lipid phages will soon be developed.

The major findings with respect to all lipid phages have been the determination of the lipid composition in which PG predominates and the demonstration that phage lipids are derived from material synthesized both before and at the time of assembly, indicating that the lipids are picked up from the host plasma membrane. Of particular interest is the finding of the nonstructural protein P12 that is involved in the formation of the membrane of $\phi 6$.

The PM2 reconstitution experiments of Schäfer and Franklin have demonstrated the simple nature of the PM2 construction and that the phage structure can be arrived at without special cleavage or assembly reactions. The asymmetry of the PM2 lipid distribution is also striking.

The pathway for the assembly of $\phi 6$ is likely to involve the filling of a capsid structure with dsRNA to form the complete nucleocapsid and then the envelopment of this structure with a membrane. The other lipid phages do not have an easily disrupted membrane. It is likely that the assembly of these may proceed through the early formation of a lipid-containing vesicle structure, as in the case of the poxviruses (Dales and Mosbach, 1968). The electron micrographs of developing PM2 suggest such a pathway (Brewer, 1976; Dahlberg and Franklin, 1970). Therefore, $\phi 6$ differs from the other lipid phages both in construction and in the role of the lipid structure in morphogenesis.

ACKNOWLEDGMENTS

I wish to express my gratitude to Ms. Annabel Howard for her help in preparing this review and my thanks to Drs. J. Lehman and R. M. Franklin for their useful comments on its content. I also appreciate the opportunity to read prepublication manuscripts of Drs. G. Brewer, J. Sands, J. Van Etten, and J. Partridge.

6. REFERENCES

Acs, G., Klett, H., Schonberg, M., Christman, J., Levin, D. H., and Silverstein, S. C., 1971, Mechanism of reovirus double-stranded ribonucleic acid synthesis *in vivo* and *in vitro*, *J. Virol.* **8**:684–689.

Ball, L. A., and Kaesberg, P., 1973, A polarity gradient in the expression of the replicase gene of RNA bacteriophage Qβ, *J. Mol. Biol.* **74**:547–562.

Bamford, D. H., Palva, E. T., and Lounatmaa, K., 1976, Ultrastructure and life cycle of the lipid-containing bacteriophage $\phi 6$, *J. Gen. Virol.* **32**:249–259.

Bell, R. M., Mavis, R. D., Osborn, M. J., and Vagelos, P. R., 1971, Enzymes of phospholipid metabolism: Localization in the cytoplasmic and outer membrane of the cell envelope of *Escherichia coli* and *Salmonella typhimurium, Biochim. Biophys. Acta* **249**:628–635.

Bowman, B. U., Jr., Newman, H. A. I., Moritz, J. M., and Koehler, R. M., 1973, Properties of mycobacteriophage DS6A. II. Lipid composition, *Am. Rev. Resp. Dis.* **107**:42–49.

Bradley, D. E., 1976, Adsorption of the R-specific bacteriophage PR4 to pili determined by a drug resistance plasmid of the W compatibility group, *J. Gen. Microbiol.* **95**:181–185.

Bradley, D. E., and Rutherford, E. L., 1975, Basic characterization of lipid-containing bacteriophage specific for plasmids of the P, N, and W compatibility groups, *Can. J. Microbiol.* **21**:152–163.

Brady, G. W., Fein, D. B., and Brumberger, H., 1976, X-ray diffraction studies of circular superhelical DNA at 300–10,000 Å resolution, *Nature (London)* **264**:231–234.

Braunstein, S., and Franklin, R. M., 1971, Structure and synthesis of a lipid-containing bacteriophage. V. Phospholipids of the host BAL-31 and of the bacteriophage PM2, *Virology* **43**:685–695.

Bretscher, M. S., 1973, Membrane structure: Some general principles, *Science* **181**:622–629.

Brewer, G. J., 1976, Control of membrane morphogenesis in bacteriophage PM2, *J. Supramol. Struct.* **5**:73–79.

Brewer, G. J., and Singer, S. J., 1974, On the disposition of the proteins of the membrane-containing bacteriophage PM2, *Biochemistry* **13**:3580–3588.

Cadden, S. P., and Sands, J. A., 1976, Proteins of a lipid containing bacteriophage which replicates in *Escherichia coli*: Phage PR4, *Abst. Annu. Meet. Am. Soc. Microbiol.*, S20, p. 207.

Camerini-Otero, R. D., and Franklin, R. M., 1972, Structure and synthesis of a lipid-containing bacteriophage. XII. The fatty acids and lipid content of bacteriophage PM2, *Virology* **49**:385–393.

Camerini-Otero, R. D., and Franklin, R. M., 1975, Structure and synthesis of a lipid-containing bacteriophage. XVII. The molecular weight and other physical properties of bacteriophage PM2, *Eur. J. Biochem.* **53**:343–348.

Camerini-Otero, R. D., Datta, A., and Franklin, R. M., 1972, Structure and synthesis of a lipid-containing bacteriophage. XI. Studies on the structural glycoprotein of the virus particle, *Virology* **49**:522–536.

Capaldi, R. A., and Vanderkooi, G., 1972, The low polarity of many membrane proteins, *Proc. Natl. Acad. Sci. U.S.A.* **69**:930–932.

Casjens, S., and King, J., 1975, Virus assembly, *Annu. Rev. Biochem.* **44**:555–611.

Coplin, D. L., Van Etten, J. L., Koski, R. K., and Vidaver, A. K., 1975, Intermediates in the biosynthesis of double-stranded ribonucleic acids of bacteriophage $\phi6$, *Proc. Natl. Acad. Sci. U.S.A.* **72**:849–853.

Coplin, D. L., Van Etten, J. L., and Vidaver, A. K., 1976, Synthesis of bacteriophage $\phi6$ double-stranded ribonucleic acid. *J. Gen. Virol.* **33**:509–512.

Cota-Robles, E., Espejo, R. T., and Haywood, P. W., 1968, Ultrastructure of bacterial cells infected with bacteriophage PM2, a lipid-containing bacterial virus, *J. Virol.* **2**:56–68.

Dahlberg, J. E., and Franklin, R. M., 1970, Structure and synthesis of a lipid-containing bacteriophage. IV. Electron microscopic studies of PM2-infected *Pseudomonas* BAL-31, *Virology* **42**:1073–1086.

Dales, S., and Mosbach, E. H., 1968, Vaccinia as a model for membrane biogenesis, *Virology* **35**:564–583.

Datta, A., Camerini-Otero, R. D., Braunstein, S. N., and Franklin, R. M., 1971, Structure and synthesis of a lipid-containing bacteriophage. VII. Structural proteins of bacteriophage PM2, *Virology* **45**:232–239.

Diedrich, D. L., and Cota-Robles, E. H., 1974, Heterogeneity in lipid composition of the outer membrane and cytoplasmic membrane of *Pseudomonas* BAL-31, *J. Bacteriol.* **119**:1006–1018.

Diedrich, D. L., and Cota-Robles, E. H., 1976, Phospholipid metabolism in *Pseudomonas* BAL-31 infected with lipid-containing bacteriophage PM2, *J. Virol.* **19**:446–456.

Edgar, R. A., Denhardt, G. H., and Epstein, R. H., 1964, A comparative genetic study of conditional lethal mutations of bacteriophage T4D, *Genetics* **49**:635–658.

Ellis, L. F., and Schlegel, R. A., 1974, Electron microscopy of *Pseudomonas* ϕ6 bacteriophage, *J. Virol.* **14**:1547–1551.

Espejo, R. T., and Canelo, E. S., 1968a, Properties of bacteriophage PM2: A lipid-containing bacterial virus, *Virology* **34**:738–747.

Espejo, R. T., and Canelo, E. S., 1968b, Origin of phospholipid in bacteriophage PM2, *J. Virol.* **2**:1235–1240.

Espejo, R. T., and Canelo, E. S., 1968c, Properties and characterization of the host bacterium of bacteriophage PM2, *J. Bacteriol.* **95**:1887–1891.

Espejo, R. T., and Canelo, E. S., 1969, The DNA of bacteriophage PM2: Ultracentrifugal evidence for a circular structure, *Virology* **37**:495–498.

Espejo, R. T., Canelo, E. S., and Sinsheimer, R. L., 1969, DNA of bacteriophage PM2: A closed circular double-stranded molecule, *Proc. Natl. Acad. Sci. U.S.A.* **63**:1164–1168.

Espejo, R. T., Canelo, E. S., and Sinsheimer, R. L., 1971a, Replication of bacteriophage PM2 deoxyribonucleic acid: A closed circular double-stranded molecule, *J. Mol. Biol.* **56**:597–621.

Espejo, R. T., Espejo-Canelo, E., and Sinsheimer, R. L., 1971b A difference between intracellular and viral supercoiled PM2 DNA, *J. Mol. Biol.* **56**:623–626.

Fields, B. N., and Joklik, W. K., 1969, Isolation and preliminary genetic and biochemical characterization of temperature-sensitive mutants of reovirus, *Virology* **37**:335–342.

Fiers, W., 1975, Chemical structure and biological activity of bacteriophage MS2 RNA, in: *RNA Phages* (N. D. Zinder, ed.), pp. 353–396, Cold Spring Harbor Laboratory, Cold Spring Harbor.

Franklin, R. M., 1974, Structure and synthesis of bacteriophage PM2, with particular emphasis on the viral lipid bilayer. *Curr. Top. Microbiol. Immunol.* **68**:108–159.

Franklin, R. M., 1976, PM2 bacteriophage as a model for the structure and synthesis of lipid membranes, *Cell Surface Rev.* (in press).

Franklin, R. M., Salditt, M., and Silbert, J. A., 1969, Structure and synthesis of a lipid-containing bacteriophage. I. Growth of bacteriophage PM2 and alterations in nucleic acid metabolism in the infected cell, *Virology* **38**:627–640.

Franklin, R. M., Hinnen, R., Schäfer, R., and Tsukagoshi, N., 1976, Structure and assembly of lipid-containing viruses, with special reference to bacteriophage PM2 as one type of model system, *Phil. Trans. R. Soc. London. Ser. B* **276**:63–80.

Gilbert, W., and Dressler, D., 1968, DNA replication: The rolling circle model, *Cold Spring Harbor Symp. Quant. Biol.* **33**:473–484.

Gonzalez, C. F., Langenberg, W. G., Van Etten, J. L., and Vidaver, A. K., 1976, Ultrastructure of bacteriophage $\phi6$: Arrangement of dsRNA and lipid envelope, *Abstr. Annu. Proc. Am. Phytopathol. Soc. Abstr.*, No. 223.

Grodzicker, T., and Zipser, D., 1968, A mutation which creates a new site for the re-initiation of polypeptide synthesis in the *z* gene of *lac* operon of *Escherichia coli, J. Mol. Biol.* **38**:305–314.

Happe, M., and Jockusch, H., 1975, Phage $Q\beta$ replicase: Cell-free synthesis of the phage-specific subunit and its assembly with host subunits to form active enzyme, *Eur. J. Biochem.* **58**:359–366.

Harrison, S. C., Caspar, D. L. D., Camerini-Otero, R. D., and Franklin, R. M., 1971, Lipid and protein arrangement in bacteriophage PM2, *Nature (London) New Biol.* **229**:197–201.

Hinnen, R., Schäfer, R., and Franklin, R. M., 1974, Structure and synthesis of a lipid-containing bacteriophage: Preparation of virus and localization of the structural proteins, *Eur. J. Biochem.* **50**:1–14.

Hinnen, R., Chassin, R., Schäfer, R., Franklin, R. M., Hitz, H., and Schäfer, D., 1976, Structure and synthesis of a lipid-containing bacteriophage: Purification, chemical composition, and partial sequences of the structural proteins, *Eur. J. Biochem.* **68**:139–152.

Israelachvili, J. N., 1973, Theoretical considerations on the asymmetric distribution of charged phospholipid molecules on the inner and outer layers of curved bilayer membranes, *Biochim. Biophys. Acta* **323**:659–663.

Laval, F., 1974, Endonuclease activity associated with purified PM2 bacteriophages, *Proc. Natl. Acad. Sci. U.S.A.* **71**:4965–4969.

Linney, E., and Hayashi, M., 1974, Intragenic regulation of the synthesis of ϕX174 gene A proteins, *Nature (London)* **249**:345–348.

Marvin, D. H., and Hohn, B., 1969, Filamentous bacterial viruses, *Bacteriol. Rev.* **33**:172–209.

McDowell, M. J., Joklik, W. K., Villa-Komaroff, L., and Lodish, H. J., 1972, Translation of reovirus messenger RNAs synthesized *in vitro* into reovirus polypeptides by several mammalian cell-free extracts, *Proc. Natl. Acad. Sci. U.S.A.* **69**:2649–2653.

Michaelson, D. M., Horwitz, A. F., and Klein, M. P., 1973, Transbilayer asymmetry and surface homogeneity of mixed phospholipids in cosonicated vesicles, *Biochemistry* **12**:2637–2645.

Mindich, L., Cohen, J., and Weisburd, M., 1976a. Isolation of nonsense suppressor mutants in *Pseudomonas, J. Bacteriol.* **126**:177–182.

Mindich, L., Sinclair, J. F., Levine, D., and Cohen, J., 1976b. Genetic studies of temperature-sensitive and nonsense mutants of bacteriophage $\phi6$, *Virology* **75**:218–223.

Mindich, L., Sinclair, J. F., and Cohen, J., 1976c. The morphogenesis of bacteriophage $\phi6$: Particles formed by nonsense mutants, *Virology* **75**:224–231.

Morgan, E. M., and Zweerink, H. J., 1975, Characterization of transcriptase and replicase particles isolated from reovirus-infected cells, *Virology* **68**:455–466.

Nagy, E., Prágai, B., and Ivanovics, G., 1976, Characteristics of phage AP50, an RNA phage containing phospholipids, *J. Gen. Virol.* **32**:129–132.

Newton, A., 1969, Re-initiation of polypeptide synthesis and polarity in the *lac* operon of *Escherichia coli, J. Mol. Biol.* **41**:329–339.

Nuss, D. L., Oppermann, H., and Koch, G., 1975, Selective blockage of initiation of host protein synthesis in RNA-virus-infected cells, *Proc. Natl. Acad. Sci. U.S.A.* **72**:1252–1262.

Olsen, R. H., Siak, J. S., and Gray, R. H., 1974, Characteristics of PRD1, a plasmid-dependent broad host range DNA bacteriophage, *J. Virol.* **14**:689–699.

Osborne, M. J., Gander, J. E., Parisi, E., and Carson, J., 1972. Mechanism of assembly of the outer membrane of *Salmonella typhimurium*: Isolation and characterization of cytoplasmic and outer membrane, *J. Biol. Chem.* **247**:3962–3972.

Pons, M. W., 1976, A reexamination of influenza single- and double-stranded RNAs by gel electrophoresis, *Virology* **69**:789–792.

Sakaki, Y., and Oshima, T., 1976, A new lipid-containing phage infecting acidophilic thermophilic bacteria, *Virology* **75**:256–259.

Sands, J. A., 1973, The phospholipid composition of bacteriophage φ6, *Biochem. Biophys. Res. Commun.* **55**:111–116.

Sands, J. A., 1976, Studies on the origin of the phospholipids of a lipid-containing virus which replicates in *Escherichia coli*: Bacteriophage PR4, *J. Virol.* **19**:296–301.

Sands, J. A., and Cadden, S. P., 1975, Phospholipids in an *Escherichia coli* bacteriophage, *FEBS Lett.* **58**:43–46.

Sands, J. A., and Lowlicht, R. A., 1976. Temporal origin of viral phospholipids of the enveloped bacteriophage φ6, *Can. J. Microbiol.* **22**:154–158.

Sands, J. A., Cupp, J., Keith, A., and Snipes, W., 1974, Temperature sensitivity of the assembly process of the enveloped bacteriophage φ6, *Biochim. Biophys. Acta* **373**:277–285.

Scandella, C. J., Schindler, H., Franklin, R. M., and Seelig, J., 1974, Structure and synthesis of a lipid-containing bacteriophage. Acyl-chain motion in the PM2 virus membrane, *Eur. J. Biochem.* **50**:29–32.

Schäfer, R., and Franklin, R. M. 1975a, Structure and synthesis of a lipid-containing bacteriophage: A polynucleotide-dependent polynucleotide–pyrophosphorylase activity in bacteriophage PM2, *Eur. J. Biochem.* **58**:81–85.

Schäfer, R., and Franklin, R. M., 1975b, Structure and synthesis of a lipid-containing bacteriophage. XIX. Reconstitution of bacteriophage PM2 *in vitro, J. Mol. Biol.* **97**:21–34.

Schäfer, R., Hinnen, R., and Franklin, R. M., 1974a, Structure and synthesis of a lipid-containing bacteriophage: Properties of the structural proteins and distribution of the phospholipid, *Eur. J. Biochem.* **50**:15–27.

Schäfer, R., Hinnen, R., and Franklin, R. M., 1974b, Further observations on the structure of the lipid-containing bacteriophage PM2, *Nature (London)* **248**:681–682.

Schäfer, R., Huber, U., Franklin, R. M., and Seelig, J., 1975, Structure and synthesis of a lipid-containing bacteriophage. XXI. Chemical modifications of bacteriophage PM2 and the resulting alterations in acyl-chain motion in the PM2 membrane, *Eur. J. Biochem.* **58**:291–296.

Schonberg, M., Silverstein, S. C., Levin, D. H., and Acs, G., 1971, Asynchronous synthesis of the complementary strands of the reovirus genome, *Proc. Natl. Acad. Sci. U.S.A.* **68**:505–508.

Semancik, J. S., Vidaver, A. K., and Van Etten, J. L., 1973, Characterization of a segmented double-helical RNA from bacteriophage φ6, *J. Mol. Biol.* **78**:617–625.

Silbert, J. A., Salditt, M., and Franklin, R. M., 1969, Structure and synthesis of a lipid-containing bacteriophage. III. Purification of bacteriophage PM2 and some structural studies on the virion, *Virology* **39**:666–681.

Simpson, R. W., and Hirst, G. K., 1968, Temperature-sensitive mutants of influenza A-virus: Isolation of mutants and preliminary observations on genetic recombination and complementation, *Virology* **35**:41–49.

Sinclair, J. F., and Mindich, L., 1976, RNA synthesis during infection with bacteriophage φ6, *Virology* **75**:209–217.

Sinclair, J. F., Tzagoloff, A., Levine, D., and Mindich, L., 1975, Proteins of bacteriophage φ6, *J. Virol.* **16**:685–695.

Sinclair, J. F., Cohen, J., and Mindich, L., 1976, The isolation of suppressible nonsense mutants of bacteriophage φ6. *Virology* **75**:198–208.

Snipes, W., Douthwright, J., Sands, J., and Keith, A., 1974, Control of phospholipid synthesis and viral assembly by bacteriophage PM2, *Biochim. Biophys. Acta* **363**:340–350.

Spear, P. G., and Roizman, B., 1972, Proteins specified by herpes simplex virus. V. Purification and structural proteins of the herpes virion. *J. Virol.* **9**:143–159.

Spencer, R., 1963, Bacterial viruses in the sea, *Symposium on Marine Microbiology* (C. H. Oppenheimer, ed.), pp. 350–365, Thomas, Springfield, Ill.

Stanisich, V., 1974, The properties and host range of male-specific bacteriophages of *Pseudomonas aeruginosa, J. Gen. Microbiol.* **84**:332–342.

Stern, W., and Dales, S., 1974, Biogenesis of vaccinia: Concerning the origin of the envelope phospholipids, *Virology* **62**:293–306.

Strauss, J. H., Jr., Burge, B. W., Pfefferkorn, E. R., and Darnell, J. E., Jr., 1968, Identification of the membrane protein and "core" protein of Sindbis virus, *Proc. Natl. Acad. Sci. U.S.A.* **59**:533–537.

Studier, F. W., 1973a, Genetic analysis of non-essential bacteriophage T7 genes, *J. Mol. Biol.* **79**:227–236.

Studier, F. W., 1973b, Analysis of bacteriophage T7 early RNA's and proteins on slab gels, *J. Mol. Biol.* **79**:237–248.

Tsukagoshi, N., and Franklin, R. M., 1974, Structure and synthesis of a lipid-containing bacteriophage. XIII. Studies on the origin of the viral phospholipids, *Virology* **59**:408–417.

Tsukagoshi, N., Petersen, M. H., and Franklin, R. M., 1975, Effect of unsaturated fatty acids on the lipid composition of bacteriophage PM2, *Nature (London)* **253**:125–126.

Tsukagoshi, N., Kania, M. N., and Franklin, R. M., 1976a, Identification of acyl phosphatidylglycerol as a minor phospholipid of *Pseudomonas* BAL-31, *Biochim. Biophys. Acta* **450**:131–136.

Tsukagoshi, N., Petersen, M. H., Huber, U., Franklin, R. M., and Seelig, J., 1976b, Phase transitions in the membrane of a marine bacterium, *Pseudomonas* BAL-31, *Eur. J. Biochem.* **62**:257–262.

Van der Schans, G. P., Weyermans, J. P., and Bleichrodt, J. F., 1971, Infection of spheroplasts of *Pseudomonas* with DNA of bacteriophage PM2, *Mol. Gen. Genet.* **110**:263–271.

Van Dieijen, G., Van Knippenberg, P. H., and Van Duin, J., 1976, The specific role of ribosomal protein S1 in the recognition of native phage RNA, *Eur. J. Biochem.* **64**:511–518.

Van Etten, J. L., Vidaver, A. K., Koski, R. K., and Semancik, J. S., 1973, RNA polymerase activity associated with bacteriophage φ6, *J. Virol.* **12**:464–471.

Van Etten, J. L., Vidaver, A. K., Koski, R. K., and Burnett, J. P., 1974, Base composition and hybridization studies of the three double-stranded RNA segments of bacteriophage φ6, *J. Virol.* **13**:1254–1262.

Van Etten, J., Lane, L., Gonzalez, C., Partridge, J., and Vidaver, A., 1976, Comparative properties of bacteriophage φ6 and φ6 nucleocapsid, *J. Virol.* **18**:652–658.

Vidaver, A. K., Koski, R. K., and Van Etten, J. L., 1973, Bacteriophage φ6: A lipid-containing virus of *Pseudomonas phaseolicola*, *J. Virol.* **11**:799–805.

Wehrli, W., and Staehelin, M., 1971, Actions of the rifamycins, *Bacteriol. Rev.* **35**:290–309.

Weiner, A. M., and Weber, K., 1971, Natural read-through at the UGA termination signal of Qβ coat protein cistron, *Nature (London) New Biol.* **234**:206–209.

Wirtz, K. W. A., 1974, Transfer of phospholipids between membranes, *Biochim. Biophys. Acta* **344**:95–117.

Yamamoto, K. R., and Alberts, B. M., 1970, Rapid bacteriophage sedimentation in the presence of polyethylene glycol and its application to large-scale virus purification, *Virology* **40**:734–744.

Yanofsky, C., and Ito, J., 1966. Nonsense codons and polarity in the tryptophan operon, *J. Mol. Biol.* **21**:313–334.

Yanofsky, C., Horn, V., Bonner, M., and Stasioski, S., 1971, Polarity and enzyme functions in mutants of the first three genes of the tryptophan operon of *Escherichia coli*, *Genetics* **69**:409–433.

Zipser, D., Zabell, S., Rothman, J., Grodzicker, T., and Wenk, H., 1970, Fine structure of the gradient of polarity in the *z* gene of the *lac* operon of *Escherichia coli*, *J. Mol. Biol.* **49**:251–254.

Index